Advanced Nanocomposites Materials Based on Graphene Oxide/Reduced Graphene Oxide: Potential Applications and Perspectives

Advanced Nanocomposites Materials Based on Graphene Oxide/Reduced Graphene Oxide: Potential Applications and Perspectives

Guest Editors

Angela Longo
Mariano Palomba

MDPI

Basel • Beijing • Wuhan • Barcelona • Belgrade • Novi Sad • Cluj • Manchester

Guest Editors

Angela Longo
Institute for Polymers,
Composites, and Biomaterials
National Research Council,
CNR
Naples
Italy

Mariano Palomba
Institute for Polymers,
Composites, and Biomaterials
National Research Council,
CNR
Naples
Italy

Editorial Office
MDPI AG
Grosspeteranlage 5
4052 Basel, Switzerland

This is a reprint of the Special Issue, published open access by the journal *Materials* (ISSN 1996-1944), freely accessible at: https://www.mdpi.com/journal/materials/special_issues/adv_nanocomposites.

For citation purposes, cite each article independently as indicated on the article page online and as indicated below:

Lastname, A.A.; Lastname, B.B. Article Title. *Journal Name* **Year**, *Volume Number*, Page Range.

ISBN 978-3-7258-5237-6 (Hbk)
ISBN 978-3-7258-5238-3 (PDF)
https://doi.org/10.3390/books978-3-7258-5238-3

Contents

About the Editors

Angela Longo

Angela Longo, PhD in *Materials and Structures Engineering* and graduate in Chemical Engineering, is currently a researcher at the Institute for Polymers, Composites and Biomaterials of the National Research Council (IPCB-CNR) in Naples. Her research focuses on the development of innovative methods for producing nanostructured materials, particularly polymer-based systems incorporating inorganic nanostructures.

She is also involved in the design and fabrication of advanced devices for applications in electronics, optoelectronics, and biomedicine, leveraging the unique properties of nanostructured materials. Her work includes the exploration of novel chemical synthesis routes and the morphological and structural characterization of nanomaterials using techniques such as transmission and scanning electron microscopy (TEM and SEM), X-ray powder diffraction, and optical spectroscopy (UV–visible absorption and fluorescence). These analyses aim to correlate the physicochemical properties of nanomaterials with their composition, size, and morphology.

Her research interests span a wide range of topics, including graphene and graphene oxide, metal nanoparticles, inorganic/polymer nanocomposites, and luminescent nanomaterials. She particularly focuses on optical characterization through ultraviolet–visible absorption and fluorescence spectroscopy, as well as morphological analysis using electron microscopy and structural investigation via X-ray diffraction techniques.

She is the author of numerous scientific publications in international journals and has participated in several international conferences.

Mariano Palomba

Mariano Palomba has served as a researcher at the Institute for Polymers, Composites and Biomaterials (IPCB) of the National Research Council (CNR), at the secondary office in Naples/Portici, since 2014. He holds a degree in Chemical Engineering (2002) and a PhD in Materials and Structural Engineering (2013), both from the University of Naples "Federico II". He carried out postgraduate contractual research at the Department of Materials and Structural Engineering (DIMP) of the University of Naples "Federico II" from 2005 to 2006, and at the Institute for Composite and Biomedical Materials (IMCB) of the CNR from 2007 to 2014. He is the author and co-author of numerous publications in international journals in the field of materials science and engineering, and of several presentations at international conferences (H index of 11, as evaluated on the Scopus database in June 2025). His current research interests include the following: functional nanocomposite materials, metal–polymer nanocomposites, polymer nanostructures, nanocoatings of graphite supported on polymers, new electrical conductors, chemical sensors, and the chemical–physical characterization of these materials with particular reference to microscopic techniques of morphological characterization (scanning electron microscopy and transmission electron microscopy) and calorimetric techniques (thermogravimetric analysis, differential scanning calorimetry, and dynamic mechanical analysis).

Preface

This Reprint, titled "*Advanced Nanocomposite Materials Based on Graphene Oxide/Reduced Graphene Oxide: Potential Applications and Perspectives*", brings together a curated selection of original research articles and reviews that reflect the current state of the art in the field of graphene-based nanocomposites. The central subject of this collection is the synthesis, functionalization, and application of graphene oxide (GO) and reduced graphene oxide (r-GO), two materials that have emerged as key components in the development of advanced nanostructured systems.

The scope of this Reprint spans a wide range of topics, including the integration of GO/r-GO with polymers, inorganic nanoparticles, and other carbon-based nanomaterials. These hybrid systems exhibit enhanced mechanical, thermal, electrical, and chemical properties, making them suitable for applications in energy storage, catalysis, environmental remediation, sensing, and biomedicine.

The aim of this Reprint is to provide a comprehensive overview of recent advances in the field, highlighting both experimental and theoretical approaches to the design of GO/r-GO-based nanocomposites. Special attention is given to novel synthesis strategies, green and sustainable processing methods, and the development of multifunctional materials with tailored properties.

The motivation behind this work stems from the growing interest in graphene-derived materials and their potential to address pressing technological and environmental challenges. By compiling contributions from leading researchers, this Reprint serves as a valuable resource for scientists, engineers, and professionals working in materials science, nanotechnology, chemistry, and related disciplines.

We hope that this collection will inspire further research and foster collaboration between academia and industry, ultimately contributing to the advancement of innovative solutions based on GO and r-GO nanocomposites.

Angela Longo and Mariano Palomba
Guest Editors

materials

MDPI

Editorial

Special Issue: Advanced Nanocomposite Materials Based on Graphene Oxide/Reduced Graphene Oxide: Potential Applications and Perspectives

Mariano Palomba * and Angela Longo *

Institute for Polymers, Composites, and Biomaterials, National Research Council, 80055 Portici, Italy
* Correspondence: mariano.palomba@cnr.it (M.P.); angela.longo@cnr.it (A.L.); Tel.: +39-081-7758824 (A.L.)

In recent years, graphene oxide (GO) and reduced graphene oxide (r-GO) have received much attention as precursors of graphene-like 2D nanomaterials. GO consists of a layered material based on a carbon skeleton functionalized by different oxygen-containing groups (typically with a C/O atomic ratio of less than 3), and r-GO is obtained via the almost complete removal of oxygen functional groups [1–3]. These materials have unique intrinsic physical and chemical properties, including a large surface area, functional groups, good conductivity, and good biocompatibility. For this reason, many experimental studies have been conducted to improve the preparation methods of GO and r-GO, and to analyze their possible applications [4,5]. Recently, a growing number of studies have been published concerning the preparation and characterization of new nanocomposites, which integrate GO and/or r-GO (GO/r-GO) with polymers [6], and inorganic nanoparticles (metal, metal oxide, etc.) [7].

For example, GO/r-GO-based polymer nanocomposites are receiving remarkable interest due to their excellent mechanical, thermal, and electrical properties. GO/r-GO can be used as nanometric fillers embedded in a polymeric matrix to enhance the structural, morphological, and functional properties of the composite material [3]. In addition, these materials are suitable for electronic and energy storage applications in the form of polymer–graphene composites [8]. Furthermore, nanocomposites based on GO/r-GO and inorganic nanoparticles such as Au, Ag, and Pt have attracted much attention due to their applications as catalysts, photocatalysts, electrodes, sensors, substrates for surface-enhanced Raman spectroscopy, and in biomedicine [7].

This Special Issue contains unique articles and reviews that reflect the current state-of-the-art, focusing on the performance peculiarities of nanocomposite materials based on GO and/or r-GO in specific fields of application.

Jiang and co-workers studied the in situ growth of Fe_3O_4 nanoparticles as homogeneous clusters on reduced graphene oxide [8]. In fact, the results showed that Fe_3O_4 nanoparticles, anchored on r-GO surfaces, formed nanoclusters without aggregation, and they expanded the interlayer spacing between r-GO sheets. The improved electrochemical properties of the Fe_3O_4/r-GO nanocomposite compared with pure Fe_3O_4 can be attributed to high electron transport, increased interfaces, and positive synergistic effects between Fe_3O_4 and r-GO. These nanocomposites represent an effective strategy to develop new advanced supercapacitor electrodes for energy storage devices [8].

Ullah et al. described the nanocomposite synthesis of magnetic nanoparticle (MNPs) and r-GO in polymethylmethacrylate (PMMA) [9]. The combination of the high performance of PMMA (due to its mechanical and physical properties) with active adsorbents, such as GO magnetized using $FeCl_3$ and $FeSO_4$ salts, was proven to be an interesting approach for the removal of hazardous Cr(VI) from tannery wastewater through batch- and continuous-mode adsorption [9].

Regarding new materials for storing solar energy, which have received a great deal of attention, Yang et al. [10] studied a photothermal conversion material obtained by attaching trifluoromethylated azobenzene (AzoF) to reduce GO. This system exhibited remarkable energy storage performance, as well as an excellent storage life span, and it is equipped with the ability to release heat at low temperatures [10].

Due to the SARS-CoV-2 pandemic, there has been an increase in the search for affordable healthcare devices for mass testing and rapid diagnosis. In this context, Zaccariotto et al. [11] described a new methodology for SARS-CoV-2 detection based on an impedimetric immunosensor developed using the advantageous immobilization of antibodies in the r-GO. An electrochemical immunoassay was considered for the detection of the SARS-CoV-2 spike protein RBD using a impedimetric immunosensor and redox couple ($[(Fe(CN)_6)]^{3-/4-}$) as a probe [11].

This Special Issue also presents three interesting reviews that could act as good starting points for the development of innovative materials based on GO, using green processes characterized by good feasibility, cost-effectiveness, and sustainability.

Leve et al. [12] report on several literature results pertaining to the promising functionalization of GO and r-GO surfaces with metal oxide, for enhanced performance in selectivity and sensitivity in gas sensing. In this review, the functionalization of graphene, the synthesis of heterostructured nanohybrids, and the assessment of their collaborative performance towards gas-sensing applications are discussed [12].

Wei et al. [13] studied graphene-based composite aerogels (GCAs), which are a solid porous substance formed by graphene or its derivatives, graphene oxide (GO) and reduced graphene oxide (rGO), with inorganic materials and polymers. This review demonstrates the super-high adsorption, separation, electrical properties, and sensitivity of GCA, which could have great potential for applications in super-strong adsorption and separation materials, long-life fast-charging batteries, and flexible sensing materials [13].

Palomba et al. [14] analyzed papers from the literature concerning the most important approach to GO reduction based on the use of L-ascorbic acid. The results were organized according to two important approaches: reduction in the liquid-phase and in the gel-phase. The achieved r-GO quality enabled its technological exploitation in various forms; for example, it could be used as a coating, self-supported, or embedded in a polymer. Knowledge of all aspects of the synthesis and properties of r-GO obtained using the L-aa reduction technique is critical in bringing this process into mass production [14].

As demonstrated in this Special Issue, the design and development of nanocomposites based on GO/r-GO with tailored properties are essential in order to expand their range of potential applications. We hope this will stimulate further development and new ideas by prompting fruitful discussions between academic and industry experts who work in the field of graphene-related materials.

Author Contributions: Conceptualization, M.P. and A.L.; writing—original draft preparation, M.P. and A.L.; writing—review and editing, M.P. and A.L.; visualization, A.L.; supervision, A.L. All authors have read and agreed to the published version of the manuscript.

Funding: This research received no external funding.

Acknowledgments: The Guest Editors would like to thank: the featured authors for submitting their research to the present Special Issue and for assisting in its successful completion; the reviewers for enhancing the quality and impact of the submitted papers; and the editorial staff for their support during the publication of this Special Issue. Finally, we offer special thanks to Karena Tang for her valuable support during the entire process of developing and publishing this Special Issue.

Conflicts of Interest: The authors declare no conflict of interest.

References

1. Lerf, A.; He, H.; Forster, M.; Klinowski, J. Structure of graphite oxide revisited. *J. Phys. Chem. B* **1998**, *102*, 4477–4482. [CrossRef]
2. He, H.; Klinowski, J.; Forster, M.; Lerf, A. A new structural model for graphite oxide. *Chem. Phys. Lett.* **1998**, *287*, 53–56. [CrossRef]
3. Smith, A.T.; LaChance, A.M.; Zeng, S.; Liu, B.; Sun, L. Synthesis, properties, and applications of graphene oxide/reduced grapheneoxide and their nanocomposites. *Nano Mater. Sci.* **2019**, *1*, 31–47. [CrossRef]

4. Ray, S. *Applications of Graphene and Graphene-Oxide Based Nanomaterials Micro and Nano Technologies*; Andrew, W., Ed.; Elsevier: Oxford, UK, 2015.
5. Rowley-Neale, S.J.; Randviir, E.P.; Abo Dena, A.S.; Banks, C.E. An overview of recent applications of reduced graphene oxide as a basis of electroanalytical sensing platforms. *Appl. Mater. Today* **2018**, *10*, 218–226. [CrossRef]
6. Shahryari, Z.; Yeganeh, M.; Gheisari, K.; Ramezanzadeh, B. A brief review of the graphene oxide-based polymer nanocomposite coatings: Preparation, characterization, and properties. *J. Coat. Technol. Res.* **2021**, *18*, 945–969. [CrossRef]
7. Khan, A.A.P.; Khan, A.; Asiri, A.M.; Ashraf, G.M.; Alhogbia, B.G. Graphene oxide based metallic nanoparticles and their some biological and environmental application. *Curr. Drug Metab.* **2017**, *18*, 1020–1029. [CrossRef] [PubMed]
8. Jiang, Y.; Han, J.; Wei, X.; Zhang, H.; Zhang, Z.; Ren, L. Magnetite Nanoparticles In-Situ Grown and Clustered on Reduced Graphene Oxide for Supercapacitor Electrodes. *Materials* **2022**, *15*, 5371. [CrossRef] [PubMed]
9. Ullah, R.; Ahmad, W.; Yaseen, M.; Khan, M.; Iqbal Khattak, M.; Mohamed Jan, B.; Ikram, R.; Kenanakis, G. Fabrication of MNPs/rGO/PMMA Composite for the Removal of Hazardous Cr(VI) from Tannery Wastewater through Batch and Continuous Mode Adsorption. *Materials* **2021**, *14*, 6923. [CrossRef]
10. Yang, X.; Li, S.; Zhang, J.; Wang, X.; Wang, Y.; Zhao, J. A Low-Temperature Heat Output Photoactive Material-Based High-Performance Thermal Energy Storage Closed System. *Materials* **2021**, *14*, 1434. [CrossRef]
11. Zaccariotto, G.C.; Silva, M.K.L.; Rocha, G.S.; Cesarino, I. A Novel Method for the Detection of SARS-CoV-2 Based on Graphene-Impedimetric Immunosensor. *Materials* **2021**, *14*, 4230. [CrossRef] [PubMed]
12. Leve, Z.D.; Iwuoha, E.I.; Ross, N. The Synergistic Properties and Gas Sensing Performance of Functionalized Graphene-Based Sensors. *Materials* **2022**, *15*, 1326. [CrossRef]
13. Wei, D.; Liu, X.; Lv, S.; Liu, L.; Wu, L.; Li, Z.; Hou, Y. Fabrication, Structure, Performance, and Application of Graphene-Based Composite Aerogel. *Materials* **2022**, *15*, 299. [CrossRef]
14. Palomba, M.; Carotenuto, G.; Longo, A. A Brief Review: The Use of L-Ascorbic Acid as a Green Reducing Agent of Graphene Oxide. *Materials* **2022**, *15*, 6456. [CrossRef] [PubMed]

materials

MDPI

Article

A Low-Temperature Heat Output Photoactive Material-Based High-Performance Thermal Energy Storage Closed System

Xiangyu Yang [1,2], Shijie Li [2], Jin Zhang [2], Xiaomin Wang [1], Yongzhen Wang [1,*] and Jianguo Zhao [1,2,*]

[1] College of Materials Science and Engineering, Taiyuan University of Technology, Yingze West Street, Taiyuan 030024, China; yangxiangyu0039@link.tyut.edu.cn (X.Y.); wangxm62@126.com (X.W.)
[2] Institute of Carbon Materials Science, Shanxi Datong University, Xingyun Street, Datong 037009, China; li841974@sina.com (S.L.); zhangjin50@hrbeu.edu.cn (J.Z.)
* Correspondence: wangyz62@163.com (Y.W.); zhaojianguo@sxdtdx.edu.cn (J.Z.)

Abstract: Designing and synthesizing photothermal conversion materials with better storage capacity, long-term stability as well as low temperature energy output capability is still a huge challenge in the area of photothermal storage. In this work, we report a brand new photothermal conversion material obtained by attaching trifluoromethylated azobenzene (Azo_F) to reduced graphene oxide (rGO). Azo_F-rGO exhibits outstanding heat storage density and power density up to 386.1 $kJ \cdot kg^{-1}$ and 890.6 $W \cdot kg^{-1}$, respectively, with a long half-life (87.7 h) because of the H-bonds based on high attachment density. Azo_F-rGO also exhibits excellent cycling stability and is equipped with low-temperature energy output capability, which achieves the reversible cycle of photothermal conversion within a closed system. This novel Azo_F-rGO complex, which on the one hand exhibits remarkable energy storage performance as well as excellent storage life span, and on the other hand is equipped with the ability to release heat at low temperatures, shows broad prospects in the practical application of actual photothermal storage.

Keywords: photothermal conversion material; outstanding heat storage density; long-term storage; low temperature energy output; closed system

1. Introduction

With the fast development of society, people's demand for energy is increasing and the energy issue has now become one of the major problems that human beings need to deal with [1]. Solar energy has the advantages of sufficient reserves, no pollution and economical availability. Efficiently converting and storing solar energy has become an important way to overcome the current energy shortage crisis [2–5]. Recently, photothermal conversion materials have attracted extensive attention as a new method for storing solar energy [6]. Photothermal conversion materials can store solar energy in chemical bonds through photo-isomerization of units and then releasing the stored energy as thermal energy when exposed to different external stimulus, achieving photothermal conversion within a closed system. Such materials are able to effectively convert light energy into its own chemical bonds and release its stored energy while avoiding the emission of additional greenhouse gases, with the potential to achieve low-cost and large-scale industrial solar storage [7]. However, photothermal conversion materials still have the shortcomings of short storage time, low energy density and inability to achieve energy release under low temperatures, which are key factors limiting its practical application in solar thermal energy storage [8,9].

Owing to its special photoisomerization ability, good structural stability and controllable configuration recoverability, azobenzene and its derivatives with numerous applications [10,11] has received extensive research interest as a kind of photothermal conversion material [12,13]. However, due to the disadvantages of poor storage performance and storage half-life ($\tau_{1/2}$) arising from low isomerization enthalpy (ΔH), azobenzene did not exert its full potential in terms of photothermal conversion and storage [14]. To override the above hurdles, great efforts have been made on the basis of molecular design by introducing different substituents and increasing the interaction between molecules [15–17]. Grossman et al. [18] reported azobenzene derivatives with bulky aromatic groups as photoactive chemical heat storage materials. Owing to the introduction of bulky phenyl groups, the solid-state azobenzene derivatives not only improve the energy density but also improve the corresponding thermal stability. Bléger et al. [19] reported *o*-Fluoroazobenzenes and derivatives which exhibit an unprecedented long half-life owing to the ortho-fluorine substituent which reduces electron density around the –N=N– double bond. Despite great efforts having been made, it is still an intractable problem to apply azobenzene photothermal conversion material to practical energy storage.

Different from freely dispersed azobenzene, many azobenzene carbon materials were formed by introducing azobenzene into high-strength carbon nanomaterials forms many azobenzene carbon nanocomposites [7,20,21] accompanied by a more closely ordered structure, which have excellent storage capacity and life cycle. The templated, structure modified azobenzene enhance the intermolecular interactions while obtaining a more stable and tightly ordered structure, which jointly improved the storage capacity of azobenzene carbon materials [22,23]. In addition, because of the unique 2D structure and broad surface of graphene with numerous applications [24,25] which contributes to high attachment density, the templated azobenzene/graphene nanomaterials show broad prospects in photothermal storage [26]. Unfortunately, azobenzene carbon nanomaterials still have problems such as difficulty in releasing storage heat at low temperatures and the inability to balance energy density and half-life, which limits their further practical application [27,28]. Therefore, how to simultaneously achieve the improvement of storage capacity and life cycle with low-temperature energy output capability is still a key issue in current research.

In this work, we report a novel photothermal conversion material by attaching trifluoromethylated azobenzene (Azo_F) to reduced graphene oxide (rGO). The storage capacity and storage life span as well as the cycling stability performance of Azo_F-rGO has made great progress. Azo_F-rGO exhibits great development potential in recyclable and long term photothermal storage.

2. Materials and Methods

2.1. Materials

3-amino-5-(trifluoromethyl)benzoic acid (99%), 3,5-dimethoxyaniline (99%), sodium nitrite (97%), Na_2CO_3 (97%) and $NaBH_4$ (97%) were purchased from Aladdin Reagent (Shanghai, China).

2.2. Detailed Synthesis Steps

- 3-amino-5-(trifluoromethyl)benzoic acid (1.025 g) was dissolved in the HCl solution (50 mL, 0.5 mol·L^{-1}), then $NaNO_2$ (0.380 g) was added and reacted at ice bath for 80 min. After dissolving 3,5-dimethoxyaniline (0.765 g) in water, we slowly added the above mixture to it, adjusted the pH to 7 and reacted it in an ice bath for 4 h. Azo_F was obtained after further purification (1.255 g, 68%).
- GO was synthesized according to the literature reports [29]. First, we used NaOH (1 mol·L^{-1}) solution to change the pH of the GO aqueous solution (300 mL, 0.5 mg·mL^{-1}) to 10, then we reacted it at 90 °C for 4 h with $NaBH_4$ (180 mg) under N_2 atmosphere. When the reaction was complete, rGO was obtained by washing the mixture with water multiple times.
- Azo_F (0.738 g) was dissolved in the HCl solution (60 mL, 0.5·mol L^{-1}), then $NaNO_2$ (0.141 g) was slowly added and reacted in an ice bath for 80 min, and the above mixture was slowly added to the rGO solution (62 mL, 1 mg·mL^{-1}). The mixture was first reacted at 0 °C for 4 h and then at 30 °C for 16 h. Azo_F-rGO was obtained by purifying the mixture with water and DMF multiple times.

2.3. Characterizations

The FT-IR was gathered from Vertex 70 (Bruker, Karlsruhe, Germany). The XRD was gathered from X'Pert Pro MPD (PANalytical, Almelo, Holland). Raman spectrum was gathered from LabRAM Aramis (HORIBA, Paris, France). The XPS was gathered from ESCALAB 250Xi (ThermoFisher, Waltham, MA, USA) using C1 s = 284.8 eV for energy calibration procedures, Operation Mode:CAE:Pass Energy 100.0 Ev, software:Thermo Avantage 5.976 and hemispherical energy analyzer were used for the test, the test vacuum was 5×10^{-9} Torr, the sample was fixed on the sample stage with conductive glue, the background was buckled through the smart method, and the energy calibration was performed with gold, silver and copper. The TGA was performed on STA449F5 (NETZSCH, Bavaria, Germany). TEM was gathered from Tecnai F20 (FEI, Hillsboro, Oregon, USA). SEM were gathered from SU8010 (Hitachi, Tokyo, Japan). The UV–Vis absorption spectra was performed on SPECORD 50 PLUS (ANALYTIK JENA, Jena, Germany) in the range of 250~550 nm with the resolution of 0.1 nm. The *trans* $\rightarrow$ *cis* transition was introduced by a multiband LED lamp at 365 nm. The *cis* $\rightarrow$ *trans* transition was introduced by a multiband LED lamp at 540 nm. The light intensity was gathered from an optical power meter (PL-MW2000, Bofeilai Technology, Beijing, China). The heat storage density was determined through differential scanning calorimetry (DSC, 214 Polyma, NETZSCH, Bavaria, Germany) under N_2.

3. Results and Discussion

3.1. Chemical Structure

As shown in Figure 1a, the low-resolution TEM image of rGO exhibited a smooth structure and its electron diffraction exhibited a hexagonal lattice according to Fast Fourier Transform (FFT) patterns within Figure 1b, demonstrating its good crystallinity. Figure 1c shows that the surface of the material became rough, and the electron diffraction spot of Azo_F-rGO (Figure 1d) has become a closed loop attributed to the adhesion of Azo_F on rGO [30,31]. Furthermore, the SEM of Azo_F-rGO (Figure 1f) shows a stacking phenomenon compared with rGO (Figure 1e). This phenomenon not only reduced the distance between adjacent graphene layers but also enhanced the intermolecular interaction, resulting in a growth in the storage capacity as well as $\tau_{1/2}$ of Azo_F-rGO [21]. In addition, it can also be concluded that the distance between layers was reduced based on the XRD results (Figure S2). After the reduction of GO, the (0 0 1) diffraction peak at 11.3° disappeared [32] and was replaced by the (0 0 2) diffraction peak at 22.9° of rGO, and the corresponding grain size was 25.51 nm based on Scherrer formula [33]. After attaching Azo_F onto rGO, the 2θ of Azo_F-rGO has become to 25.2° with the grain size of 22.63 nm, which is consistent with the SEM observation (Figure 1f) [34].

The Azo_F-rGO had new peaks of –N=N– (1430 cm^{-1}) and $–CF_3$ (1140 cm^{-1}) compared to rGO [35] according to Figure 2a. Moreover, the FT-IR spectra of Azo_F-rGO and Azo_F also showed peaks derived from -OH (3298 cm^{-1}) and –C=O (1640 cm^{-1}). It can also be seen from Figure 2a that the wavenumbers of -OH and –C=O of Azo_F-rGO show a significant red shift compared to that of Azo_F (3204 cm^{-1} and 1700 cm^{-1}), confirming the formation H-bond of Azo_F on rGO [36]. XPS results also proven the successful grafting of Azo_F on rGO. In addition, the characteristic peaks of Azo_F at 287.5 eV and 292.5 eV corresponding to C–N and C–F bond also appeared in Azo_F-rGO (Figure S3) [35]. Additionally, the fact that there were characteristic peaks of –N=N– (400.3 eV) and $–CF_3$ (688.3 eV) in Azo_F-rGO also confirmed the successful bonding between Azo_F and rGO [35].

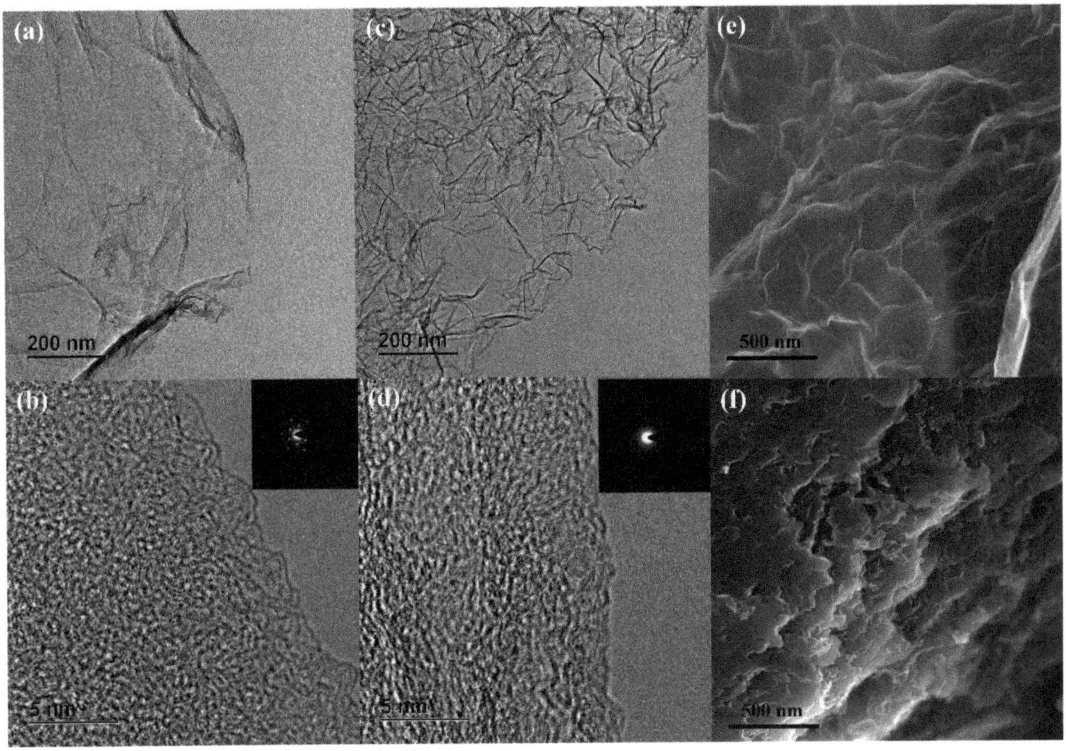

Figure 1. (**a**,**c**) Low resolution TEM images of rGO and Azo_F–rGO, (**b**,**d**) high resolution TEM images of rGO and Azo_F-rGO with FFTs, and SEM images of (**e**) rGO and (**f**) Azo_F–rGO.

The high-density adhesion of Azo_F onto rGO nanosheets is inextricably linked to the improvement of the performance of Azo_F-rGO. The decomposition of rGO during the whole heating process was linear according to Figure 2d, and its weight loss mainly attributed to the disappearance of oxygen-containing groups [37]. The Azo_F was stable before 185 °C, and its weight loss was attributed to self-decomposition. Additionally, the weight loss of Azo_F-rGO was caused by the weight loss of Azo_F and rGO [27]. Therefore, the attachment density (A_d) of Azo_F on rGO after different time reactions can be obtained based on Equation (1) [38].

$$D_g = \frac{R_p - R}{R_p - R_a} \times 100\% \quad (1)$$

where *Ra* is the residual weight percentage of Azo_F, *R* is the residual weight percentage of Azo_F-rGO, *Rp* is the residual weight percentage of rGO.

Table 1 shows that the attachment density (A_d) was 1/40 after the first reaction and increased to 1/16 after the third reaction. The attachment density can also be obtained based on XPS [39]. It can also be seen from Table 1 that the results obtained by XPS and TGA were almost identical. From the above results, it can be concluded that almost every 16 carbon atoms of rGO correspond to one Azo_F after the third reaction, which is better than previous research [21,40]. High adhesion density on the one hand helps to form intermolecular hydrogen bonds, while on the other hand it also enhances intermolecular interactions, which improves the storage performance of Azo_F-rGO [41]. In addition, Raman spectroscopy also proved this result. It can also be seen from Figure S4 that the I_D/I_G value of Azo_F-rGO-1 (1.14) and Azo_F-rGO-3 (1.18) was much larger than rGO

(1.08), which indicates that the crystal structure of rGO has changed after attachment [31], proving the remarkable attachment density of Azo_F on rGO.

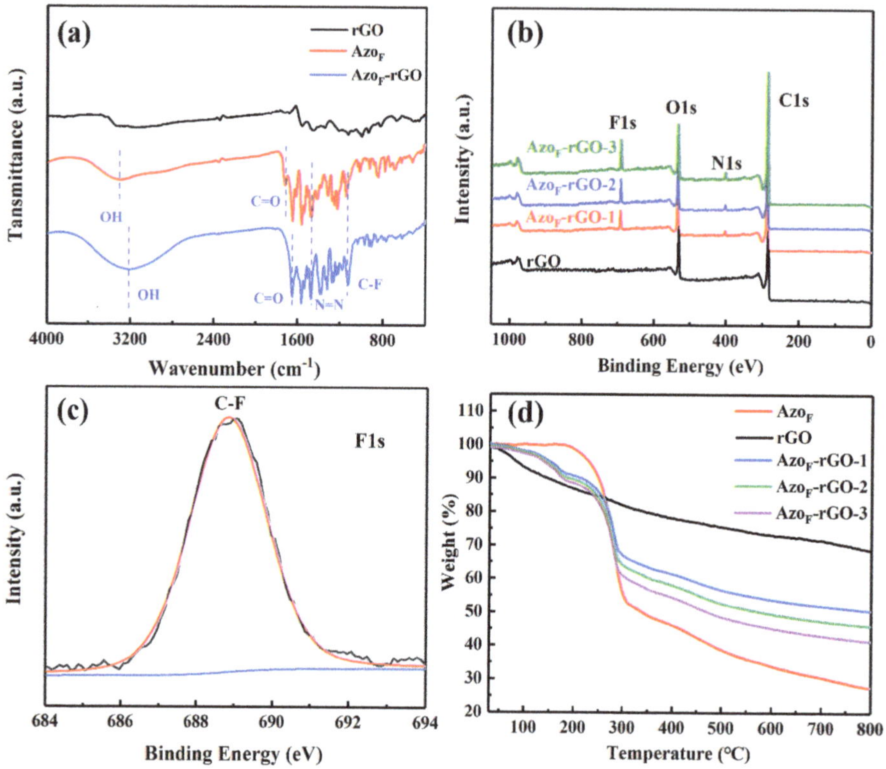

Figure 2. (**a**) FT-IR spectra of rGO, Azo_F and Azo_F-rGO. (**b**) XPS spectra of rGO and Azo_F-rGO. (**c**) F1s XPS spectra of Azo_F-rGO. (**d**) TGA spectra of rGO, Azo_F and Azo_F-rGO.

Table 1. A_d of Azo_F on rGO.

Reaction Times	TGA		XPS			
	Dg (%) [a]	A_d	Element Content (%)			A_d
			C	F	O	
Azo_F-rGO-1	43.41	1:40.1	77.42	4.13	15.71	1:40.2
Azo_F-rGO-2	52.95	1:27.3	74.13	5.09	17.39	1:27.7
Azo_F-rGO-3	65.73	1:16.0	71.07	6.64	17.90	1:16.1

[a] Dg is the average weight percentage of Azo_F in Azo_F-rGO at 600 °C, 700 °C and 800 °C.

3.2. Cycling Stability and Storage Performance

The optical properties performance of Azo_F and Azo_F-rGO was investigated through time-evolved absorption spectra. It can be seen from Figure 3 that Azo_F-rGO went through a *trans* → *cis* isomerization process under 365 nm ultraviolet light irradiation. Compared with Azo_F ($\tau_{1/2}$: 195.2 min), Azo_F-rGO ($\tau_{1/2}$: 87.7 h) takes more time to complete the isomerization process from *cis*-isomer to *trans*-isomer, indicating that Azo_F-rGO has better thermal stability than pristine Azo_F. The same conclusion can be drawn from the fact that the first-order reversion rate constant (K_{rev}) of Azo_F-rGO ($3.29 \times 10^{-6} \cdot s^{-1}$) was much smaller than that of Azo_F ($1.20 \times 10^{-4} \cdot s^{-1}$) under dark conditions derived from Equation (2) [21].

$$\ln\left(\frac{A_t - A_\infty}{A_0 - A_\infty}\right) = -k_{rev}t \quad (2)$$

where A_0 is the absorption intensity of Azo_F-rGO and Azo_F at metastable state (*cis*-rich) irradiated by UV light, A_t is the absorbance of Azo_F-rGO and Azo_F reversing for "t" time and A_∞ is the absorption intensity of Azo_F-rGO and Azo_F after complete *cis*-to-*trans* reversion. Moreover, compared to pristine Azo_F (Figure S5), Azo_F-rGO exhibited a lower isomerization degree owing to the intermolecular H-bonds and steric hindrance owing to high attachment density, resulting in a better storage performance of this material. Furthermore, the ΔEa value of the *cis*-isomer of Azo_F-rGO (1.05 eV) was higher than that of Azo_F (0.94 eV) according to Equation (3) [42], which again proves the formation of intermolecular hydrogen bonds [43].

$$E_a = -\mathrm{RT}\ln\frac{h\ln 2}{\tau_{1/2}k_B T} \quad (3)$$

where E_a is the activation barrier for *cis*-to-*trans* isomerization process, T represents the temperature and $\tau_{1/2}$ represents the half-life. k_B, R and h are the Boltzman, universal gas and Plank constants. Additionally, the optical band gap of Azo_F-rGO complex was estimated to be ~1.8 eV based on the Tauc formula (Figure S6) [44]. The increase in the stability of the *cis*-isomer means extension of the life cycle of Azo_F-rGO, which is directly related to the large-scale promotion of photoactive chemical heat storage materials.

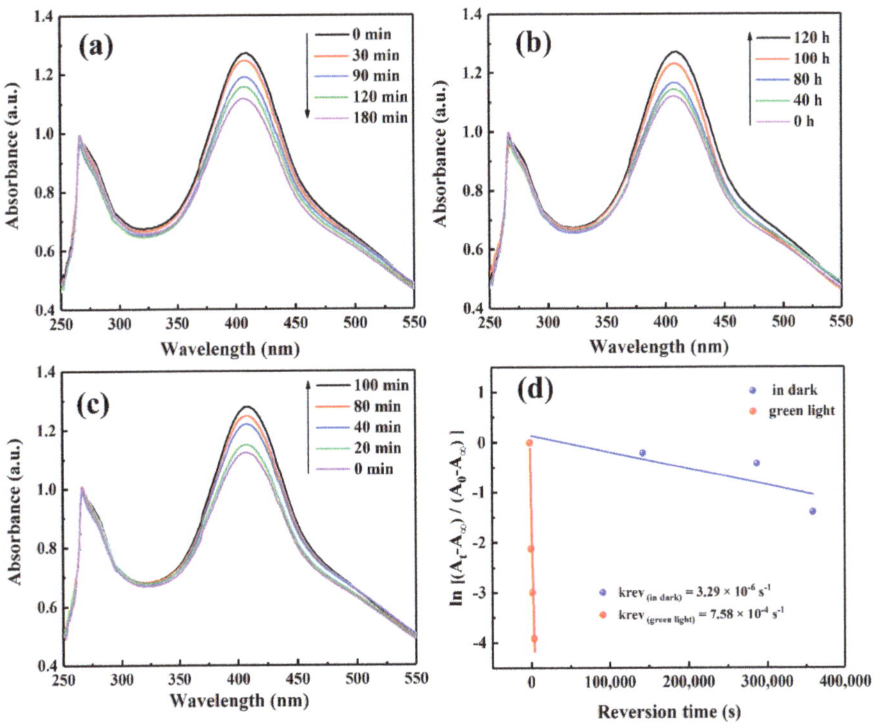

Figure 3. UV–Vis absorption spectra of Azo_F-rGO-3 (**a**) under UV irradiation, (**b**) in dark conditions, (**c**) under visible light irradiation, (**d**) reversion rates curves of Azo_F-rGO in different environments.

Similar to the length of the life cycle, whether the controllable heat release under external stimuli can be achieved is critical to the future application value of Azo_F-rGO. Figure 3c showed that compared with dark conditions, the irradiation of green light (540 nm) significantly accelerated the recovery process of Azo_F-rGO from *cis* -isomer to *trans*-isomer. Compared with dark conditions, the result that K_{rev} ($7.58 \times 10^{-4} \cdot s^{-1}$) was significantly larger under green light irradiation also confirmed the conclusion of faster reversion. The same effect can also be achieved by absorbing heat from the external environment according to DSC. The reason for this phenomenon is that the cis-isomer can absorb energy from external stimuli to overcome the energy barrier of configuration reversion isomerization, thereby achieving the purpose of accelerating energy output [45,46]. The above results show that Azo_F-rGO has successfully possessed the controllable heat output capability.

The stability of repeated *cis* ↔ *trans* configuration transformations of Azo_F-rGO and Azo_F has also been studied. It can be seen from Figure 4 that both Azo_F-rGO and Azo_F have no significant decrease in the absorption intensity at 407 nm after repeated irradiation of ultraviolet light (365 nm) and visible light (540 nm) for 50 times, which shows that they have outstanding isomerization stability. The Azo_F-rGO can not only be stored for a long time under the premise of ensuring the storage effect, but also can control the output of the stored energy, which is essential for actual photothermal conversion.

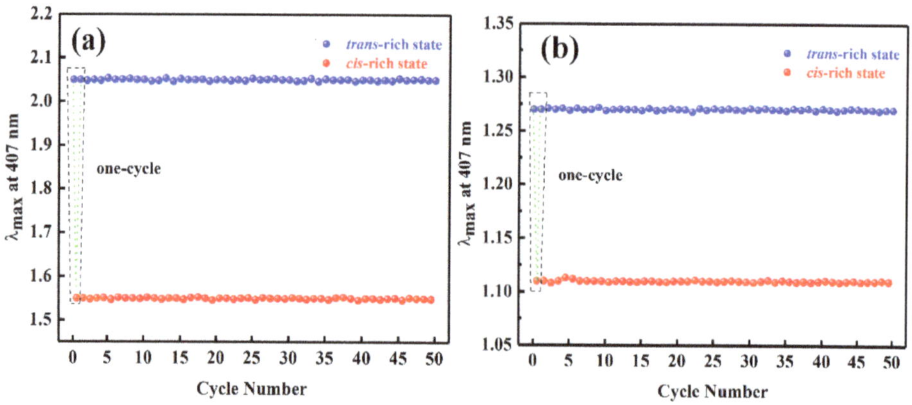

Figure 4. Stability performance of (**a**) Azo_F and (**b**) Azo_F–rGO-3 for 50 times.

The photothermal storage capacity of Azo_F and Azo_F-rGO was investigated through DSC [7]. All objects were stable between 10–140 °C based on TGA. Azo_F and Azo_F-rGO released significant heat under the first round of heating stimulation, but no heat was released during the second round according to Figure 5. The above results prove that the research subjects have released all the energy stored through the configuration transformation in the form of heat. Furthermore, most photothermal storage materials start to release the stored energy after 100 °C, while this kind of heat storage material can start energy output at 35 °C, which is a milestone in achieving fast energy output at lower temperatures [7].

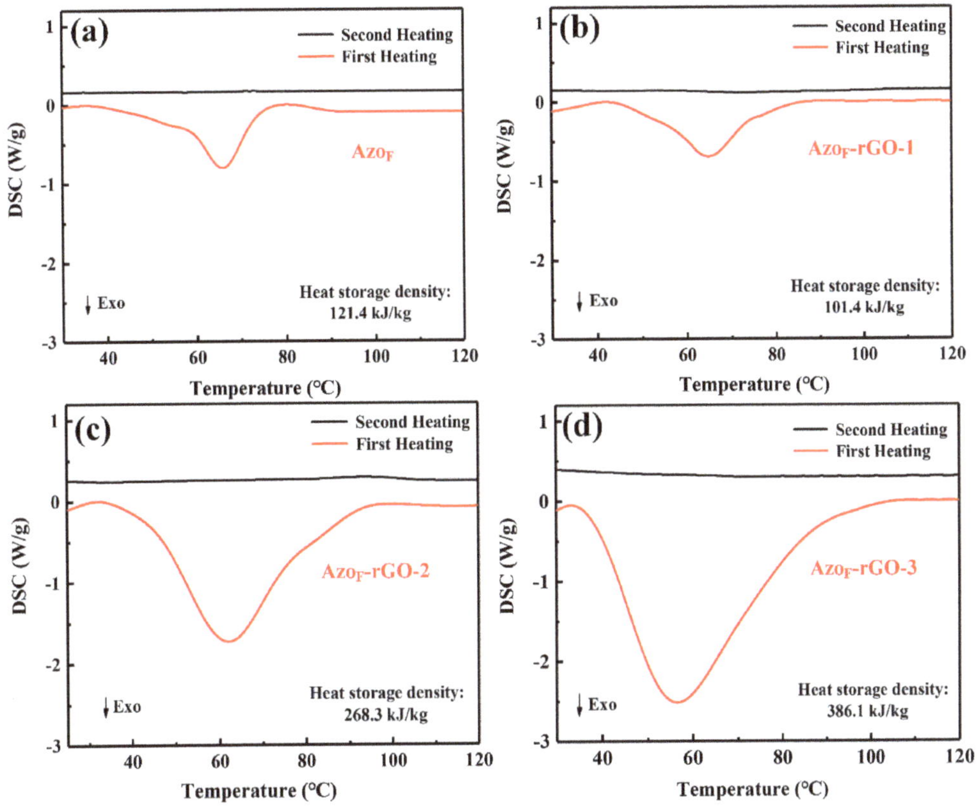

Figure 5. DSC (differential scanning calorimetry) traces of (**a**) Azo$_F$ and (**b–d**) Azo$_F$-rGO after 1, 2 and 3-times reaction.

It can be seen from Figure 5 that the heat storage density of Azo$_F$-rGO-3 has reached to 386.1 kJ kg^{-1}, which shows a significant increase over Azo$_F$ (121.4 kJ kg^{-1}). This is because of the close-packed orderly distribute of Azo$_F$ on rGO as a result of high attachment density, which strengthens the intermolecular interaction [23]. In addition, high attachment density also enhances the steric hindrance and promotes the formation of H-bonds, which further increases the photothermal storage capacity [47]. The reason for Azo$_F$-rGO-1 showing less effectiveness compared to the Azo$_F$ is the low attachment density, which leads to weak intermolecular interaction and therefore relatively low energy density. Moreover, the heat storage density of Azo$_F$-rGO-3 was also higher than Azo$_F$-rGO-1 and Azo$_F$-rGO-2, which shows that the attachment density was positively correlated with great storage performance.

Similar to heat storage density, power density is also a key element to measure the possibility of practical application of Azo$_F$-rGO. It can be seen from Figure 6 that the power density of Azo$_F$-rGO-3 was 890.6 W kg^{-1}, which shows a huge improvement compared to Azo$_F$ (448.6 W kg^{-1}). Furthermore, the power density of Azo$_F$-rGO-3 was also higher than Azo$_F$-rGO-1 and Azo$_F$-rGO-2, which shows that the attachment density is directly related to the heat output performance. It is worth noting that high power density means fast output of energy, which further increases the feasibility of practical application of Azo$_F$-rGO. As shown in Table 2, the performance of Azo$_F$-rGO in many aspects has been greatly improved compared to other similar materials [7,15,48,49]. The above results demonstrate that Azo$_F$-rGO, which not only exhibits remarkable photothermal capacity but also equipped with low temperature energy output capability, has shown great development potential in achieving the goal of efficient photothermal storage.

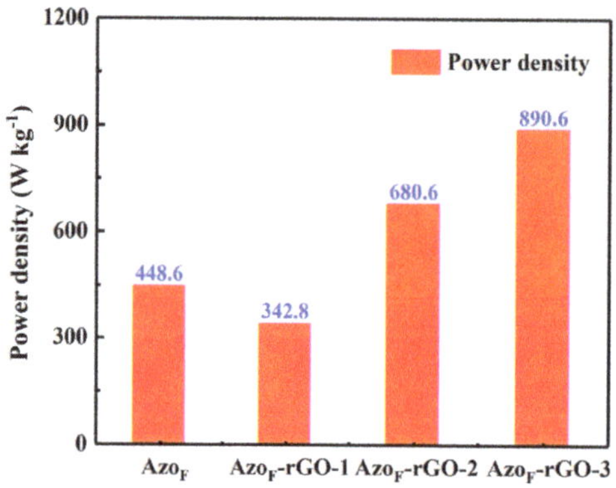

Figure 6. Power density of Azo$_F$ and Azo$_F$-rGO after 1, 2 and 3-times reaction.

Table 2. Performance of different photothermal conversion materials.

Photothermal Conversion Material	Energy Density (kJ mol^{-1})	Power Density (W mol^{-1})	Half-Life (h)	Ref.
Azo-diacetylene polymer	176.2	1289.5	27.8	[48]
Azo-SWCNT complex	92.0	457.1	0.5	[7]
Azo-PCM complex	79.3	–	–	[15]
Azo-alkyl polymer	89.0	148.6	55	[49]
Azo$_F$-rGO-3 complex	367.7	848.6	87.7	This paper

4. Conclusions

In summary, Azo$_F$-rGO with good photothermal storage performance, outstanding storage lifespan and low-temperature energy output capability has been proven to be a great photothermal conversion material. The formation of hydrogen bonds and the enhancement of intermolecular interactions owing to the high attachment density has simultaneously achieved the improvement of the heat storage density (max. 386.1 kJ kg^{-1}), power density (max. 890.6 W kg^{-1}) and half-life (up to 87.7 h) of Azo$_F$-rGO for photothermal storage. Azo$_F$-rGO also exhibits exceptional cycling stability, which realizes long-term recyclability and efficient and pollution-free utilization of solar energy in a closed system. Furthermore, Azo$_F$-rGO can start energy output at 35 °C, which shows that the goal of low-temperature energy output has been achieved. The above results indicate that Azo$_F$-rGO, combining outstanding photothermal capacity with a long-life cycle as well as low-temperature energy output capability, is a prominent photothermal conversion material with great practical application value.

Supplementary Materials: The following are available online at https://www.mdpi.com/1996-1944/14/6/1434/s1, Figure S1: ^{1}H NMR, ^{13}C NMR and ^{19}F NMR spectra of Azo$_F$, Figure S2: XRD patterns of GO, rGO, Azo$_F$-rGO, Figure S3: C1s region in XPS spectra of (a) rGO, (b) Azo$_F$-rGO and (c) N1s region in XPS spectra, Figure S4: Raman spectra, Figure S5: Time-evolved absorption spectra of Azo$_F$ and Figure S6: UV-Vis absorption spectra of Azo$_F$-rGO powder at room temperature (25 °C).

Author Contributions: Conceptualization, X.Y. and S.L.; methodology, X.Y.; software, X.Y.; validation, X.W. and Y.W.; formal analysis, X.Y.; investigation, X.Y.; resources, J.Z. (Jin Zhang); data curation, X.Y. and S.L.; writing—original draft preparation, X.Y.; writing—review and editing, S.L.; visualization, J.Z. (Jianguo Zhao); supervision, J.Z. (Jin Zhang); project administration, J.Z. (Jianguo

Zhao); funding acquisition, J.Z. (Jin Zhang). All authors have read and agreed to the published version of the manuscript.

Funding: This research was funded by National Natural Science Foundation of China (Grant No. 52071192), Shanxi "1331 project" foundation for the construction of collaborative innovation center of graphene industrial application, Key Research and Development Project of Shanxi Province (Grant No. 201803D121122), Research and development projects in key areas of Guangdong Province (Grant No. 2020B0202010004), The Program for Scientific and Technological Innovation of Higher Education Institutions in Shanxi (Grant No. 2020L0477).

Institutional Review Board Statement: Not applicable.

Informed Consent Statement: Not applicable.

Data Availability Statement: The data presented in this study are available in [insert article or Supplementary Material].

Conflicts of Interest: The authors declare no conflict of interest.

References

1. Armaroli, N.; Balzani, V. The future of energy supply: Challenges and opportunities. *Angew. Chem. Int. Ed. Engl.* **2007**, *46*, 52–66. [CrossRef] [PubMed]
2. Zhang, X.; Peng, T.; Song, S. Recent advances in dye-sensitized semiconductor systems for photocatalytic hydrogen production. *J. Mater. Chem. A* **2016**, *4*, 2365–2402. [CrossRef]
3. Aguirre, M.E.; Zhou, R.; Eugene, A.J.; Guzman, M.I.; Grela, M.A. Cu2O/TiO2 heterostructures for CO2 reduction through a direct Z-scheme: Protecting Cu2O from photocorrosion. *Appl. Catal. B Environ.* **2017**, *217*, 485. [CrossRef]
4. Zhou, R.; Guzman, M.I. CO2 Reduction under Periodic Illumination of ZnS. *J. Phys. Chem. C* **2015**, *118*, 11649–11656. [CrossRef]
5. Moth-Poulsen, K.; Ćoso, D.; Börjesson, K.; Vinokurov, N.; Meier, S.K.; Majumdar, A.; Vollhardt, K.P.C.; Segalman, R.A. Molecular solar thermal (MOST) energy storage and release system. *Energy Environ. Sci.* **2012**, *5*, 8534–8537. [CrossRef]
6. Yoshida, Z.-I. New molecular energy storage systems. *J. Photochem.* **1985**, *29*, 27–40. [CrossRef]
7. Kucharski, T.J.; Ferralis, N.; Kolpak, A.M.; Zheng, J.O.; Nocera, D.G.; Grossman, J.C. Templated assembly of photoswitches significantly increases the energy-storage capacity of solar thermal fuels. *Nature Chem.* **2014**, *6*, 441–447. [CrossRef] [PubMed]
8. Kanai, Y.; Srinivasan, V.; Meier, S.K.; Vollhardt, K.P.; Grossman, J.C. Mechanism of thermal reversal of the (fulvalene)tetracarbonyldiruthenium photoisomerization: Toward molecular solar-thermal energy storage. *Angew. Chem. Int. Ed. Engl.* **2010**, *49*, 8926–8929. [CrossRef]
9. Berthiller, F.; Dall'asta, C.; Corradini, R.; Marchelli, R.; Sulyok, M.; Krska, R.; Adam, G.; Schuhmacher, R. Occurrence of deoxynivalenol and its 3-beta-D-glucoside in wheat and maize. *Food Addit. Contam. Part A Chem. Anal. Control. Expo. Risk Assess* **2009**, *26*, 507–511. [CrossRef]
10. Lee, K.M.; Smith, M.L.; Koerner, H.; Tabiryan, N.; Vaia, R.A.; Bunning, T.J.; White, T.J. Photodriven, Flexural-Torsional Oscillation of Glassy Azobenzene Liquid Crystal Polymer Networks. *Adv. Funct. Mater.* **2011**, *21*, 2913–2918. [CrossRef]
11. Liu, J.; Bu, W.; Pan, L.; Shi, J. NIR-triggered anticancer drug delivery by upconverting nanoparticles with integrated azobenzene-modified mesoporous silica. *Angew. Chem. Int. Ed. Engl.* **2013**, *52*, 4375–4379. [CrossRef] [PubMed]
12. Sun, L.; Gao, F.; Shen, D.; Liu, Z.; Yao, Y.; Lin, S. Rationally designed hyperbranched azopolymer with temperature, photo and pH responsive behavior. *Polym. Chem.* **2018**, *9*, 2977–2983. [CrossRef]
13. Wang, G.; Yuan, D.; Yuan, T.; Dong, J.; Feng, N.; Han, G. A visible light responsive azobenzene-functionalized polymer: Synthesis, self-assembly, and photoresponsive properties. *J. Polym. Sci. Part A Polym. Chem.* **2015**, *53*, 2768–2775. [CrossRef]
14. Iii, J.O.; Lawrence, J.; Yee, G.G. Photochemical storage potential of azobenzenes. *Sol. Energy* **1983**, *30*, 271–274.
15. Han, G.G.D.; Li, H.; Grossman, J.C. Optically-controlled long-term storage and release of thermal energy in phase-change materials. *Nat. Commun.* **2017**, *8*, 1446. [CrossRef]
16. Joshi, D.K.; Mitchell, M.J.; Bruce, D.; Lough, A.J.; Yan, H. Synthesis of cyclic azobenzene analogues. *Tetrahedron* **2012**, *68*, 8670–8676. [CrossRef]
17. Weston, C.E.; Richardson, R.D.; Haycock, P.R.; White, A.J.; Fuchter, M.J. Arylazopyrazoles: Azoheteroarene photoswitches offering quantitative isomerization and long thermal half-lives. *J. Am. Chem. Soc.* **2014**, *136*, 11878–11881. [CrossRef]
18. Cho, E.N.; Zhitomirsky, D.; Han, G.G.; Liu, Y.; Grossman, J.C. Molecularly Engineered Azobenzene Derivatives for High Energy Density Solid-State Solar Thermal Fuels. *ACS Appl. Mater. Interfaces* **2017**, *9*, 8679–8687. [CrossRef]
19. Bleger, D.; Schwarz, J.; Brouwer, A.M.; Hecht, S. o-Fluoroazobenzenes as readily synthesized photoswitches offering nearly quantitative two-way isomerization with visible light. *J. Am. Chem. Soc.* **2012**, *134*, 20597–20600. [CrossRef]
20. Lennartson, A.; Roffey, A.; Moth-Poulsen, K. Designing photoswitches for molecular solar thermal energy storage. *Tetrahedron Lett.* **2015**, *56*, 1457–1465. [CrossRef]
21. Feng, Y.; Liu, H.; Luo, W.; Liu, E.; Zhao, N.; Yoshino, K.; Feng, W. Covalent functionalization of graphene by azobenzene with molecular hydrogen bonds for long-term solar thermal storage. *Sci. Rep.* **2013**, *3*, 3260. [CrossRef]

22. Kurihara, S.; Nomiyama, S.; Nonaka, T. Photochemical control of the macrostructure of cholesteric liquid crystals by means of photoisomerization of chiral azobenzene molecules. *Chem. Mater.* **2001**, *13*, 1992–1997. [CrossRef]
23. Kolpak, A.M.; Grossman, J.C. Hybrid chromophore/template nanostructures: A customizable platform material for solar energy storage and conversion. *J. Chem. Phys.* **2013**, *138*, 034303. [CrossRef]
24. Wu, J.; Yang, Y.; Qu, Y.; Jia, L.; Zhang, Y.; Xu, X.; Chu, S.T.; Little, B.E.; Morandotti, R.; Jia, B.; et al. 2D Layered Graphene Oxide Films Integrated with Micro-Ring Resonators for Enhanced Nonlinear Optics. *Small* **2020**, *16*, 1906563. [CrossRef]
25. Wu, J.; Yang, Y.; Qu, Y.; Xu, X.; Liang, Y.; Chu, S.T.; Little, B.E.; Morandotti, R.; Jia, B.; Moss, D.J. Graphene Oxide Waveguide and Micro-Ring Resonator Polarizers. *Laser Photonics Rev.* **2019**, *13*, 1900056. [CrossRef]
26. Wu, S.; Butt, H.-J. Solar-Thermal Energy Conversion and Storage Using Photoresponsive Azobenzene-Containing Polymers. *Macromol. Rapid Commun.* **2020**, *41*, 1900413. [CrossRef] [PubMed]
27. Zhang, X.; Feng, Y.; Huang, D.; Li, Y.; Feng, W. Investigation of optical modulated conductance effects based on a graphene oxide–azobenzene hybrid. *Carbon* **2010**, *48*, 3236–3241. [CrossRef]
28. Wang, Z.; Li, Z.-x.; Liu, Z. Photostimulated reversible attachment of gold nanoparticles on multiwalled carbon nanotubes. *J. Phys. Chem. C* **2009**, *113*, 3899–3902. [CrossRef]
29. Hummers, W.S.; Offeman, R.E. Preparation of Graphitic Oxide. *J. Am. Chem. Soc.* **1958**, *80*, 1339. [CrossRef]
30. An, X.; Simmons, T.J.; Shah, R.; Wolfe, C.S.; Lewis, K.M.; Washington, M.; Nayak, S.K.; Talapatra, S.; Kar, S. Stable Aqueous Dispersions of Noncovalently Functionalized Graphene from Graphite and their Multifunctional High-Performance Applications. *Nano Lett.* **2010**, *10*, 4295–4301. [CrossRef]
31. Englert, J.M.; Dotzer, C.; Yang, G.; Schmid, M.; Papp, C.; Gottfried, J.M.; Steinruck, H.; Spiecker, E.; Hauke, F.; Hirsch, A. Covalent bulk functionalization of graphene. *Nat. Chem.* **2011**, *3*, 279–286. [CrossRef] [PubMed]
32. Marcano, D.C.; Kosynkin, D.V.; Berlin, J.M.; Sinitskii, A.; Sun, Z.; Slesarev, A.; Alemany, L.B.; Lu, W.; Tour, J.M. Improved Synthesis of Graphene Oxide. *ACS Nano* **2010**, *4*, 4806–4814. [CrossRef] [PubMed]
33. Lim, D.J.; Marks, N.A.; Rowles, M.R. Universal Scherrer equation for graphene fragments. *Carbon* **2020**, *162*, 475–480. [CrossRef]
34. Li, D.; Muller, M.B.; Gilje, S.; Kaner, R.B.; Wallace, G.G. Processable aqueous dispersions of graphene nanosheets. *Nat. Nanotechnol.* **2008**, *3*, 101–105. [CrossRef]
35. Min, M.; Bang, G.S.; Lee, H.; Yu, B. A photoswitchable methylene-spaced fluorinated aryl azobenzene monolayer grafted on silicon. *Chem. Commun.* **2010**, *46*, 5232–5234. [CrossRef]
36. Gearba, R.I.; Lehmann, M.; Levin, J.; Ivanov, D.A.; Koch, M.H.J.; Barbera, J.; Debije, M.G.; Piris, J.; Geerts, Y. Tailoring discotic mesophases: Columnar order enforced with hydrogen bonds. *Adv. Mater.* **2003**, *15*, 1614–1618. [CrossRef]
37. Zhang, X.; Feng, Y.; Lv, P.; Shen, Y.; Feng, W. Enhanced reversible photoswitching of azobenzene-functionalized graphene oxide hybrids. *Langmuir* **2010**, *26*, 18508–18511. [CrossRef]
38. Zhang, B.; Zhang, Y.; Peng, C.; Yu, M.; Li, L.; Deng, B.; Hu, P.; Fan, C.; Li, J.; Huang, Q. Preparation of polymer decorated graphene oxide by γ-ray induced graft polymerization. *Nanoscale* **2012**, *4*, 1742–1748. [CrossRef]
39. Yu, D.S.; Kuila, T.; Kim, N.H.; Khanra, P.; Lee, J.H. Effects of covalent surface modifications on the electrical and electrochemical properties of graphene using sodium 4-aminoazobenzene-4′-sulfonate. *Carbon* **2013**, *54*, 310–322. [CrossRef]
40. Kim, M.; Safron, N.S.; Huang, C.; Arnold, M.S.; Gopalan, P. Light-driven reversible modulation of doping in graphene. *Nano Lett.* **2012**, *12*, 182–187. [CrossRef] [PubMed]
41. Bandara, H.M.D.; Burdette, S.C. Photoisomerization in different classes of azobenzene. *Chem. Soc. Rev.* **2012**, *41*, 1809–1825. [CrossRef]
42. Samanta, S.; Beharry, A.A.; Sadovski, O.; Mccormick, T.M.; Babalhavaeji, A.; Tropepe, V.; Woolley, G.A. Photoswitching azo compounds in vivo with red light. *J. Am. Chem. Soc.* **2013**, *135*, 9777–9784. [CrossRef] [PubMed]
43. Kolpak, A.M.; Grossman, J.C. Azobenzene-Functionalized Carbon Nanotubes As High-Energy Density Solar Thermal Fuels. *Nano Lett.* **2011**, *11*, 3156–3162. [CrossRef] [PubMed]
44. Zhang, G.; Amani, M.; Chaturvedi, A.; Tan, C.; Bullock, J.; Song, X.; Kim, H.; Lien, D.H.; Scott, M.C.; Zhang, H. Optical and electrical properties of two-dimensional palladium diselenide. *Appl. Phys. Lett.* **2019**, *114*, 253102. [CrossRef]
45. Yang, Y.; Hughes, R.P.; Aprahamian, I. Visible Light Switching of a BF2-Coordinated Azo Compound. *J. Am. Chem. Soc.* **2012**, *134*, 15221–15224. [CrossRef]
46. Siewertsen, R.; Neumann, H.; Buchheimstehn, B.; Herges, R.; Nather, C.; Renth, F.; Temps, F. Highly Efficient Reversible Z−E Photoisomerization of a Bridged Azobenzene with Visible Light through Resolved S1(nπ*) Absorption Bands. *J. Am. Chem. Soc.* **2009**, *131*, 15594–15595. [CrossRef]
47. Liu, Y.; Grossman, J.C. Accelerating the Design of Solar Thermal Fuel Materials through High Throughput Simulations. *Nano Lett.* **2014**, *14*, 7046–7050. [CrossRef] [PubMed]
48. Han, G.D.; Park, S.S.; Liu, Y.; Zhitomirsky, D.; Cho, E.; Dincă, M.; Grossman, J.C. Photon energy storage materials with high energy densities based on diacetylene–azobenzene derivatives. *J. Mater. Chem. A* **2016**, *4*, 16157–16165. [CrossRef]
49. Zhitomirsky, D.; Cho, E.; Grossman, J.C. Solid-State Solar Thermal Fuels for Heat Release Applications. *Adv. Energy Mater.* **2016**, *6*, 1502006. [CrossRef]

materials

MDPI

Article

A Novel Method for the Detection of SARS-CoV-2 Based on Graphene-Impedimetric Immunosensor

Gabriel C. Zaccariotto , Martin K. L. Silva , Giovanna S. Rocha and Ivana Cesarino *

Department of Bioprocesses and Biotechnology, School of Agriculture, São Paulo State University (UNESP), Botucatu 18610-034, Brazil; g.zaccariotto@unesp.br (G.C.Z.); martin.leme@unesp.br (M.K.L.S.); giovanna.s.rocha@unesp.br (G.S.R.)

* Correspondence: ivana.cesarino@unesp.br; Tel.: +55-14-3880-7404

Abstract: Due to the SARS-CoV-2 pandemic, there has been an increase in the search for affordable healthcare devices for mass testing and rapid diagnosis. In this context, this work described a new methodology for SARS-CoV-2 detection based on an impedimetric immunosensor developed using the advantageous immobilization of antibodies in the reduced graphene oxide (rGO). The rGO was obtained by chemical synthesis from the commercial graphene oxide (GO), and the materials were morphologically, electrochemically and visually characterized. The cyclic voltammetry (CV) and electrochemical impedance spectroscopy (EIS) techniques were used to evaluate the fabrication steps of the immunosensor. The electrochemical immunoassay was considered for SARS-CoV-2 spike protein RBD detection using a impedimetric immunosensor and redox couple ($[(Fe(CN)_6)]^{3-/4-}$) as a probe. The immunosensor was effectively developed and applied in the detection of SARS-CoV-2 spike protein RBD in saliva samples.

Keywords: electrochemical sensor; reduced graphene oxide; COVID-19; SARS-CoV-2

1. Introduction

The development of rapid tests, with significant reliability, easy applicability and low cost is essential for the context of the SARS-CoV-2 pandemic [1,2]. In Brazil, specifically, the development of these devices based on the national sample of patients appears as an essential and necessary opportunity for the development of national technology with low cost and accessibility, in order to minimize the bringing down effect on the health system. Hence, developing novel electrochemical biosensors based on the detection of antigen-antibody interactions or membrane proteins of SARS-CoV-2 appears as an adequate and accessible alternative of test devices to those based on reverse transcription-polymerase chain reaction (RT-PCR) to supply the world's current demand [3–6].

Yakoh et al. [7] developed a paper-based electrochemical platform as a screening tool to detect SARS-CoV-2 immunoglobulins. The electrochemical sensor reached a limit of detection (LOD) of 1 ng/mL for SARS-CoV-2 antibodies; however, the detection limit of the protein antigen of SARS-CoV-2 has not yet achieved the detection level in real nasal swab specimens. Zhao et al. [8] reported an electrochemical detection technology using calixarene functionalized graphene oxide for targeting RNA of SARS-CoV-2. The super sandwich-type electrochemical sensor presented an LOD of the clinical specimen of 200 copies/mL. Raziq et al. [9] developed a MIP-based electrochemical sensor for detection of SARS-CoV-2 nucleoprotein. The resulting nucleoprotein sensor showed a detection and quantification limit of 15 fM and 50 fM, respectively. Torres et al. [10] prepared a low-cost biosensor for rapid detection of SARS-CoV-2 modified with human receptor angiotensin-converting enzyme-2. According to the authors, the miniaturized biosensor can detect SARS-CoV-2 using 10 μL of sample within 4 min. Other works [11] highlight electrochemical sensors as important tools in the analysis of COVID-19 summarizing the current state-of-the-art approaches to viral electrochemical biosensors, but these technologies have not yet been developed or are under the development phase.

Different studies focusing on graphene-based electrochemical biosensors have been developed in the past decade [12,13]. Taniselass et al. [14] conducted a review highlighting the development of graphene-biosensing devices for monitoring noncommunicable disease biomarkers. The graphene research for the effective immobilization of enzymes and the accurate detection of biomolecules was discussed in another review work [15]. Our research group used the reduced graphene oxide (rGO) to immobilize enzymes for the preparation of enzymatic biosensors to monitor glucose during the second-generation ethanol production [16], to analyze neurotransmitters in urine and plasmatic serum samples [17] and to determine pesticides in food [18]. rGO was also employed as a platform in the development of an electrochemical biosensor for detection of staphylococcal enterotoxin A in milk samples [19].

In this work, the advantageous immobilization of antibodies in rGO coupled with the sensitivity of the faradic impedimetric immunosensor model was used to develop a SARS-CoV-2 antigen diagnostic device. Electrochemical characterization by cyclic voltammetry (CV) and electrochemical impedance spectroscopy (EIS) techniques evaluated the fabrication steps of the immunosensor. The antigen-antibody binding on glassy carbon (GC)/rGO platform was successfully detected by EIS and CV contributing to advances on the SARS-CoV-2 electrochemical biosensing field.

2. Materials and Methods

2.1. Reagents and Solutions

Anti-SARS-CoV-2 Spike Glycoprotein S1 antibody (monoclonal) and Recombinant human coronavirus SARS-CoV-2 Spike Glycoprotein RBD (Active) were purchased from Abcam PLC (Cambridge, UK). Graphene Oxide and Bovine Serum Albumin (BSA—lyophilized powder) were purchased from Sigma-Aldrich (São Paulo, Brazil). Lauryl sulfate sodium salt (SDS), Sodium Tetrahydridoborate ($NaBH_4$), ethyl alcohol, Monopotassium phosphate (KH_2PO_4), Disodium hydrogen phosphate (Na_2HPO_4), Potassium Hexacyanoferrate (II) and (III) ($K_4[Fe(CN)_6)]$ and $K_3[Fe(CN)_6)]$) were analytical grade. Solutions and dilution steps were carried out by using ultra-pure water (resistivity $\geq$ 18 MΩ cm^{-1}) of a PURELAB Option-Q—ELGA–VEOLIA system (São Paulo, Brazil).

2.2. Production of Reduced Graphene Oxide

Reduced Graphene Oxide (rGO) was produced through a chemical reduction method. In a reaction flask, 20 mg of graphene oxide (5 mL of stock solution) was mixed with 15 mL of ethyl alcohol and 16.0 mg of SDS; then, the mixture was subjected to a sonication step for 30 min (75% amplitude). An amount of 8.0 mg of $NaBH_4$ added into the reaction promoted the reduction of GO functionalities. The mixture was sonicated for a further 30 min. In order to eliminate any residuals reagents and clean the nanomaterial, the rGO was centrifuged three times with ethanol pure grade. After the cleaning step, the rGO was dried (60 °C overnight) and subsequently dispersed in ultra-pure water at 1.0 mg/mL prior to the immunosensor confection.

2.3. SARS-CoV-2 Immunosensor Fabrication

Before the surface modification, the glassy carbon (GC) electrode was polished with alumina slurries (Al_2O_3) and cleaned in an ultrasonic bath with ethyl alcohol for 5 min, followed by ultra-pure water for a further 5 min. Next, an aliquot 10 µL of rGO (25 µg/mL) was pipetted on the surface of the GC electrode, dried at 50 °C and the electrode was incubated on 1 mL of EDC-NHS (5 mM and 8 mM, respectively) for 1 h at room temperature. After, 10 µL of Anti-SARS-CoV-2 Spike Glycoprotein S1 antibody solution (2.5 µg/mL) was added on the surface of the GC/rGO/EDC-NHS electrode and incubated for 1 h, followed by a blocking step with BSA (1%) for 30 min. The electrode surface was softly rinsed with a phosphate buffer solution (PBS) (0.2 mol L^{-1}, pH 7.4) three times during 10 s after each incubation time. Finally, the electrode was ready to the measurements of the spike protein RBD (antigen) solution.

2.4. Scanning Electron Microscopy

The morphology of GO and rGO were both characterized using a scanning electron microscopy (FEG–SEM) using a model Quanta 200 (FEI Company, Hillsboro, OR, USA) localized in the Electron Microscope Center of the Institute of Biosciences of Botucatu, UNESP (CME-IBB-UNESP).

2.5. Electrochemical Measurements

Electrochemical experiments took place in a PGSTAT-128N Autolab (Metrohm) potentiostat equipped with NOVA2.1.4 software, and the electrodes were set as follow: a glassy carbon (GC) as a working electrode, a platinum plate as an auxiliary electrode and Ag/AgCl (3.0 mol L^{-1}) as the reference electrode. Cyclic Voltammetry (CV) and Electrochemical Impedance (EIS) were carried out in a 0.2 mol L^{-1} PBS (pH 7.4) solution having 0.1 mol L^{-1} of KCl and 5.0 mmol L^{-1} of the redox couple $Fe(CN)_6]^{3-/4-}$. CV scans were recorded in the potential range of −0.5 to 1.0 V vs. Ag/AgCl, with a scan rate of 50 mV s^{-1}. An open circuit potential (OCP) setup was employed for EIS measurements. The experimental conditions of EIS were: 10 points per decade, frequency range of 10^7 and 10^{-2} Hz and amplitude of 10 mV. Equivalent circuit and fitting results were applied and obtained using the Electrochemical Circle Fit tool. The charge transfer resistance (R_{ct}) and other components of the adjusted electrical circuit obtained during the analysis were used to obtain the quantitative signal of the RBD peak protein concentration in the assay.

2.6. Analysis of Spike Protein RBD in Saliva Samples

The saliva analysis was conducted by diluting the sputum-collected sample with PBS 7.4 (1:1) prior to incubation. Sputum samples from the oral cavity were collected on an empty stomach and before morning oral hygiene to avoid contamination by toothpaste and remnants of food or coffee. No complementary extraction or purification procedures were employed. In order to evaluate the immunoassay response, electrodes were incubated with only PBS, sample and PBS (1:1) and sample and PBS (1:1) contaminated with 2.5 μg/mL of spike protein RBD. Then, the immunosensor was rinsed carefully with PBS and the electrochemical measurements were recorded.

3. Results and Discussion

3.1. Morphological and Electrochemical Characterization of the rGO and the SARS-CoV-2 Immunosensor

Before preparing the immunosensor, the rGO was morphologically, electrochemically and visually characterized, as shown in Figure 1. Different colors can be observed for the GO and rGO suspensions. The characteristic color of the GO suspension is yellowish, while after chemical reduction, the GO presents a darker color. This is a way to visually characterize the structural changes of the graphene [20]. In the microscopic analysis, it can be seen that the GO material consists of a mixture of single layers and multilayer graphene sheets and the rGO displays a wrinkled structure with plenty of defects and corrugations. The electrochemical characterization performed by CV shows the voltammetric profiles with well-defined oxidation and reduction peaks for the GO and rGO modified the GC electrodes. This behavior is due to the Fe^{3+}/Fe^{2+} redox couple. The GC/rGO electrode showed a 1.6-fold increase in the peak current compared to the GC/GO electrode. This increase is due to the presence of defects introduced in its structure, as wellas fewer oxygen atoms increasing the electron transport [21]. In accordance with the CV experiments, the study by EIS showed a lower value of R_{ct} to rGO, indicating the improvement in the electronic transfer of this material. Therefore, the rGO obtained by the chemical method was successfully characterized showing that GO was reduced.

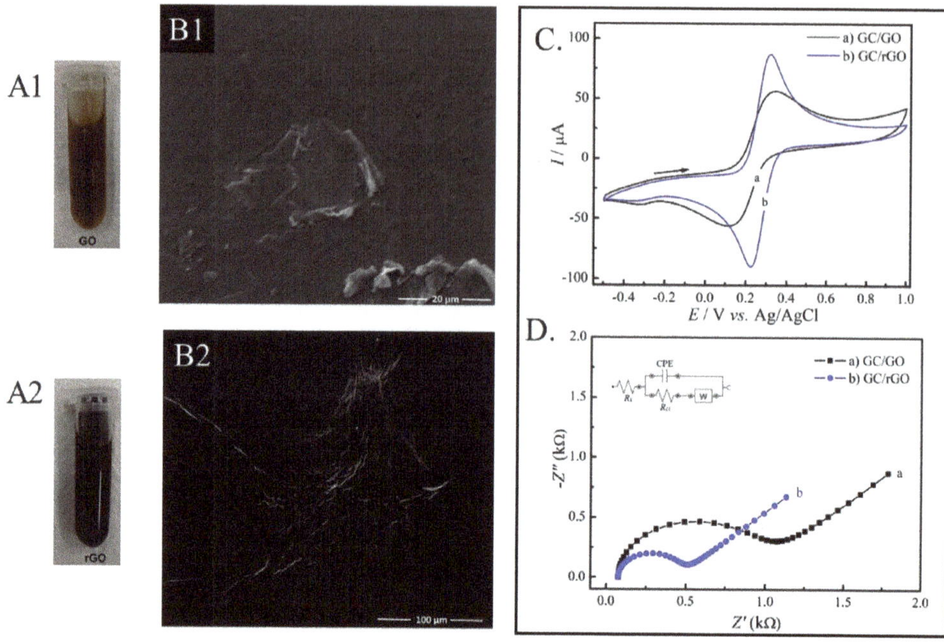

Figure 1. Visual characterization of GO (**A1**) and rGO (**A2**) suspensions. SEM images of GO (**B1**) and rGO (**B2**). Electrochemical characterization of electrodes modified with GO and rGO materials using CV (**C**) and EIS (**D**).

EIS and CV experiments were also used to monitor the single steps of the SARS-CoV-2 immunosensor assembly process as presented in Figure 2. The cyclic voltammograms obtained for the GC and GC/rGO electrodes showed well-defined oxidation and reduction peaks due to the Fe^{3+}/Fe^{2+} redox couple. Hence, an average increase of 11.5% in the peak currents were observed when comparing the GC/rGO electrode (curve b) with the bare GC electrode (curve a). This increase can be attributed to the high electron transfer properties of chemically reduced graphene. As the immunosensor fabrication was carried on, with the steps of antibody immobilization (curve c), followed by BSA surface blocking (curve d) and then the SARS-CoV-2 spike protein RBD (antigen) incubation step (curve e), a decrease in the anodic and cathodic peak currents of the redox couple was observed. This occurs because the biomolecules act as an obstacle to the electron transfer at the electrode-solution interface. These results indicate that the SARS-CoV-2 antibody and antigen are bonded to the electrode surface. Moreover, ΔE_p of 237 mV for the GC/rGO-EDC-NHS/Ab electrode and 358 mV after the SARS-CoV-2 antigen is incubated are observed. This increase in the ΔE_p was also observed in the antigen-antibody binding procedures of different types of electrochemical immunosensors [19,22]. In addition, Figure 2C showed that the decrease of currents generated by the $[Fe(CN)_6]^{4-/3-}$ system observed in the cyclic voltammograms after SARS-CoV-2 antigen binding at the GC/rGO-EDC-NHS/Ab/BSA electrode has a clear correlation with the concentration of the antigen.

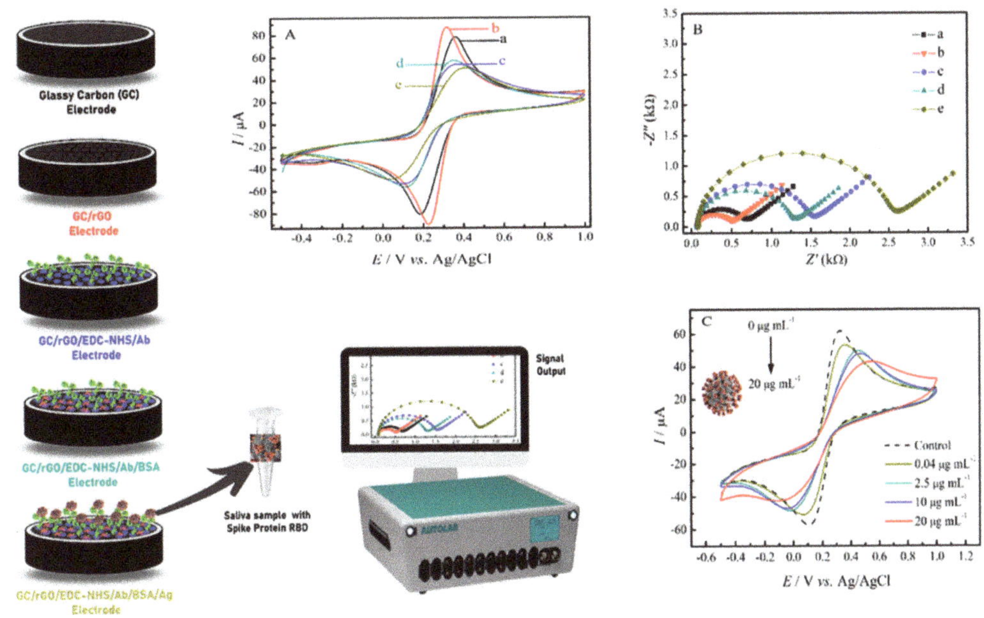

Figure 2. CV (**A**) and EIS (**B**) electrochemical characterization of the immunosensor steps fabrication in 0.2 mol L^{-1} PBS pH 7.4, 0.1 mol L^{-1} KCl containing 5.0 mmol L^{-1} of $[Fe(CN_6)]^{3-}/^{4-}$ for the electrodes: (a) GC, (b) GC/rGO, (c) GC/rGO-EDC-NHS/Ab, (d) GC/rGO-EDC-NHS/Ab/BSA and (e) GC/rGO-EDC-NHS/Ab/BSA/Ag. (**C**) CV experiments in for the GC/rGO-EDC-NHS/Ab/BSA electrode (control) and after the incubation of different antigen concentration.

Figure 2B also presented the Nyquist plots for bare GC (■), GC/rGO (▼), GC/rGO-EDC-NHS/Ab (•), GC/rGO-EDC-NHS/Ab/BSA (▲) and GC/rGO-EDC-NHS/Ab/BSA/Ag (♦) electrodes. The EIS results were quantitatively optimized in a Randles equivalent circuit (inset Figure 2B) in order to calculate the charge-transfer resistance (R_{ct}), the electrolyte ohmic resistance (R_s), the constant phase element (CPE) and the surface roughness (α). The EIS experimental values obtained are summarized in Table 1. As expected and in agreement with CV experiments, a lower value of R_{ct} for rGO was observed, indicating the improvement in electron transfer of when the GC electrode is modified with this material. However, when biomolecules, such as proteins and enzymes, that have poor electrical conductivity at low frequencies are immobilized on the electrodes surface, the electron transfer process between the solution-based mediators and the electrode surface is impeded. Thus, an R_{ct} value of 1464.5 Ω for the GC/rGO-EDC-NHS/Ab electrode (curve c) was found, and after the SARS-CoV-2 antigen binding at the GC/rGO-EDC-NHS/Ab/BSA electrode (curve e), the R_{ct} value was 2398.8 Ω. This behavior of increase in R_{ct} as the deposition of the biomaterial occurs on the biosensor surface is reported in several studies [23,24]. Leva-Bueno et al. [25] did a general scheme of EIS for each step of biosensor construction, showing that the impedance increases as the deposition over the surface electrode increases. Therefore, the CV and EIS experiments indicated that the SARS-CoV-2 immunosensor was effectively prepared.

Table 1. Resume of fitted parameters of EIS experiments.

Steps of Immunosensor Fabrication	R_{ct} (Ω)	R_s (Ω)	CPE (μF sα^{-1})	α
GC	720	69.5	1.49	0.95
GC/rGO	550	91	1.30	0.97
GC/rGO-EDC-NHS/Ab	1464.5	67.6	0.79	0.97
GC/rGO-EDC-NHS/Ab/BSA	1241.7	60	0.81	0.94
GC/rGO-EDC-NHS/Ab/BSA/Ag	2398.8	81.7	0.75	1.00

3.2. Optimization and Stability of the Impedimetric Immunosensor for SARS-CoV-2 Spike Protein RBD

The optimization of the impedimetric immunosensor was carried out with the antibody and antigen aliquots diluted from stock solutions with filtered PBS pH 7.4. The optimization experiments were conducted by diluting the antibody (Ab) at 1:1600, 1:800, 1:400, 1:200 and 1:100 (stock solution: 1.0 mg mL^{-1}), and corresponding antigen (Ag) dilutions at 1:10, 1:20, 1:40, 1:80 and 1:160 (stock solution: 0.2 mg mL^{-1}). The results presented in Table 2 and Figure 3 represent the effect of Ab and Ag concentration on the R_{ct} value during the immunoassay. It is possible to observe that the highest increase in the R_{ct} value was obtained for 1:40 Ab and 1:20 Ag dilutions. Therefore, this proportion was considered as an optimal value and used in the next studies.

Table 2. Effect of antibody and antigen dilution on the immunosensor response.

Dilution		R_{ct} (Ω)
Ab	**Ag**	
1:1600	1:10	727.81
1:800	1:20	2078.5
1:400	1:40	3470.4
1:200	1:80	468.63
1:100	1:160	340.36

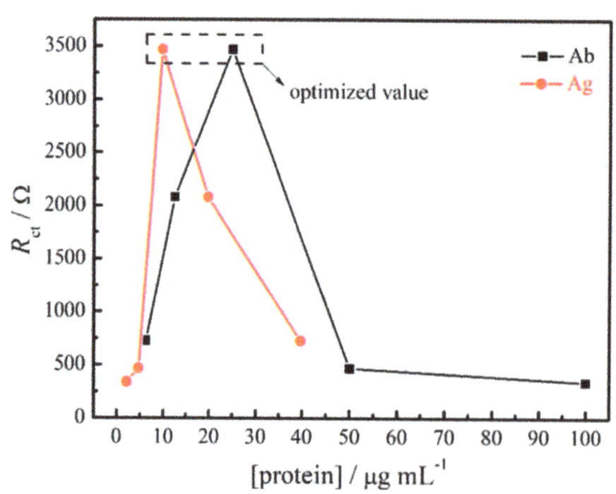

Figure 3. Optimization of antibody and antigen proteins concentrations at the immunosensor surface by EIS experiments.

To evaluate the stability of the proposed immunosensor, EIS and CV experiments were performed for the GC/rGO-EDC-NHS/Ab and GC/rGO-EDC-NHS/Ab/BSA/Ag electrodes. Five sequential experiments were carried out for the GC/rGO-EDC-NHS/Ab electrode, and it is observed that the I_{pa}/I_{pc} and R_{ct} values did not show significant differences between the measurements presenting an average value of 1.44 ± 0.03 μA (n = 5) and 1350.02 ± 70.60 Ω (n = 5), respectively, thus demonstrating that antibody proteins were effectively immobilized on the GC/rGO-EDC-NHS surface (Figure not shown).

3.3. Analytical Performance of the SARS-CoV-2 Immunosensor

The analytical performance of SARS-CoV-2 was evaluated by using the Nyquist plots obtained from the EIS experiments at different concentrations of SARS-CoV-2 spike protein RBD. As shown in Figure 4, R_{ct} values were enhanced with the increase of the antigen concentration, indicating a clear dependence on target concentration. The resulting calibration plots presented a good liner relationship between ΔR_{ct} (subtraction of electrode's R_{ct} before and after spike protein RBD incubation) and the logarithm concentrations of the antigen. In addition, two linear segments were obtained with different slopes. The first segment of the analytical curve is linear for a protein concentration range of 0.16 to 1.25 μg/mL (•). Meanwhile, the second segment of the calibration curve is also linear for a range of 2.5 to 40 μg/mL (■) RBD S protein concentration. The detection limit (calculated as LOD = 3SDblank/Slope) obtained for the lowest antigen concentrations was 150 ng/mL.

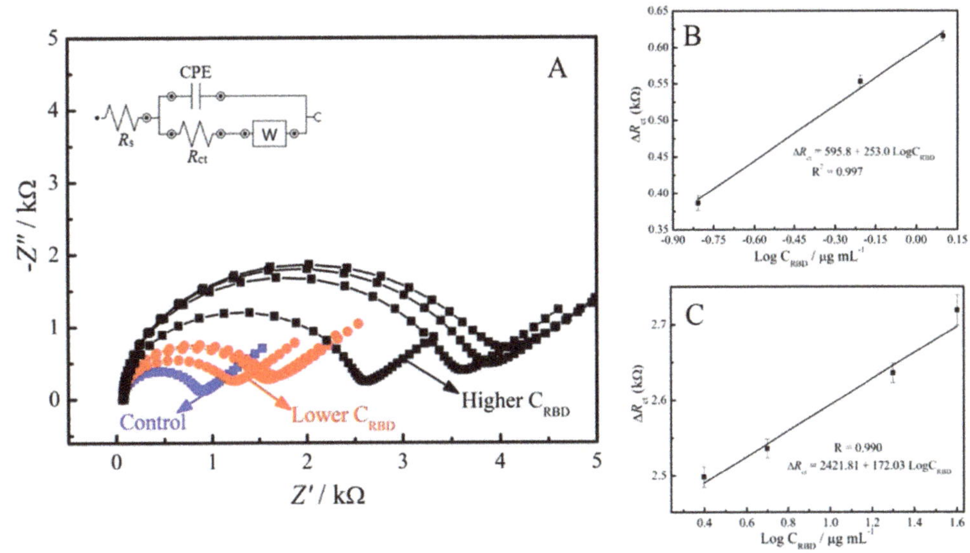

Figure 4. (**A**) EIS responses of the impedimetric immunosensor with different concentrations of the SARS-CoV-2 spike protein. The respective calibration curves plotted between the ΔR_{ct} and logarithmic concentration of SARS-CoV-2 spike protein from (**B**) 0.16 to 1.25 μg/mL, and (**C**) 2.5 to 40 μg/mL.

The sensitivity of the analytical device is a crucial point for the detection of the disease at the beginning of the infection. It is known that PCR test, mainly in saliva samples, does not detect the virus in the first days of the infection. Therefore, low cost and high sensitivity analytical methods are very important. The diagnostic platform developed in this work can be used for SARS-CoV-2 detection using other voltammetric techniques, such as square wave voltammetry (SWV), which increases the sensitivity of the proposed diagnostic. Figure 5 shows the square wave voltammograms obtained for the control (GC/rGO-EDC-NHS/Ab/BSA electrode) and after the binding of different concentrations of SARS-CoV-2 antigen. Using the SWV technique, the proposed sensor detected a concentration of 2.40 ng/mL of the virus. This study shows the potential of the technique and the immunosensor proposed in the SARS-CoV-2 spike protein analysis at low concentrations [26–28].

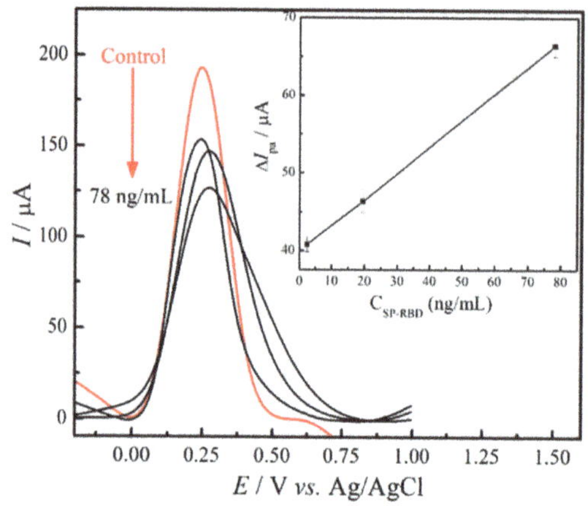

Figure 5. SWV data of the immunosensor in the absence (control) and presence of spike protein RBD concentrations ranging from 2.44 to 78 ng/mL; the inset shows the relationship between the ΔI_{pa} and concentration of the SARS-CoV-2 spike protein.

3.4. SARS-CoV-2 Spike Protein RBD Analysis in Saliva Samples

The analysis of the SARS-CoV-2 spike protein RBD in saliva samples was performed by EIS experiments in triplicate. The samples preparation is described in Section 2.6 and the obtained results are presented in Figure 6. As expected, it is possible to observe the increase in the R_{ct} values when the saliva samples were spiked with virus. The mean R_{ct} values found in the presence of 2.5 µg/mL of SARS-CoV-2 spike protein RBD were 3283.2 ± 451.5 Ω (n = 3), and in the absence of the antigen were 2316.2 ± 345.1 Ω (n = 3). The immunosensor showed a good response towards SARS-CoV-2 determination in the saliva samples. The proposed immunosensor is an effective tool towards early COVID-19 diagnosis. The European Union (EU) has stated that antibody-based kits have limitations in detecting SARS-CoV-2 infections because antibodies only became detectable within several days after infection [29,30]. In addition, the saliva samples are much easier and less invasive method, and depending on age, it can be done even by self-collection [31–33].

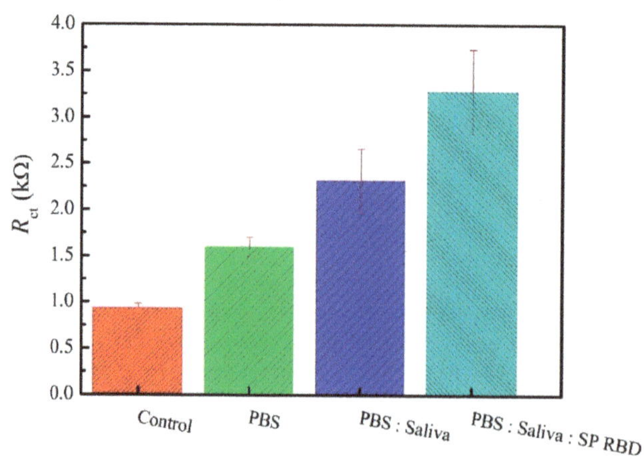

Figure 6. Response of the proposed immunosensor towards detecting SARS-CoV-2 spike protein RBD in real saliva sample.

4. Conclusions

A new methodology based on an electrochemical immunosensor developed with reduced graphene oxide for SARS-CoV-2 determination was successfully described in this work, presenting a low-cost technology by employing glassy carbon electrodes modified with rGO, a graphene material derivative through electrochemical reduction, which has an inexpensive, easy, fast and green way of obtention if compared with other materials also employed as biosensor surface modifiers, such as gold and silver. The large surface area of this material allows the coupling of interested biomolecules, and its conductivity properties can be enhanced, functionalizing it with immobilization agents, as EDC/NHS, and this type of surface modification can easily be transposed to printed carbon electrodes, which enables the integration of this immunosensor in point-of-care (POC) devices.

The immunosensor was characterized and optimized by electrochemical techniques and successfully applied to the determination of the SARS-CoV-2 spike protein RBD in saliva samples. Compared with other diagnostic methods and developed biosensors aiming the detection of SARS-CoV-2, this work combines feasibility and reliability, without any complex steps of building it, and is less reactive and time consuming compared to RT-PCR, having a great potential for large-scale production of a diagnostic tool with medical care capability and not needing specialized personnel in its management, contributing to a more effective control of the spread of SARS-CoV-2. In addition, the immunosensor demonstrated robustness towards SARS-CoV-2 analysis, showing good reproducibility and contributing to advances in the SARS-CoV-2 electrochemical biosensing area.

Author Contributions: G.C.Z. and I.C. conceived and planned the study. G.C.Z., M.K.L.S. and G.S.R. carried out the experiments. I.C. and M.K.L.S. writing—review and editing the manuscript. I.C., M.K.L.S. and G.S.R. revised the final version manuscript. I.C. was responsible for supervision, project administration and funding acquisition. All authors provided critical feedback and helped shape the research, analysis. All authors have read and agreed to the published version of the manuscript.

Funding: This research was funded by CNPq grant 440158/2020-6 and FAPESP grant 2017/24274-3.

Institutional Review Board Statement: Not applicable.

Informed Consent Statement: Not applicable.

Data Availability Statement: The data is available under the request to the correspondence.

Acknowledgments: The authors would like to acknowledge the research grant (301192/2019-7) awarded by CNPq.

Conflicts of Interest: The authors declare no conflict of interest.

References

1. Chauhan, D.S.; Prasad, R.; Srivastava, R.; Jaggi, M.; Chauhan, S.C.; Yallapu, M.M. Comprehensive Review on Current Interventions, Diagnostics, and Nanotechnology Perspectives against SARS-CoV-2. *Bioconjug. Chem.* **2020**, *31*, 2021–2045. [CrossRef] [PubMed]
2. Udugama, B.; Kadhiresan, P.; Kozlowski, H.N.; Malekjahani, A.; Osborne, M.; Li, V.Y.C.; Chen, H.; Mubareka, S.; Gubbay, J.B.; Chan, W.C.W. Diagnosing COVID-19: The Disease and Tools for Detection. *ACS Nano* **2020**, *14*, 3822–3835. [CrossRef] [PubMed]
3. Kudr, J.; Michalek, P.; Ilieva, L.; Adam, V.; Zitka, O. COVID-19: A challenge for electrochemical biosensors. *TrAC Trends Anal. Chem.* **2021**, *136*, 116192. [CrossRef] [PubMed]
4. Ranjan, P.; Singhal, A.; Yadav, S.; Kumar, N.; Murali, S.; Sanghi, S.K.; Khan, R. Rapid diagnosis of SARS-CoV-2 using potential point-of-care electrochemical immunosensor: Toward the future prospects. *Int. Rev. Immunol.* **2021**, *40*, 126–142. [CrossRef]
5. Chandra, P. Miniaturized label-free smartphone assisted electrochemical sensing approach for personalized COVID-19 diagnosis. *Sens. Int.* **2020**, *1*, 100019. [CrossRef]
6. Choi, J.R. Development of Point-of-Care Biosensors for COVID-19. *Front. Chem.* **2020**, *8*, 517. [CrossRef]
7. Yakoh, A.; Pimpitak, U.; Rengpipat, S.; Hirankarn, N.; Chailapakul, O.; Chaiyo, S. Paper-based electrochemical biosensor for diagnosing COVID-19: Detection of SARS-CoV-2 antibodies and antigen. *Biosens. Bioelectron.* **2021**, *176*, 112912. [CrossRef]
8. Zhao, H.; Liu, F.; Xie, W.; Zhou, T.C.; OuYang, J.; Jin, L.; Li, H.; Zhao, C.Y.; Zhang, L.; Wei, J.; et al. Ultrasensitive supersandwich-type electrochemical sensor for SARS-CoV-2 from the infected COVID-19 patients using a smartphone. *Sens. Actuators B Chem.* **2021**, *327*, 128899. [CrossRef]

9. Raziq, A.; Kidakova, A.; Boroznjak, R.; Reut, J.; Öpik, A.; Syritski, V. Development of a portable MIP-based electrochemical sensor for detection of SARS-CoV-2 antigen. *Biosens. Bioelectron.* **2021**, *178*, 113029. [CrossRef]
10. Torres, M.D.T.; de Araujo, W.R.; de Lima, L.F.; Ferreira, A.L.; de la Fuente-Nunez, C. Low-cost Biosensor for Rapid Detection of SARS-CoV-2 at the Point-of-Care. *Matter* **2021**, *4*, 2403–2416. [CrossRef]
11. Bandala, E.R.; Kruger, B.R.; Cesarino, I.; Leao, A.L.; Wijesiri, B.; Goonetilleke, A. Impacts of COVID-19 pandemic on the wastewater pathway into surface water: A review. *Sci. Total Environ.* **2021**, *774*, 145586. [CrossRef]
12. Bahadir, E.B.; Sezgintürk, M.K. Applications of graphene in electrochemical sensing and biosensing. *TrAC Trends Anal. Chem.* **2016**, *76*, 1–14. [CrossRef]
13. Prattis, I.; Hui, E.; Gubeljak, P.; Kaminski Schierle, G.S.; Lombardo, A.; Occhipinti, L.G. Graphene for Biosensing Applications in Point-of-Care Testing. *Trends Biotechnol.* **2021**. [CrossRef] [PubMed]
14. Taniselass, S.; Arshad, M.K.M.; Gopinath, S.C.B. Graphene-based electrochemical biosensors for monitoring noncommunicable disease biomarkers. *Biosens. Bioelectron.* **2019**, *130*, 276–292. [CrossRef] [PubMed]
15. Zhang, Y.; Shen, J.; Li, H.; Wang, L.; Cao, D.; Feng, X.; Liu, Y.; Ma, Y.; Wang, L. Recent Progress on Graphene-based Electrochemical Biosensors. *Chem. Rec.* **2016**, *16*, 273–294. [CrossRef] [PubMed]
16. Donini, C.A.; Silva, M.K.L.; Bronzato, G.R.; Leão, A.L.; Cesarino, I. Evaluation of a biosensor based on reduced graphene oxide and glucose oxidase enzyme on the monitoring of second-generation ethanol production. *J. Solid State Electrochem.* **2019**, 1–8. [CrossRef]
17. Kohori, N.A.; da Silva, M.K.L.; Cesarino, I. Evaluation of graphene oxide and reduced graphene oxide in the immobilization of laccase enzyme and its application in the determination of dopamine. *J. Solid State Electrochem.* **2018**, *22*, 141–148. [CrossRef]
18. da Silva, M.K.L.; Vanzela, H.C.; Defavari, L.M.; Cesarino, I. Determination of carbamate pesticide in food using a biosensor based on reduced graphene oxide and acetylcholinesterase enzyme. *Sens. Actuators B Chem.* **2018**, *277*, 555–561. [CrossRef]
19. Rocha, G.S.; Silva, M.K.L.; Cesarino, I. Reduced Graphene Oxide-Based Impedimetric Immunosensor for Detection of Enterotoxin A in Milk Samples. *Materials* **2020**, *13*, 1751. [CrossRef] [PubMed]
20. Silva, M.K.L.; Cesarino, I. Graphene functionalization and nanopolymers. In *Carbon Nanostructures*; Springer International Publishing: Berlin/Heidelberg, Germany, 2019; pp. 157–178, ISBN 978-981-32-9057-0.
21. Pei, S.; Cheng, H.-M. The reduction of graphene oxide. *Carbon* **2012**, *50*, 3210–3228. [CrossRef]
22. Zhuo, Y.; Yuan, P.X.; Yuan, R.; Chai, Y.Q.; Hong, C.L. Bienzyme functionalized three-layer composite magnetic nanoparticles for electrochemical immunosensors. *Biomaterials* **2009**, *30*, 2284–2290. [CrossRef]
23. Yu, L.; Zhang, Y.; Hu, C.; Wu, H.; Yang, Y.; Huang, C.; Jia, N. Highly sensitive electrochemical impedance spectroscopy immunosensor for the detection of AFB1 in olive oil. *Food Chem.* **2015**, *176*, 22–26. [CrossRef] [PubMed]
24. Li, H.; Hu, Y.; Li, A.; Wang, X.; Hou, P.; Wang, C.; Chen, K.; Zhao, C. A highly sensitive electrochemical impedance immunosensor for indole-3-acetic acid and its determination in sunflowers under salt stress. *RSC Adv.* **2017**, *7*, 54416–54421. [CrossRef]
25. Leva-Bueno, J.; Peyman, S.A.; Millner, P.A. A review on impedimetric immunosensors for pathogen and biomarker detection. *Med. Microbiol. Immunol.* **2020**, *209*, 343–362. [CrossRef]
26. Da Silva, M.K.L.; Plana Simões, R.; Cesarino, I. Evaluation of Reduced Graphene Oxide Modified with Antimony and Copper Nanoparticles for Levofloxacin Oxidation. *Electroanalysis* **2018**, *30*, 2066–2076. [CrossRef]
27. Mojsoska, B.; Larsen, S.; Olsen, D.A.; Madsen, J.S.; Brandslund, I.; Alatraktchi, F.A. Rapid SARS-CoV-2 Detection Using Electrochemical Immunosensor. *Sensors* **2021**, *21*, 390. [CrossRef]
28. Liu, X.; Duckworth, P.A.; Wong, D.K.Y. Square wave voltammetry versus electrochemical impedance spectroscopy as a rapid detection technique at electrochemical immunosensors. *Biosens. Bioelectron.* **2010**, *25*, 1467–1473. [CrossRef] [PubMed]
29. European Union Communication from the Commission—Guidelines on COVID-19 In Vitro Diagnostic Tests and Their Performance. Available online: https://op.europa.eu/en/publication-detail/-/publication/9d8c5572-7f12-11ea-aea8-01aa75ed71a1/language-en (accessed on 16 May 2021).
30. Shen, Y.; Anwar, T.B.; Mulchandani, A. Current status, advances, challenges and perspectives on biosensors for COVID-19 diagnosis in resource-limited settings. *Sens. Actuators Rep.* **2021**, *3*, 100025. [CrossRef]
31. Wyllie, A.L.; Fournier, J.; Casanovas-Massana, A.; Campbell, M.; Tokuyama, M.; Vijayakumar, P.; Warren, J.L.; Geng, B.; Muenker, M.C.; Moore, A.J.; et al. Saliva or Nasopharyngeal Swab Specimens for Detection of SARS-CoV-2. *N. Engl. J. Med.* **2020**, *383*, 1283–1286. [CrossRef]
32. Fabiani, L.; Saroglia, M.; Galatà, G.; De Santis, R.; Fillo, S.; Luca, V.; Faggioni, G.; D'Amore, N.; Regalbuto, E.; Salvatori, P.; et al. Magnetic beads combined with carbon black-based screen-printed electrodes for COVID-19: A reliable and miniaturized electrochemical immunosensor for SARS-CoV-2 detection in saliva. *Biosens. Bioelectron.* **2021**, *171*, 112686. [CrossRef] [PubMed]
33. Wyllie, A.L.; Fournier, J.; Casanovas-Massana, A.; Campbell, M.; Tokuyama, M.; Vijayakumar, P.; Geng, B.; Muenker, M.C.; Moore A.J.; Vogels, C.B.F.; et al. Saliva is more sensitive for SARS-CoV-2 detection in COVID-19 patients than nasopharyngeal swabs. *medRxiv* **2020**, *3*. [CrossRef]

materials

MDPI

Article

Fabrication of MNPs/rGO/PMMA Composite for the Removal of Hazardous Cr(VI) from Tannery Wastewater through Batch and Continuous Mode Adsorption

Rahman Ullah [1], Waqas Ahmad [1,*], Muhammad Yaseen [1], Mansoor Khan [2], Mehmood Iqbal Khattak [3], Badrul Mohamed Jan [4,*], Rabia Ikram [4,*] and George Kenanakis [5]

1 Institute of Chemical Sciences, University of Peshawar, Peshawar 25120, Khyber Pakhtunkhwa, Pakistan; rahmandawar@uop.edu.pk (R.U.); myyousafzai@gmail.com (M.Y.)
2 Department of Chemistry, Kohat University of Science and Technology, Kohat 26000, Khyber Pakhtunkhwa, Pakistan; mansoor009988@gmail.com
3 Material Science Center (PCSIR) Laboratories Complex, Peshawar 25120, Khyber Pakhtunkhwa, Pakistan; mahmood2162002@gmail.com
4 Department of Chemical Engineering, University of Malaya, Kuala Lumpur 50603, Malaysia
5 Institute of Electronic Structure and Laser, Foundation for Research and Technology-Hellas, N. Plastira 100, Vasilika Vouton, GR-70013 Heraklion, Crete, Greece; gkenanak@iesl.forth.gr
* Correspondence: waqasahmad@uop.edu.pk (W.A.); badrules@um.edu.my (B.M.J.); raab@um.edu.my (R.I.)

Abstract: Herein, we report the synthesis of magnetic nanoparticle (MNP)-reduced graphene oxide (rGO) and polymethylmethacrylate (PMMA) composite (MNPs/rGO/PMMA) as adsorbent via an in situ fabrication strategy and, in turn, the application for adsorptive removal and recovery of Cr(VI) from tannery wastewater. The composite material was characterized via XRD, FTIR and SEM analyses. Under batch mode experiments, the composite achieved maximum adsorption of the Cr(VI) ion (99.53 ± 1.4%, i.e., 1636.49 mg of Cr(VI)/150 mg of adsorbent) at pH 2, adsorbent dose of 150 mg/10 mL of solution and 30 min of contact time. The adsorption process was endothermic, feasible and spontaneous and followed a pseudo-2nd order kinetic model. The Cr ions were completely desorbed (99.32 ± 2%) from the composite using 30 mL of NaOH solution (2M); hence, the composite exhibited high efficiency for five consecutive cycles without prominent loss in activity. The adsorbent was washed with distilled water and diluted HCl (0.1M), then dried under vacuum at 60 °C for reuse. The XRD analysis confirmed the synthesis and incorporation of magnetic iron oxide at 2θ of 30.38°, 35.5°, 43.22° and 57.36°, respectively, and graphene oxide (GO) at 25.5°. The FTIR analysids revealed that the composite retained the configurations of the individual components, whereas the SEM analysis indicated that the magnetic Fe_3O_4–NPs (MNPs) dispersed on the surface of the PMMA/rGO sheets. To anticipate the behavior of breakthrough, the Thomas and Yoon–Nelson models were applied to fixed-bed column data, which indicated good agreement with the experimental data. This study evaluates useful reference information for designing a cost-effective and easy-to-use adsorbent for the efficient removal of Cr(VI) from wastewater. Therefore, it can be envisioned as an alternative approach for a variety of unexplored industrial-level operations.

Keywords: reduced graphene oxide; MNPs/rGO/PMMA composite; hexavalent chromium; tannery wastewater; fixed bed adsorption

1. Introduction

Over the years, enormous studies have been investigated for wastewater treatments with the breakthrough of research using value-added nanomaterials, including NPs and nanocomposites [1,2]. Tanning is one of the global water-and-soil polluting industries due to its high environmental risks [3]. The usage of chemicals during the tanning process produces a large amount of carcinogenic mobilization of toxic metals, which is the vital concern of the current era [4]. Chrome tanning relies on chromium salts and liquors to minimize hazardous environmental impacts. On the other hand, vegetable tanning is a natural process of tree tannins and water [5]. Among these common tanning procedures, chrome tanning is highly favorable due to its low cost and high productivity [6].

There is a continuous shift of production wastewater, which is highly polluted, containing both inorganic and organic pollutants from the tanneries [7]. Notably, the presence of organic pollutants comprises blood particles, flesh, soluble proteins and hair [8]. Conversely, the inorganic pollutants used as tanning agents are typically salts of various metals, such as Chromium (Cr-III) sulphate and chrome alum [9]. Cr, in aqueous solution, exists in two stable oxidation states (Cr-III and Cr-VI), where both have diverse biological, chemical and environmental possessions [10]. Cr(III) is a moderately insoluble and useful micronutrient for animals, plants and humans. As compared to Cr(III), Cr(VI) is 100 times more contaminant and hazardous and 1000 times more mutagenic and carcinogenic for living organisms due to its oxidizing ability [11]. The occurrence of Cr(VI) in water beyond permissible limits causes liver, cancer and skin disturbance, bringing about ulcer development, diarrhea, hemorrhage, vomiting and damage to the kidneys [12]. According to the World Health Organization (WHO), the acceptable amount of Cr in drinking water is 0.1 $mg{\cdot}L^{-1}$, while, in industrial wastewater, it is 0.05 $mg{\cdot}L^{-1}$ and 5 $mg{\cdot}L^{-1}$ for Cr(VI) and Cr(III), respectively [13]. Therefore, it is a substantial task to discover facile and cost-effective ways to eliminate Cr(VI) from environmental water.

Various techniques have been applied for the effective and economical elimination of Cr(VI) from industrial effluents, including electrochemical technologies [14], use of chemical coagulation solvents [15], extraction [16], ion exchange [17], membrane filtration [18] and adsorption [19]. The remarkable advantages and disadvantages of these techniques have been critically reviewed [5,11]. It has been observed that electrochemical techniques represent low cost and high selectivity along with the fruitless recovery of treated metals for a recycled, high volume of sludge formation [9,14]. Likewise, the ion exchange and membrane filtration processes are restricted due to their selectivity, high operational cost and maintenance issues [20]. However, the adsorption process, by all accounts, has shown unrivaled outcomes due to its minimal effort, high effectiveness and selectivity [21].

Among the nanomaterials, GO has gained enormous attention due to its excellence in thermal, electronic and mechanical impendence [22]. Owing to its high surface area and tremendous adsorption capacity, GO has been utilized as an exceptional adsorbent for the adsorption of heavy metals [23,24]. Very recent studies have utilized novel GO–ionic liquid composites [25], GO/chitosan [26] and Fe_2O_3-GO-chitosan composites [27], for enhanced Cr(VI) adsorption.

Besides, MNPs have been employed for the adsorption of substantial metals from wastewater [28] MNPs demonstrate exceptional paramagnetic properties and have a rich surface, which makes their target to be easily separated from the aqueous medium using external magnetic field [29]. However, the adsorption properties of MNPs are limited due to their mobility and aggregation in aqueous media [30]. The MNPs may be combined into composites with other active adsorbents, such as GO, to eliminate the operations of filtration, centrifugation and aggregation and, eventually, to increase the adsorption capacity [23,31]. A variety of magnetized adsorbents has proved high adsorption efficiency using different types of pollutants, including Cr from water [30,32]. The incorporation of a polymer matrix to these nanomaterials may also help reducing their aggregation and enhance their adsorption capacity by offering the addition of a reactive functional group as adsorption site. The chemical and mechanical stability of the adsorbent may also be enhanced with the incorporation of a polymer matrix [33].

In this study, we fabricated a novel composite of MNPs with reduced GO (rGO) and polymethylmethacrylate (MNPs/rGO/PMMA) through in situ polymerization and applied it for the adsorptive removal and recovery of Cr(VI) from tannery wastewater. Polymethylmethacrylate (PMMA) has high performance in multipurpose products due to its outstanding mechanical and physical properties. It is resistant to oils, waxes, greases, acids, bases, ozone, weathering and water immersion, which can withstand high temperatures (160 °C). Additionally, PMMA provides multiple carboxylate functionalities for binding the Cr ions, which impart the dispersion properties to the composite and inhibit particle aggregations of rGO and MNPs. Therefore, PMMA was selected as the polymer matrix for the preparation of a composite material in this work. rGO was magnetized using $FeCl_3$ and $FeSO_4$ salts, using an in situ magnetization process. The synthesized composite was characterized using the scanning electron microscopy (SEM), Fourier transform infra-red spectroscopy (FTIR) and X-rays diffraction (XRD) techniques. The adsorption performance of MNPs/rGO/PMMA for Cr(VI) was evaluated using batch adsorption techniques and fixed-bed column-based experimental set up from model aqueous solution and tannery wastewater.

2. Experimental Procedure

2.1. List of Chemicals and Standards

In this study, all the chemicals used were of analytical grade and were used without further purification. The MMA monomer and 2,2-Azobisisobutyronitrile (AIBN) were provided by Daejung chemicals, (Siheung-si, Korea). Acetone (CH_3COCH_3), sulfuric acid (H_2SO_4, 98%), hydrogen peroxide (H_2O_2, 30%), sodium nitrate ($NaNO_3$, 99%) and hydrazine hydrate were purchased from Merk (Darmstadt, Germany). Ferric chloride ($FeCl_3$, 97%) and ferrous sulfate ($FeSO_4.6H_2O$, 97%) were provided by Scharlau (Barcelona, Spain). Ammonium hydroxide (NH_4OH, 35%) was purchased from Fisher scientific (Leicestershire, UK). The tannery wastewater was collected from the effluent discharge of a local tannery industry (Prime Tanning Industry (Pvt.) Ltv K.M, 25 GT Rd, Muridke, Sheikhupura, Punjab) Pakistan. The tannery wastewater was characterized by determining various parameters such as pH, chemical oxygen demand (COD), biological oxygen demand (BOD) and suspended solids (SS).

2.2. List of Instruments

The analysis of Cr was carried out by an Atomic Absorption Spectrometer (Perkin Elmer Analyst 700, Waltham, MA, USA). An ultrasonicator (wt-230HTD, Seoul, Korea) was used for sample dispersion. To examine the surface morphology of the adsorbent, the scanning electron microscope (SEM) JSM 5910, manufactured by JEOL JEM-2100F (Tokyo, Japan), was utilized. The samples were scanned over a 2θ range of 0–70° at the scanning rate of 1 degree per minute. The FTIR analysis was carried out with an FTIR spectrometer (Perkin Elmer 103385, Waltham, MA, USA). The structural crystallinity of the adsorbent was studied using the XRD analysis model (X'Pert^3 Powder-Malvern Panalytical, Almelo, The Netherlands). The XRD analysis was conducted using an advanced Bruker anode X-ray diffractometer with Cu Kα (λ = 1.5406 A) radiation. A glass electrode of model 422 WTW, Weilhmein, Germany, was used for the adjustment of the sample's pH.

2.3. Synthesis of MNPs/rGO/PMMA Composite

2.3.1. Synthesis of GO

GO was synthesized from powder graphite with Hummers' technique [34]. Firstly, 2 g of graphite was slowly added to the solution of 1 g of $NaNO_3$ dissolved in 92 mL of sulfuric acid (H_2SO_4) in the conical flak fitted in a bath containing ice, to maintain the temperature at 0 °C. $KMnO_4$ (12 g) was added and stirred vigorously; the paste that formed was removed from the ice bath, further stirred for two h at 35 °C and then left to settle overnight. Deionized water (184 mL) was added to the suspension under vigorous stirring. The temperature was raised to 95 °C and the flask was cooled to 45 °C on an ice bath. About 10 mL of H_2O_2 was added to the reaction mixture until a brown–yellow product of GO was formed, which was collected through filtration. Further, it was washed with distilled water and diluted HCl, then dried under vacuum at 60 °C.

2.3.2. Synthesis of rGO

The synthesized GO was then reduced to reduced graphene oxide (rGO) through the method reported in the literature [35]. In total, 1.0 g of GO was dispersed into 150 mL of deionized water and sonicated using an ultrasonic bath for 3 h under nitrogen atmosphere. About 5 mL of hydrazine hydrate solution was added to the dispersion and then stirred for 30 min till the color of the content changed from dark brown to dark black. The end product was washed with double-distilled water and ethanol for many times to evacuate undesirable contaminations.

2.3.3. Preparation of MNPs/GO

An in situ method was used to prepare MNPs/rGO, in which 1 g of rGO was dispersed in 100 mL of distilled water and sonicated using ultrasonication for 2 h. The $FeCl_3$ and $FeSO_4$ solutions were added to the suspension in a ratio of 2:1 and agitated for 2 h at 80 °C under nitrogen atmosphere. A volume of 10 mL of 35% NH_4OH solution was added dropwise for 2 h to blend. A dark-black precipitate of MNPs/rGO was obtained; the sequences of changes are indicated by the following reactions:

$$Fe^{2+} + 2Fe^{3+} + 6OH^- \rightarrow 3Fe(OH)_2$$
$$3Fe(OH)_2 \rightarrow Fe_3O_4 + 2H_2O$$

The suspension was separated from the solution using an external magnetic field. It was washed several times with distilled water and ethanol to eliminate excessive base and dissociate free ions of Fe(II) and Fe(III). Then, the final product was dried under vacuum at 30 °C. The sequence of reactions is shown in Scheme 1. The ratio of $FeCl_3$ and $FeSO_4$ to rGO is 2:1:1, respectively.

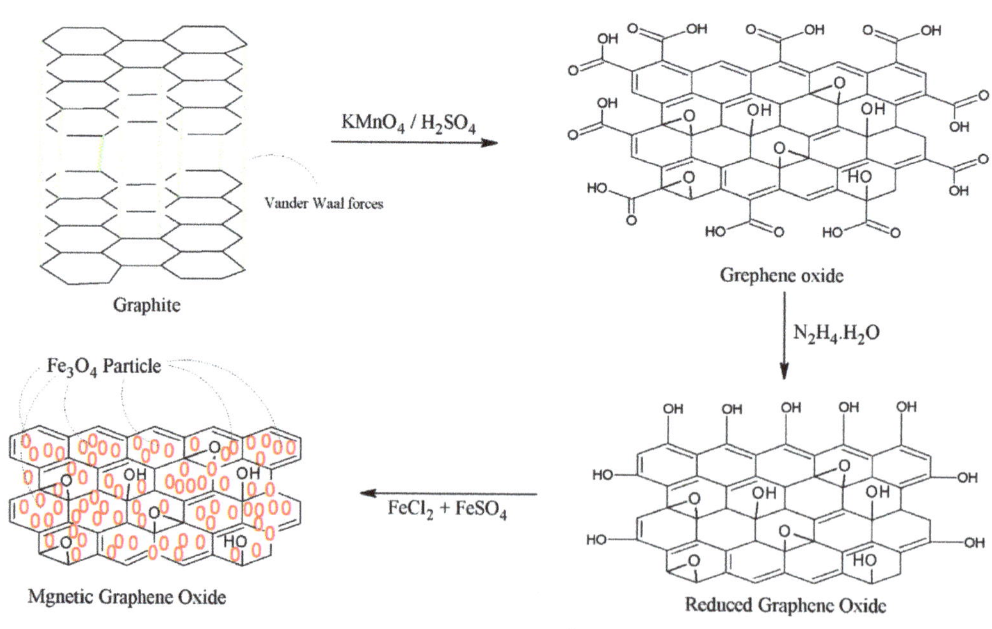

Scheme 1. Synthesis of GO, rGO and MNPs/rGO.

2.3.4. Preparation of MNPs/rGO and PMMA Composite

The MNPs/rGO/PMMA composite was fabricated using an in situ polymerization technique [36]. A total of 1 g MNPs/rGO was dispersed in 100 mL of deionized waster and 20 mL methyl methacrylate (MMA) was mixed to it. The suspension was dispersed through ultrasonication for 30 min. A total of 0.2 g of initiator 2,2-Azobisisobutyronitrile (ABBN) was added to the reaction mixture and heated to 70 °C, then stirred for 25 min under controlled condition. Then, a few mL of $(CH_3)_2CO$ was added to the final product, decanted into a petri dish and dried at room temperature. The theoretical weight ratio of MNPs:rGO:PMMA was approximately 4:1:1.8 g.

2.4. Batch Adsorption and Desorption Studies

The tannery wastewater sample was first filtered with Whatman filter paper (grade 41, 20–25 μm) to remove suspended solids. In the batch mode adsorption experiments, a wastewater sample of about 10 mL containing 1640 mg/L of Cr(VI) was taken in a 50 mL conical flask. Its pH was adjusted to 2 (with the addition of 0.1 M HCl or NaOH) and 150 mg of the MNPs/rGO–PMMA composite adsorbent was added and agitated for 30 min in a shaking bath. The adsorbent was collected from the water sample by holding a magnet along the wall of the flask, then the sample was collected and subjected to analysis. The concentration of Cr(VI) ions was determined via flame atomic absorption spectrometry (FAAS).

In the desorption experiments, the composite sample was shacked with 2 M sodium hydroxide (NaOH) for 30 min to desorb Cr(VI) ions from the surface. The composite was collected from the medium by an external magnetic field. The eluent was also analyzed via FAAS to determine the Cr(VI) concentration. The scheme of the experiment is shown below (Scheme 2).

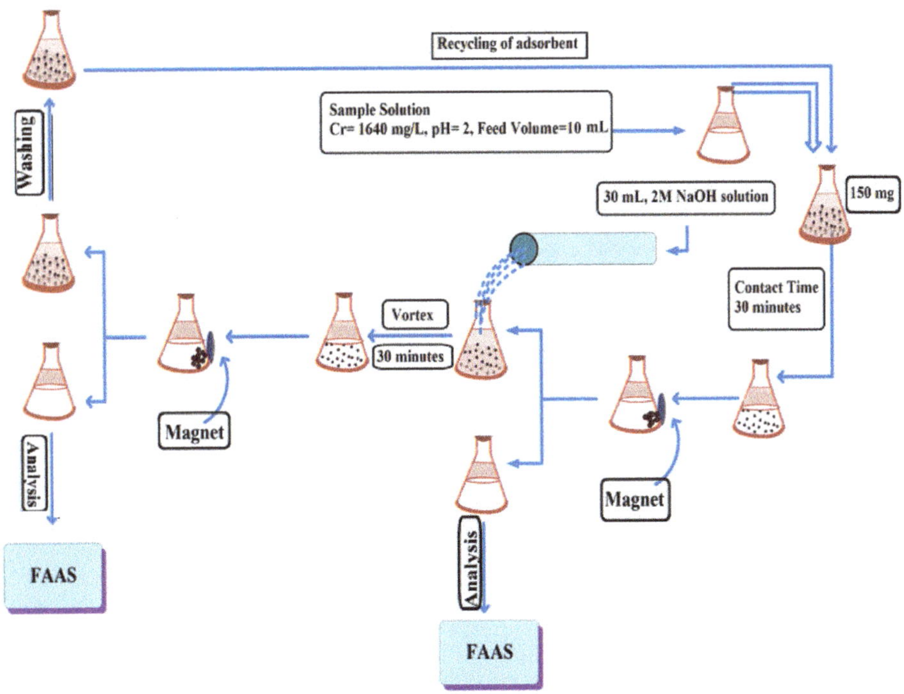

Scheme 2. Batch adsorption and desorption studies.

The batch mode adsorption of Cr(VI) onto the MNPs/rGO/PMMA was investigated in triplicate and the average values of the standard deviations were recorded. The percent adsorption of Cr was determined using Equation (1)

$$\%,\ \text{Adsorption} = \frac{C_0 - C_t}{C_0} \times 100 \tag{1}$$

where "C_0" ($mg \cdot L^{-1}$) is the original Cr concentration in the wastewater sample, while "C_t" ($mg \cdot L^{-1}$) is the final Cr concentration in the outlet sample. The adsorption capacity (q_e) (the amount of adsorbate ($mg \cdot L^{-1}$) held per mass of adsorbent (mg)) was determined using Equation (2)

$$q_e = \frac{C_0 - C_t}{m} \times v \tag{2}$$

where "m" is the adsorbent mass (g), while "V" (mL) is the volume of the sample solution. The batch experiments were repeated for the amount of adsorbent to calculate the maximum adsorbent dose. Various temperatures were applied, ranging from 293 K to 393 K, to calculate the entropy change (ΔS^o), Gibbs free energy (ΔG^o) and enthalpy change (ΔH^o). Then, the pH was optimized to describe the pH dependence of the adsorption process. Similar batch tests were directed for sample volume to investigate the maximum sample volume. The adsorption kinetics was calculated from the vortex time, running from 10 min to 60 min.

2.5. Fixed-Bed Column Experiments

For fixed-bed column adsorption studies, a glass column of 50 cm in height and 2 cm in internal diameter was used, into which a swab of glass wool was inserted. A total of 1 g of MNPs/rGO/PMMA adsorbent was loaded into the column and another glass wool swab was placed on the top of the adsorbent bed to avoid disturbance of the adsorbent surface during the flow of the sample. The diagrammatic illustration of the fixed-bed column is shown in Scheme 3. Initially, the composite packed column was rinsed with distilled water and left for a night to clarify the thorough distribution of adsorbent particles. Wastewater containing 1640 mg/g was impelled, downstream mode, into the column with the assistance of a peristaltic pump. Primarily, the collected adsorbate was discarded for one minute to ensure an unbroken stream rate. Afterward, the discharge samples were collected successively for further analysis at consistent interval of time, flow rate and adsorbent masses. This was performed from the bottom of the column to investigate the breakthrough point.

The continuous fixed-bed column process was stopped when the effluent concentration was equal to the initial concentration. The breakthrough cure was calculated from different adsorbent masses and flow rates by plotting graphs between C_t/C_0 and time. In most of the continuous fixed-bed system, the breakthrough curves are S-shaped with variable degrees of sharpness and location of the breakthrough point.

To regenerate the adsorbent and recover the Cr(VI) ion, an eluent reagent was used. The Cr(VI) ions which were loaded on different masses of adsorbent with a uniform flow rate 2 mL/min were desorbed using a 2 M NaOH solution. Similarly, the Cr(VI) analytes which were loaded on 3 g of composite with different flow rates (1, 2 and 3 m/L/min) were also desorbed through the 2M NaOH solution.

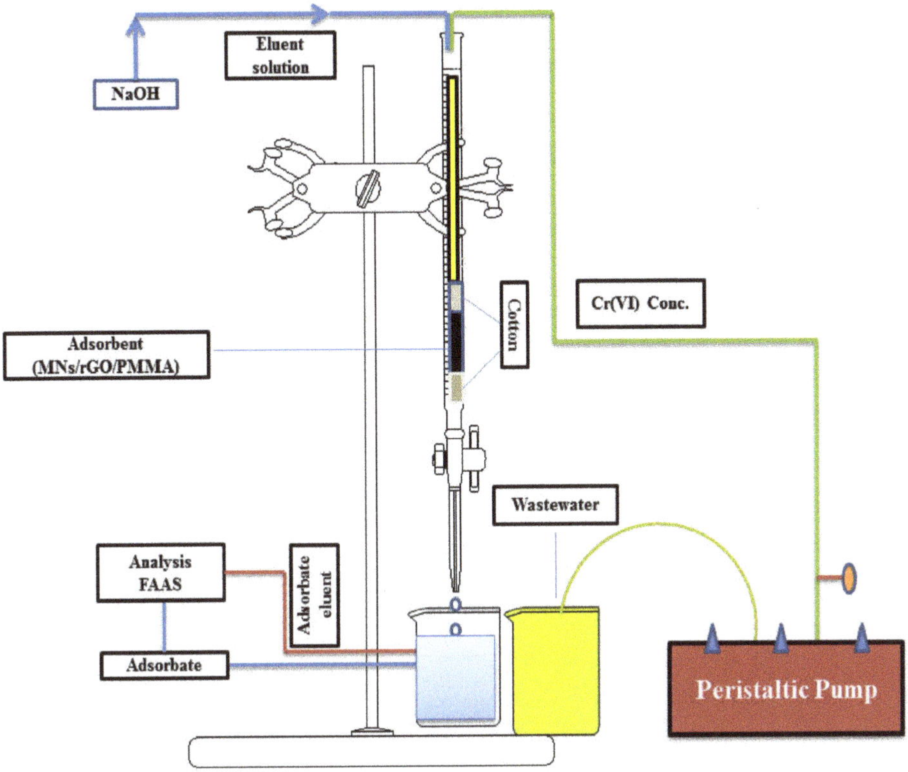

Scheme 3. Adsorption and desorption scheme of fixed-bed column.

3. Results and Discussions

3.1. Characteristics of Adsorbent

The synthesized composite (MNPs/rGO/PMMA) was confirmed via FTIR analysis, XRD analysis and SEM analysis. Figure 1 shows the FTIR spectra of rGO, MNPs/rGO, PMMA and MNPs/rGO/PMMA. The FTIR spectrum of rGO shows characteristic peaks of the C=C stretching bond of alkene and C–O stretching of the hydroxyl group at "1637 cm^{-1}" and "1061 cm^{-1}", respectively. The rGO (rGO) impregnated with magnetic Fe_3O_4 showed an additional peak at "537 cm^{-1}", which gives a clear identification of the Fe–O–Fe bond [37]. The FTIR spectrum of PMMA displays the characteristics peaks of C–H bending of the α– methyl group, C–O stretching of aliphatic ether, C–H bending of a methylene group, C=O stretching of the carboxylic functional group and C–H stretching of the methyl group at 735 cm^{-1}, 1140 cm^{-1}, 1423 cm^{-1}, 1717 cm^{-1} and 2950 cm^{-1}, respectively. The spectrum of the composite sample reflects the extra peak of Fe–O–Fe stretching at 537 cm^{-1} due to the magnetic rGO (MNPs/rGO) incorporated to PMMA, which indicates that the polymer was successfully magnetized [38].

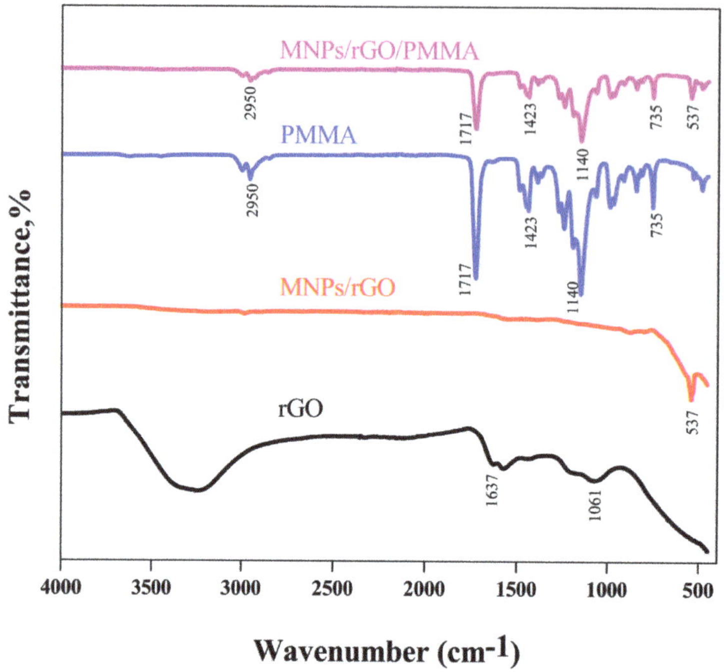

Figure 1. FTIR spectra of rGO, Fe_3O_4/rGO, PMMA and MNPs/rGO/PMMA.

The XRD configuration of rGO indicates a solid peak at 2θ of 25.8, which confirms the reduction of GO as shown in Figure 2 [39]. In the case of magnetized rGO, a peak at 2θ of 35.5 displayed that it matched with a face-centered cubic crystal of JCPD file card number 19-0629, suggesting the confirmation of impregnation of magnetic Fe_3O_4 on the rGO [37]. The XRD arrangement of pure PMMA gives no peak due to its amorphous nature. The XRD design of the composite (MNPs/rGO/PMMA), exhibits peaks corresponding to Fe_3O_4 at 2θ of 30.38° (220), 35.5° (331), 43.22° (400) and 57.36° (511), respectively, which confirms the MNPs/rGO/PMMA composite [40,41]. From the XRD data, the crystallite size of the composite was estimated to be 31 nm.

The SEM micrographs of rGO, MNPs/rGO, PMMA and MNPs/rGO/PMMA are given in Figure 3a–d. The micrograph of rGO indicates a fluffy and chipped morphology, which may be a stacked structure of graphene sheets linked by Van der Waal's forces, as shown in Figure 3a. Fe_3O_4-NPs were consistently scattered over the rGO sheets, as shown in Figure 3b [42]. The PMMA matrix is in the form of smooth uniform layers with no pores, in the form of sheets spread over each other, as can be seen in Figure 3c [43]. The micrograph of MNPs/rGO/PMMA confirms the magnetic MNPs dispersed on the surface of the PMMA/rGO sheets. MNPs/rGO/PMMA seems a uniform layer, indicating all components were properly incorporated together (see Figure 3d) [44]. From the SEM micrographs, the average size of the rGO flakes is found to be about 2000 nm, whereas that of MNPS is about 100 nm.

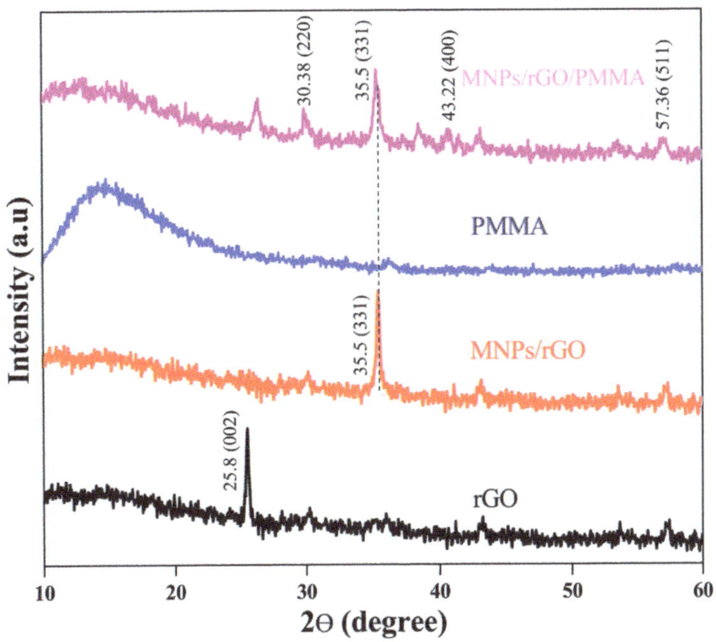

Figure 2. XRD spectra of rGO, Fe_3O_4/rGO, PMMA and MNPs/rGO/PMMA.

Figure 3. SEM of (**a**) rGO, (**b**) Fe_3O_4/rGO, (**c**) PMMA and (**d**) MNPs/rGO/PMMA.

3.2. *Physiochemical Characterization of Wastewater*

Previously, to bring the tannery wastewater into batch and column modes of adsorption, the sample was evaluated for a few limits, e.g., COD, biological oxygen demand (BOD), pH, Cr concentration and suspended solids. The Cr concentration in wastewater was found to be "1640" mg·L^{-1}, COD was "1130" mg·L^{-1}, while BOD was calculated at about "396" mg·L^{-1}. Similarly, suspended solids in the sample were calculated to be "960" mgL^{-1} and a pH of 3.17 was observed. As indicated by the Environmental Health Safety Guidelines (EHSG), the possible edge for COD and pH in wastewater are 150 mg·L^{-1} and 6–9, respectively. Likewise, the permissible level of Cr in industrial wastewater, according to the Environmental Protection Agency (EPA), is "200" mg·L^{-1}. Thus, it is obvious, from the above proof, that these limits do not follow the possible edges for wastewater, especially if the Cr concentration is extremely high.

4. Adsorption Experiment

The adsorption of Cr(VI) over the MNPs/rGO/PMMA was carried out in batch mode under different conditions, including pH, adsorption time, adsorbent dose, temperatures, etc.

4.1. *pH Investigation*

The pH of the reaction medium has an exceptional impact on the charge density on the adsorbent surface and the assurance of the change of Cr species [45]. The Cr(VI) adsorption on the composite was carried at various pH ranges, from 2 to 10, while the other parameters were kept constant. The pH of the investigated solution was adjusted using 0.1 N NaOH or HCl solution, according to necessity. It was noticed that the most extreme adsorption was found at pH 2.0 (100 ± 1), as shown in Figure 4.

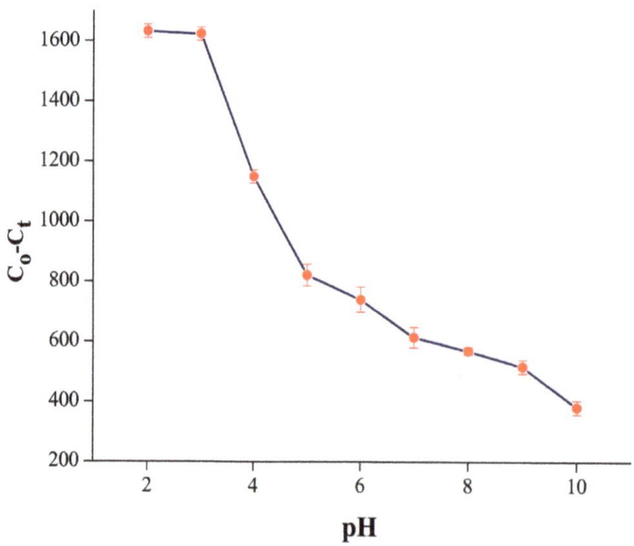

Figure 4. Effect of pH on adsorption of Cr(VI) on the MNPs/rGO/PMMA composite; adsorbent amount, 150 mg; shacking time, 30 min.

The maximum adsorption of Cr at this pH was observed due to the existence of Cr(VI) in various forms, such as dihydrogen chromate (H_2CrO_4), hydrogen chromate ($HCrO_4^-$), hydrogen dichromate ($HCr_2O_7^-$), dichromate ($Cr_2O_7^{2-}$) and chromate (CrO_4^{2-}) [46]. The dominancy of all above ions greatly depends on the pH solution, as well as on the total Cr concentration. At pH from 2 to 6, hydrogen chromate ($HCrO_4^-$) and dichromate ($Cr_2O_7^{2-}$) are the dominant species. Dihydrogen chromate (H_2CrO_4) is the main species at low pH values (less < 1) [47]. The dominant type of Cr(VI) at pH 2 is hydrogen chromate ($HCrO_4^-$), generated from the dichromate ($Cr_2O_7^{2-}$) based in reaction below [48].

$$Cr_2O_7^{2-} + H_2O \rightarrow 2HCrO_4^-$$

The $Cr_2O_7^{2-}$ particle holds two negative charges; hence, it needs two active sites for successful adsorption, while hydrogen chromate ($HCrO_4^-$) particles need one active site of the adsorbent (MNPs/rGO/PMMA) for successful adsorption [49]. Therefore, the adsorption capacity of the ($HCrO_4^-$) is twice that of ($Cr_2O_7^{2-}$). Likewise, at low pH, the surface of adsorbent became protonated due to the availability of a large number of H^+ ions, which creates an electrostatic force within a protonated functional group of adsorbent and Cr anions. The adsorption rate falls quickly and spans to 38% at pH 6, above pH 3. This is most likely because of the general change in surface charge on the adsorbent and, subsequently, adsorption decline. At pH > 8, the percent adsorption was further decreased because of the double resistance of two anions (CrO_4^{2-} and OH^-) on the surface of the adsorbent [50].

4.2. Adsorbent Dose Investigation

The percent adsorption of Cr(VI) is dependent on the amount of adsorbent dose in wastewater. To explore the impact of adsorbent (MNPs/rGO/PMMA) dose, batch adsorption experiments were conducted using dosages ranging from 20 mg to 250 mg. Results indicate that the % Cr(VI) adsorption was enhanced, as the amount of adsorbent dose was increased, and the maximum percent adsorption was found at 150 mg (99.53 ± 1), as shown in Figure 5. The significant rise in Cr(VI) adsorption with the increase in the amount of composite material was due to the availability of a larger surface region of the adsorbent and a large number of active sites [51]. A further enhancement of the amount of composite material to 240 mg showed no considerable variation in percent adsorption of Cr(VI). Hence, 150 mg was chosen as an ideal amount of (MNPs/rGO/PMMA) adsorbent to the be studied.

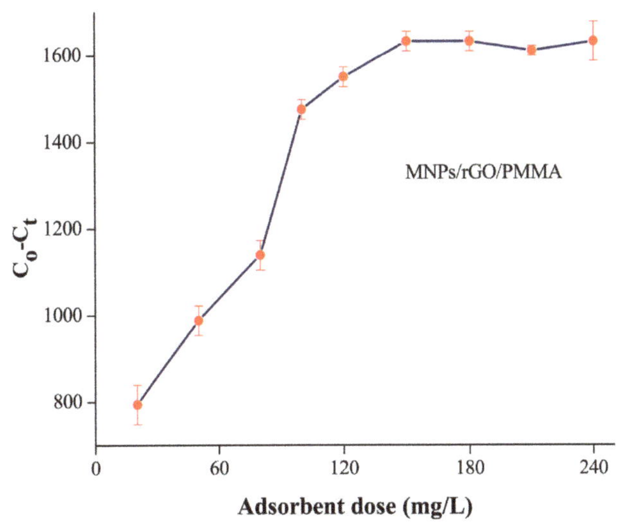

Figure 5. Effect of adsorbent amount on adsorption of Cr(VI) on MNPs/rGO/PMMA composite; pH 2; sample volume, 10 mL; shaking time, 30 min.

4.3. Kinetic Investigation

To achieve maximum adsorption over adsorbent (MNPs/rGO/PMMA) versus time, various contact periods of time, running from 10 min to 1 h, were determined. The outcomes show that the % adsorption of Cr(VI) was enhanced versus time; finally, stability was reached after 30 min, as shown in Figure 6.

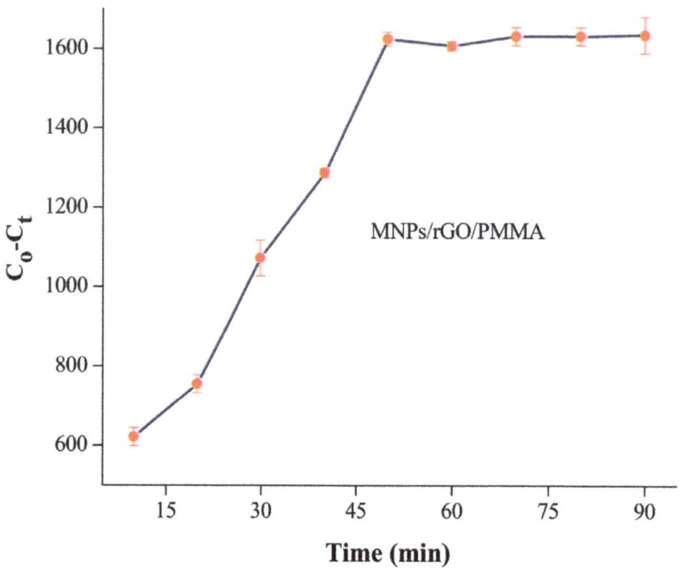

Figure 6. Effect of shaking time on adsorption of Cr(VI) on MNPs/rGO/PMMA composite; adsorbent amount, 150 mg; sample volume, 10 mL; pH 2.

A further increase in shacking time caused no considerable change in adsorption. Thus, 30 min of contact time was chosen to attain the most extreme adsorption for each batch experiment.

To clarify the systemic adsorption procedure, the shacking data were applied to pseudo-1st order and pseudo-2nd order kinetics models using Equations (3) and (4), respectively.

$$\log(q_e - q_t) = \log q_e - \frac{K_1}{2.303}t \tag{3}$$

$$\frac{t}{q_t} = \frac{t}{q_e} + \frac{1}{k_2 {q_e}^2} \tag{4}$$

where "k_1" (1/min) is the proportionality constant of pseudo-1st order model, while "K_2" ($g^{-1}\cdot mg^{-1}\cdot min^{-1}$) is the proportionality constant of pseudo-2nd order model; "q_t" ($mg\cdot g^{-1}$) is the quantity of metal ions (mg) adsorbed per unit quantity of adsorbent (g) at any time (t) and "q_e" ($mg\cdot g^{-1}$) is the quantity of metal ions (mg) adsorbed per unit quantity (g) of adsorbent at dynamic equilibrium.

Various kinetic factors of pseudo-1st order and pseudo-2nd order kinetics were determined from the slope and intercept, as displayed in Figures S1 and S2 (Supplementary Materials), respectively. The calculated and experimental results are listed in Table 1. The value of "R^2" for pseudo-1st order kinetic was calculated to be 0.86, which is smaller than those of pseudo-2nd order kinetic (0.98). The value of adsorption efficiency (q_e) for the pseudo-1st order of kinetic was calculated to be 23.93 mg g^{-1}, which is smaller than the experimental adsorption efficiency, 109.33 mg g^{-1}, whereas the calculated value of adsorption capacity (q_e) for pseudo-2nd order kinetic was 166.66 mg g^{-1}, which is higher than the experimental value of 109.33 mg g^{-1}.

Table 1. Kinetics parameters, for adsorption of Cr(VI) on MNPs/rGO/PMMA composite; pH 2; contact time, 30 min; adsorbent amount, 150 mg.

Experimental q_e ($mg \cdot g^{-1}$)	Pseudo-1st Order			Pseudo-2nd Order		
	K_1 min^{-1}	q_e ($mg \cdot g^{-1}$)	R^2	K_2 ($g.mg^{-1} \cdot min^{-1}$)	q_e ($mg \cdot g^{-1}$)	R^2
109.33	0.151	23.93	0.86	0.362	166.66	0.98

Similarly, the value of K_1 was 0.151, which is lower than that of K_2 (0.362), suggesting that the Cr(VI) adsorption process follows the pseudo-2nd order kinetic model.

4.4. Adsorption Isotherms

To calculate various adsorption parameters, the data obtained from the adsorption process were fitted to the Langmuir isotherm and Freundlich isotherm. The Langmuir adsorption isotherm is expressed by Equation (5):

$$\frac{C_e}{q_e} = \frac{1}{q_m\, k_b} + \frac{1}{q_m\, C_e} \quad (5)$$

In the above expression, "C_e" ($mg \cdot L^{-1}$) is the equilibrium adsorption concentration and "q_e" ($mg \cdot g^{-1}$) is the quantity of Cr(VI) ions adsorbed at equilibrium state. Similarly, "q_m" ($mg \cdot g^{-1}$) is the highest adsorption capacity and "K_b" (Langmuir constant) is the constant associated with energy. R_L (dimensionless isolating component) is determined from the Langmuir isotherm using Equation (6):

$$R_L = \frac{1}{1 + k_b C_e} \quad (6)$$

The projected value of "R_L" verifies whether the adsorption is favorable, unfavorable, reversible, or irreversible. If the R_L value is more than one ($1 < R_L$), the adsorption is unfavorable. Similarly, when the value of R_L is equivalent to one ($R_L = 1$), it indicates that the adsorption process is reversible. If the R_L value is equivalent to zero ($R_L = 0$), it denotes that the adsorption process is irreversible. R_L values of $0 < R_L < 1$ suggest that the adsorption process is favorable [52]. The slope and intercept values that were achieved from the Langmuir adsorption isotherm graph verified that the adsorption process may have been unfavorable or favorable, irreversible, or reversible, as indicated in Table 2. The binding constant and adsorbent capacities were calculated from the plot between C_e/q_e and C_e, as indicated in Figure S3 (Supplementary Materials). The results suggest that the adsorption capacity (q_m) of 240.96 $mg \cdot g^{-1}$ was investigated. The dimensionless separating constant (R_L) value was found to be 0.199, the coefficient correlation factor (R^2) value was 0.991, while a Langmuir constant (K_b) value of 4.021 was achieved.

Table 2. Adsorption isotherm models, for adsorption Cr(VI) on MNPs/rGO/PMMA composite.

Model Equation	R^2	Kf	1/n	K_b	q_m ($mg \cdot g^{-1}$)	R_L
Langmuir Isotherm	0.991	—	—	4.021	240.96	0.199
Freundlich Isotherm	0.972	6.85	0.351	—	—	

To examine the adsorption information through the Freundlich adsorption isotherm, Equation (7) was used:

$$\log q_e = \log Kf + \frac{1}{n} \log C_e \quad (7)$$

The slope and intercept values obtained from the Freundlich adsorption isotherm graph concluded whether the adsorption was favorable or unfavorable. It also explained whether the system was heterogeneous or not, as shown in Figure S4 (Supplementary Materials). A Freundlich isotherm parameter such as 1/n gives useful evidence about the adsorption system. A value of $1/n > 1$ suggests that the system is imperfect for low concentrations but thrives at high concentrations. Likewise, when the value of 1/n is smaller than one ($1/n < 1$), it means that the adsorption system is perfect over the whole range of concentrations. Correspondingly, if the 1/n value is equal to one ($1/n = 1$), it suggests that the adsorption system is homogenous [53].

In the current study, the value of R^2 (0.99124) showed that the Langmuir adsorption isotherm fitted better to the adsorption data than the Freundlich isotherm ($R^2 = 0.97215$). On the other hand, the R_L value was 0.1999, which means that the adsorption of Cr(VI) on MNPs/rGO/PMMA was favorable and monolayer. In addition, the value of 1/n was found to be 0.35102, which confirms that the adsorption was positive at all possible concentrations and the adsorption framework was heterogeneous.

4.5. Thermodynamic Investigation

To investigate the adsorption system versus temperature, the influence of temperature on the adsorption of Cr(VI) achieved at various temperatures, from 293 to 373 K, at optimized conditions, as illustrated in Figure 7.

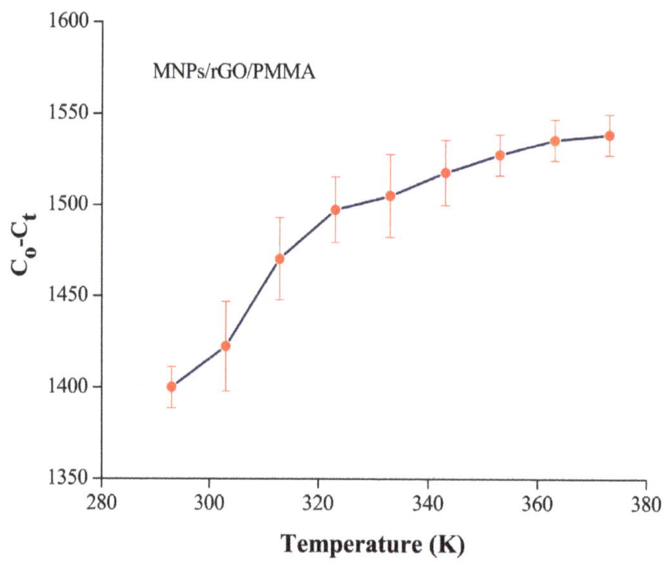

Figure 7. Effect of temperature on adsorption of Cr(VI) on MNPs/rGO/PMMA composite; adsorbent amount, 150 mg; shacking time, 30 min; pH 2.

The adsorption of Cr(VI) increased proportionally with the increase in temperature from 293 to 373 K. This can be attributed to the fact that a temperature increase leads to a corresponding increase in the system's energy, thereby providing more chances of adsorbate-active site interactions. As the temperature was increased, there might have been an increase in the number of active sites, thereby rupturing the functional group bond occurred on the surface of the adsorbent, which improves the adsorption proficiency [50]. The thermodynamic parameters, including Gibbs free energy (ΔG^o), enthalpy (ΔH^o) and entropy (ΔS^o), were determined using Equations (8)–(10), respectively.

$$\Delta G^o = -RT \ln K_D \tag{8}$$

$$\Delta H^o = R\,\frac{T_{2T_1}}{T_{2-T_1}}\ln\frac{K_2}{K_1} \tag{9}$$

$$\Delta S^o = \frac{(\Delta H^o - \Delta G^o)}{T} \tag{10}$$

The negative values of ΔG^o indicated the spontaneous nature of Cr(VI) adsorption. Similarly, ΔG^o values became more negative as the temperature increased, which shows that the adsorption process was improved as the temperature was enhanced. The ΔH^o and ΔS^o values were determined from the plot shown in Figure S5 (Supplementary Materials). The positive value of ΔH^o demonstrates that the current adsorption system was endothermic in nature. Besides, the positive value of ΔS^o confirms that the adsorption system was spontaneous, as shown in Table 3.

Table 3. Thermodynamic parameters for the adsorption of Cr(VI) on (MNPs/rGO/PMMA).

Temperature K	ΔG^o KJ.mole^{-1}	ΔH^o KJ.mole^{-1}	ΔS^o KJ.mole^{-1}
293	−14.4944		
303	−15.4132		
313	−16.5329		
323	−17.4261		
333	−18.3889	46.0074	0.1875
343	−19.4415		
353	−20.8456		
363	−22.5964		
373	−24.0239		

4.6. Volume of Eluent Optimization and Desorption Studies

The desorption of metal particles from the adsorbent is an important aspect in wastewater treatment. For this purpose, Cr(VI) was desorbed from MNPs/rGO/for 30 min with (1 and 2 M) solutions of NaOH and (1 and 2M) NH_4OH, respectively. As the desorption process was completed, the filtrate was isolated from the adsorbent utilizing the outer magnetic field and was then analyzed through the atomic absorption spectrometer. The results indicate that the greatest desorption of Cr(VI) (95 ± 1) from the adsorbent was achieved with the 2M NaOH solution, while, with NH_4OH, the maximum desorption attained was 78 ± 1, as displayed in Table 4.

Table 4. Influence of various alkaline solutions, used for desorption of Cr(VI) from NPs/rGO/PMMA.

Basic Solution	Concentration of Eluent mol·L^{-1}	%, Recovery of Cr(VI)
NaOH	1	70 ± 1
NaOH	2	95 ± 1
NH_4OH	1	65 ± 2
NH_4OH	2	78 ± 1

The desorption process for Cr(VI) from the adsorbent with alkaline solutions has been well documented [54]. The volume of NaOH solution (10–40 mL) revealed that, when the volume eluent reagent increased, the percent desorption increased and the equilibrium was established up to 30 mL. Hence, 30 mL of 2 M NaOH was suggested for the Cr(VI) desorption, as demonstrated in Figure 8.

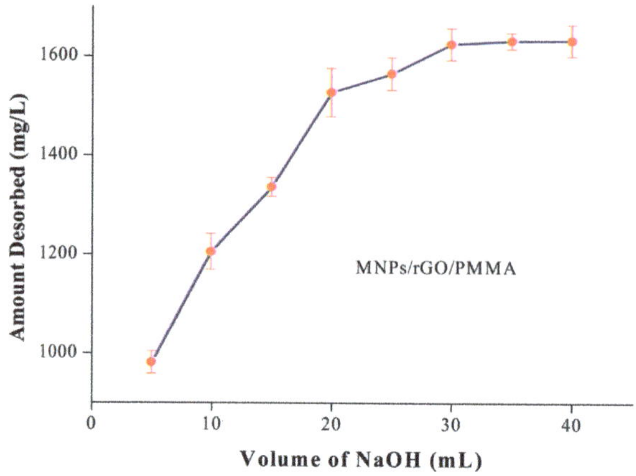

Figure 8. Effect of eluent volume on % recovery of Cr(VI) from MNPs/rGO/PMMA composite.

4.7. Recovery of Cr(VI), Recycling and Regeneration of the Adsorbent

The adsorbent's reuse and recovery of Cr(VI) are important parameters in wastewater treatment in terms of minimizing process costs. In the current examination, recycling experiments were conducted following the adsorption and desorption of Cr(VI) at optimized conditions. After adsorption, desorption was performed using 30 mL of NaOH solution (2M) shacked for 30 min. As the desorption process was finished, the MNPs/rGO/PMMA composite was washed with double distilled water as well as dilute HCl (0.1M) and dried under vacuum at 60 °C for reuse. The results of marginal loss in the adsorption of Cr(VI) after five repeated cycles indicate the highly stable nature of the newly designed MNPs/rGO/PMMA (Figure 9). During the desorption process, the recovery of Cr(VI) from the adsorbent was 95%, 94%, 96%, 95% and 96% in each cycle, respectively. The current investigation was not assisted to control the water contamination caused by Cr(VI) effluents. At the same time, it is vital to recover the industrially important Cr metal from industrial effluents which might be reused or recycled in different applications.

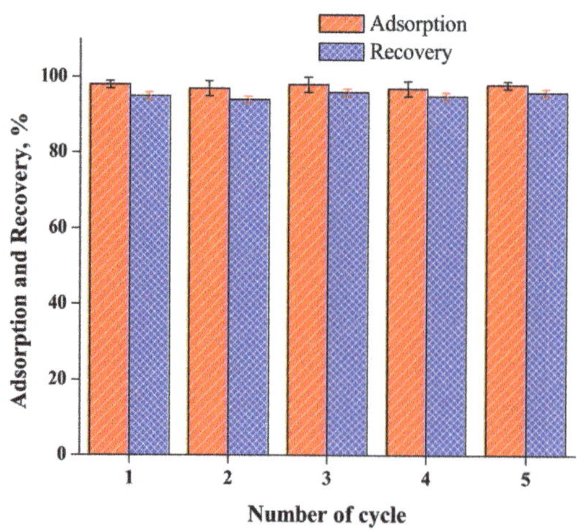

Figure 9. Recycling of MNPs/rGO/PMMA composite; eluent, 2M NaOH solution; volume, 30 mL.

The selectivity of the MNPs/rGO/PMMA composite towards the adsorption of Cr(IV) in the presence of various competing ions was investigated, in batch mode adsorption with various salts at different concentrations. The results show that, in the presence of various ions, the removal of Cr(VI) was minimally affected, indicating the high selectivity of the composite adsorbents towards Cr(VI), as indicated in Table 5.

Table 5. Effect of interfering ions on the adsorption of Cr-VI ions over the MNPs/rGO/PMMA composite.

Ions	Concentration (mg/L)	% Adsorption of Cr(VI)
Cu^{2+}	10	96 ± 4
Mn^{2+}	10	98 ± 3
Zn^{2+}	100	93 ± 1
Cd^{2+}	100	97 ± 1
Fe^{3+}	100	96 ± 1
Ni^{2+}	10	94 ± 1
Ca^{2+}	500	98 ± 1
Mg^{2+}	500	98 ± 1
CO_3^{2-}	500	96 ± 1
F^-	1000	99 ± 1
SO_4^{2-}	500	98 ± 1

5. Theoretical Analysis of Fixed-Bed Column Data

The quantity of Cr(VI) adsorbed on MNPs/rGO/PMMA was intended from the region under the breakthrough curve using Equation (11):

$$q_{total} = \frac{Q}{1000}\int_{t=0}^{t=total} C_{ad}dv \tag{11}$$

where "Q" is the flow rate (mL/min), "t" is the time of total flow (min) and "C_{ad}" is the adsorbed concentration of Cr(VI) mg/L. The adsorption capacity of the MNPs/rGO/PMMA was calculated from the breakthrough curve using Equation (12):

$$q_{e=}\frac{C_0}{m}\int_0^{vb}\left(1-\frac{C_t}{C_0}\right)dv \tag{12}$$

where "q_e" is the adsorption capacity of MNPs/rGO/PMMA at the breakthrough curve (mg/g), "m" is the mass of the adsorbent (g), "C_t" is the outlet solution concentration (mg/L) at any time (min), "C_0" is the initial concentration (mg/L) and "V" is the volume (L) of the treated solution. Similarly, Equation (13) was used to calculate the empty bed residence time of the continuous bed column:

$$\text{Empty ed residence time (EBRT)} = \frac{\text{Bed voume}}{\text{volumetric flow rate of the liquid}} \tag{13}$$

During the continuous process, the MNPs/rGO/PMMA adsorbent was frequently exhausted and was calculated using Equation (14):

$$\text{Adsorbent exhausted rate (AER)} = \frac{\text{Mass of adsorbent (g)in column}}{\text{Volume of water treated}} \tag{14}$$

The bed volume (BV) of the continuous fixed bed adsorbent in the column was obtained through Equation (15):

$$\text{Bed volume} = \frac{\text{volume of water treated at breakthrough curve (L)}}{\text{volume of adsorbent bed}} \tag{15}$$

The proficiency of the continuous fixed-bed column was studied under various flow rates and adsorbent masses. Likewise, the fixed-bed adsorption in the column was also applied to mathematical models, including the Thomas and Yoon–Nelson Models.

5.1. Effect of Mass of MNPs/rGO/PMMA Composite on the Breakthrough Curve

In order to investigate the impact of adsorbent masses on the breakthrough curve, tannery wastewater (Cr(VI), 1640 mg/L) was passed through various adsorbent masses extending from 1 g to 3 g, while other parameters, such as flow rate, pH and concentration of Cr(VI), were kept constant. The results in Figure 10 display that the breakthrough time was enhanced as the adsorbent mass was increased from 1 g to 3 g. This could be ascribed to the availability of more active sites of MNPs/rGO/PMMA for the adsorption of Cr(VI) with the increase in mass [55].

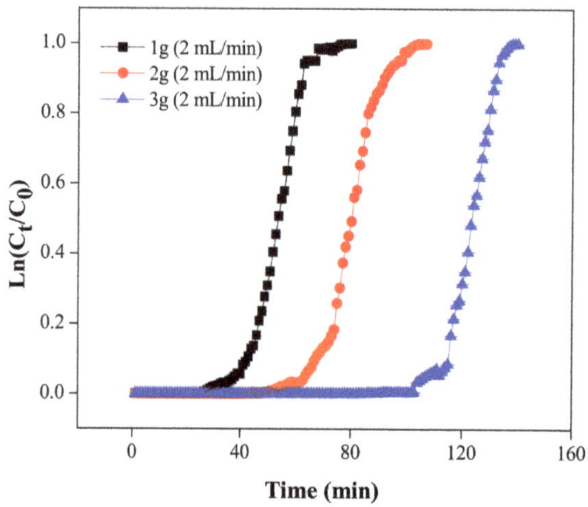

Figure 10. Cr(VI) adsorption breakthrough with different adsorbent masses.

It might be induced from the results that the breakthrough curve is dependent on the MNPs/rGO/PMMA bed mass. It was found that the former breakthrough curve was seen at lower adsorbent mass and, after the breakthrough, a sharp ascent occurs in the Cr concentration. The sharp ascent is due to the exit of the mass exchange zone; at that point, the bed has an insignificant capacity to adsorb Cr(VI) [56]. At the same time, the adsorption capacity (q_e) and adsorbent exhaustion rate (AER) were calculated from the breakthrough curve and are given in Table 6.

Table 6. Parameters for fixed-bed adsorption of Cr(VI) on MNPs/rGO/PMMA at breakthrough point.

Parameters		Adsorbent Capacity (mg/g)	AER (g/L)
Adsorbent Mass (g)	Flow Rate (mL/min)		
1	2	125.16	20.00
2	2	134.77	17.24
3	2	152.13	14.56
Flow rate (mL/min)	Adsorbent masses (g)		
1	3	120.06	11.02
2	3	111.98	10.10
3	3	75.01	9.80

The adsorption capacity was enhanced with an increase in the adsorbent mass due to the increase in the surface zone, which encouraged the availability of active sites for adsorption. Similarly, the obtained values of the adsorbent exhaustion rate (AER) declined from 20.00 g/L to 14.56 g/L with an increase in the adsorbent masses from 1 g to 3 g. The lower values of the adsorbent exhaustion rate (AER) showed better execution for adsorption [57].

5.2. *Effect of Flow Rate*

The flow rate of the wastewater stream plays a substantial role in estimating the execution of the adsorption process, especially at the industrial level, to achieve maximum treatment of the influent. Therefore, the impact of stream rate on the adsorption of Cr(VI) by MNPs/rGO/PMMA was studied at different flow rates ranging from 1 mL/min to 3 mL/min, keeping the other parameters constant. The results are shown in Figure 11 and the calculated process parameter values are recorded in Table 5. As shown in Figure 11, the breakthrough curve shortened as the flow rate was enhanced from 1 mL/min to 3 mL/min, for constant adsorbent mass.

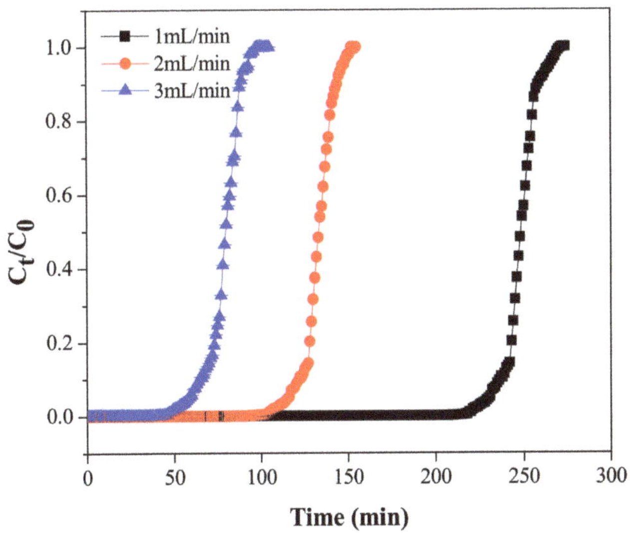

Figure 11. Cr(VI) adsorption breakthrough at different flow rates.

This is presumably due to the increased mass passage rate, which brings a reduction in the time needed to obtain the preferred breakthrough concentration [58]. Additionally, the adsorption capacity and ARE also declined from 120.06 to 75.01 mg/L and from 11.02 to 9.08 mg/L, respectively. This could be for the reason that, at a higher stream rate, the adsorbent in the fixed-bed column rapidly saturated and there was insufficient contact time for the complete adsorption of Cr on the adsorbent in the fixed-bed column [46].

5.3. *Desorption and Regeneration of the Column*

During the desorption experiments, Cr(VI) was leached from a column with various masses of the MNPs/rGO/PMMA composite at a constant flow rate (2 mL/min) of the 2 M NaOH solution. The results show that the breakthrough point for desorption was found in three distinctive times, i.e., 27, 42 and 75 min, for 1, 2 and 3 g of MNPs/rGO/PMMA masses, respectively, as shown in Figure 12. The maximum recovery of Cr(VI) attained was 79.93 ± 0.07%, 80.40 ± 07% and 76.33 ± 0.008%, when loaded with 1, 2 and 3 g MNPs/rGO/PMMA at a 2 mL/min flow rate of the 2 M NaOH solution, respectively. A further increase in time caused the concentration of leached Cr(VI) to decrease the in outlet solution due to the unsaturation of Cr(VI) on the MNPs/rGO/PMMA composite [59].

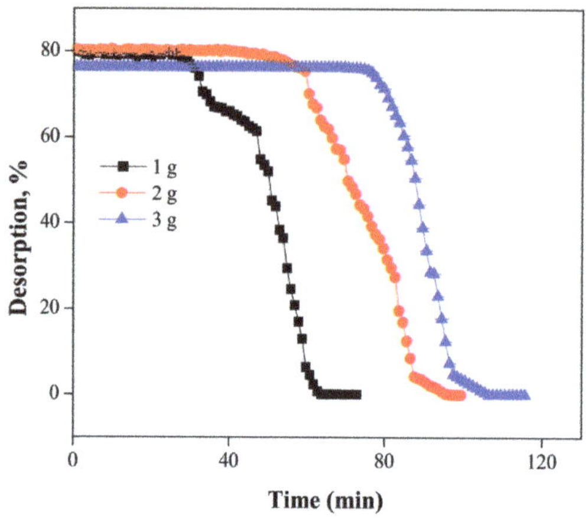

Figure 12. Desorption of Cr(VI) at various adsorbent masses using the 2 M NaOH solution.

Meanwhile, the effect of various flow rates, i.e., 1, 2 and 3 mL/min was also investigated at the constant adsorbent mass of 3 g. The results are displayed in Figure 13. We observed that, as the flow rate of the NaOH solution was increased (from 1 to 3 mL/min), the desired time for the recovery of Cr(VI) decreased from 186 to 47 min. The maximum recovery of Cr(VI) in each column was observed to be 70.77 ± 0.07, 74.82 ± 0.04 and 76.45 ± 0.001, with 1, 2 and 3 mL/min flow rates of NaOH.

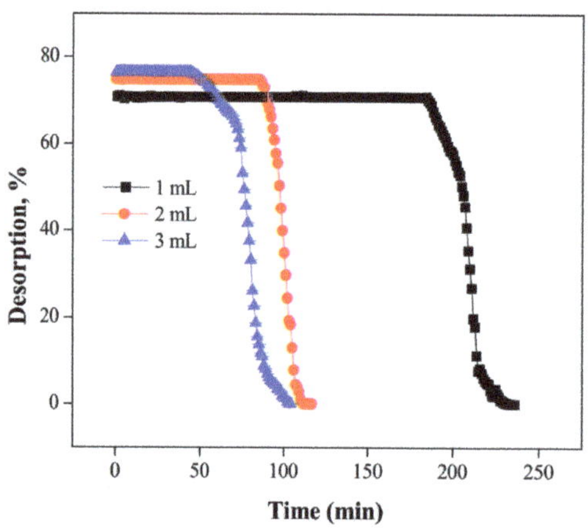

Figure 13. Desorption of Cr(VI) at various flow rates using the 2 M NaOH solution.

5.4. Theoretical Modeling of the Breakthrough Curve

The theoretical calculation of the column studies was assessed by the Thomas and Yoon–Nelson models to investigate the efficiency of the fixed-bed adsorption column.

5.4.1. Thomas Model

The Thomas model is a general theoretical model utilized for fixed-bed column data analysis and for the prediction of breakthrough points. The Thomas model is described on the supposition of pseudo-2nd order kinetic and Langmuir adsorption isotherm models. The Thomas model is expressed in Equation (16):

$$\ln \frac{C_0}{C_t - 1} = \frac{K_{TH}}{Q}(qe)(m) - K_{TH}C_0 t \quad (16)$$

where "C_0" (mL/L) and "C_t" (mL/L) are the initial and final concentration, respectively, "q_e" (mg/g) is the adsorbent capacity, "KTH" (mL/min g) is the Thomas model constant and "t" (min) is the total flow time. The values of "q_e" and "K_{TH}" were obtained from the linear plot between ($\ln \frac{C_0}{C_{t-1}}$) and time (t), as shown in Figure S6 (Supplementary Materials).

Different values of KTH and q were obtained from the Thomas model for various flow rates, as shown in Table 7, which suggests that the values of q and KTH decreased as the flow rate increased from 135.31 to 111.48 mg/g and from 11.58×10^{-2} to 8.82×10^{-2}, respectively. This may be due to the short time for salute particles to adsorb on the MNPs/rGO/PMMA composite and permitted Cr(VI) before the complete adsorption. The Thomas model exhibited good fit to the adsorption data with the R2 values of 0.999, 0.993 and 0.972 under the three flow rates, respectively. From the above results, it was concluded that there are some experimental points and prediction positions which recommend the fitting of the Thomas model on the experimental breakthrough curve. The same result was found by Shalini et al.; the adsorption capacity of Cr(VI) on chemically modified Lagerstroemia speciosa bark in the fixed-bed column decreased with the increase in the flow rate [60].

Table 7. Adsorption parameters calculated from the Thomas and Yoon–Nelson models.

Parameters				Thomas Model			Yoon–Nelson Model		
C_0 (mg/L)	Rate Flow (mL/min)	Bed Height (cm^{-1})	pH	K_{TH} (L/min gm)	q (mg/g)	R^2	K_{YN} (min^{-1})	τ (min)	R^2
1640	1	2.7	2	11.58×10^{-2}	135.31	0.999	27.19×10^{-2}	135.31	0.998
1640	2	2.7	2	9.14×10^{-2}	126.51	0.993	22.03×10^{-2}	126.51	0.997
1640	3	2.7	2	8.82×10^{-2}	111.48	0.972	43.53×10^{-2}	111.48	0.982

5.4.2. Yoon–Nelson Model

Yoon–Nelson developed a simple mathematical model for fixed-bed column adsorption and is based on the fact that the probability of adsorption rate decreases for each adsorbate molecule, which is proportional to the probability of adsorbate adsorption and the probability of adsorbate breakthrough on the adsorbent. The Yoon–Nelson model is mathematically expressed by Equation (17):

$$\ln\left(\frac{C_t}{C_0 - C_t}\right) = K_{YN}t - K_{YN}\tau \quad (17)$$

where "C_0" (mg/g) and "C_t" (mg/g) are the initial and final concentration in the outlet, "KYN" (min^{-1}) is the rate constant of Yoon–Nelson and "τ" (min) is the flow time required for the 50 percent breakthrough curve. The values of "q_e" and "K_{TH}" were obtained from the linear plot between $\left(\ln \frac{C_0}{C_{t-1}}\right)$ and time (t), as shown in Figure S7 (Supplementary Materials).

Different constants and variables were obtained from the Yoon–Nelson model at various flow rates, as displayed in Table 6, which suggests that the values of KTH were enhanced from 27.19×10^{-2} to 43.53×10^{-2}. Meanwhile, the values of τ (min) declined from 253.18 to 82.80 as the flow rates were increased. The reason is that the MNPs/rGO/PMMA composite in the column saturated more quickly to attain equilibrium as the flow rate increased. The values of R^2 achieved were 0.998, 0.997 and 0.982 for the three flow rates, respectively. This indicated that the Yoon–Nelson model fitted the experimental adsorption data well, which is consistent with results previously published in the literature [56].

6. Adsorption Mechanism

The adsorption system in the current investigation is the expected interaction between hydrogen chromate ($HCrO_4^-$) and functional groups of the adsorbent material.

The main factor involved in the mechanism of Cr(VI) adsorption is pH. The most extreme adsorption of Cr(VI) was found to be at pH 2. At this pH, $Cr_2O_7^{2-}$ was changed to $HCrO_4^-$ species, as shown in the following reaction [46]:

$$Cr_2O_7^{2-} + H_2O \rightarrow 2HCrO_4^-$$

The adsorptive capacity of the magnetic composite was due to the active sites present (organic and inorganic phases), as well as the new sites yielded from the interaction between these phases. The adsorption of $HCrO_4^-$ on the active sites of organic and inorganic phases involves an electrostatic interaction. The oxygen atom on the surface of the magnetic composite becomes protonated to a high degree at pH 2, which brings a substantial electrostatic fascination among $HCrO_4^-$ and positively charges on the adsorbent. At this state, hazardous Cr(VI) in tannery waste gets adsorbed onto the surface of MNPs via electrostatic attraction, as shown in Scheme 4.

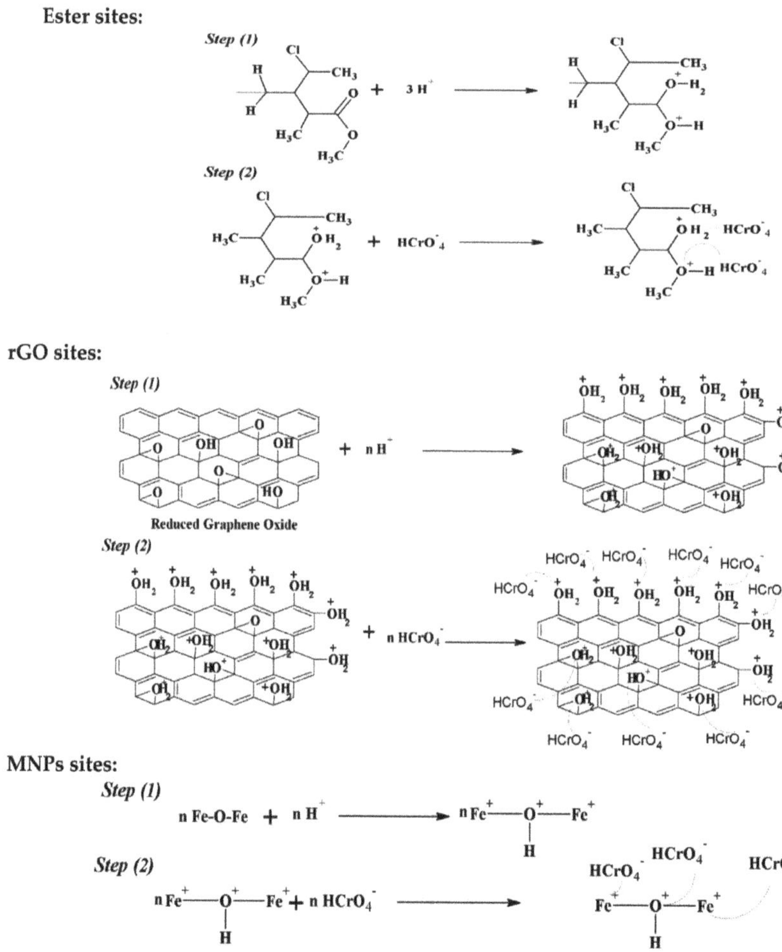

Scheme 4. Structure of adsorption mechanism.

7. Physicochemical Study of Tannery Wastewater

Tannery wastewater was collected from industry and a physicochemical study was carried out before adsorption treatment, as given in Section 3.2. Various physicochemical parameters were analyzed, including pH, COD, suspended solids, Cr concentration and BOD, as presented in Table 8. In the case of tannery wastewater, after the adsorption treatment, both in batch and column modes, the concentration of Cr, level of COD and BOD were reduced to 3.51 and 2.42, 110 and 99, 120 and 109 mg/L, respectively, whereas no suspended solids were found after adsorption, as they were removed through filtration before treatment. These findings conclude that all the parameters are well below the permissible range for the Cr(VI) in wastewater, indicating the effective role of this study in processing and cleaning tannery wastewater.

Table 8. Physicochemical characteristics of tannery wastewater before and after adsorption treatment.

Parameters	Values		
	Before Treatment	**After Batch Mode Adsorption**	**After Column Adsorption**
Cr concentration	1640 mg/L	3.51 mg/L	3.42 mg/L
pH	3.17	Variable	Variable
Chemical oxygen demand (COD)	1130 mg/L	110 mg/L	99 mg/L
Biological oxygen demand (BOD)	396 mg/L	120 mg/L	109 mg/L
Suspended solids (SS)	960 mg/L	0.00 mg/L	0.00 mg/L

To the best of our knowledge, the application of MNPs/rGO/PMMA composite materials for the adsorption and recovery of Cr(VI) from real tannery wastewater with a concentration of Cr(VI) as high as 1640 mg/L through both batch and column mode adsorption has not been reported in earlier literature. This material offers multiple functionalities for adsorption of Cr(VI) through a diverse mechanism; hence, it provides higher adsorption potential than any other conventional material, which is explained in detail in Section 6.

Owing to the hazardous nature of Cr(VI) in water bodies, the removal of Cr(VI) from wastewater streams has been extensively studied in the literature. In this regard, Table 9 presents a comparison of the adsorption efficiency of current adsorbents and various types of other materials reported in the literature, which concludes that the MNPs/rGO/PMMA composite offers superior efficiency.

Table 9. Comparison of Cr(VI) adsorption capacities of various adsorbents.

Adsorbent	Adsorption Capacity	pH	Adsorption Process	Wastewater	Ref.
Bagasse fly ash	29.07 mg/g	2–3	Batch	Synthetic	[61]
Fe_3O_4/rGO	98.1%	1	Batch	-	[41]
Non-cross-linked chitosan	80 mg/g	5	Batch	Synthetic	[62]
Polymeric based surfactant-chitosan	180 mg/g	5.3	Batch	Synthetic	[63]
Sawdust	1.74 mg/g	3	Batch	Synthetic	[64]
SWCNTs	96.9 mg/g	4	Batch	Synthetic	[65]
MNPs/rGO/PMMA	109.3/135.3 mg/g	3	Batch/column	Wastewater	Current work

8. Conclusions

A well-organized and novel MNPs/rGO/PMMA composite was successfully fabricated and applied for the adsorptive removal and recovery of Cr(VI) from tannery wastewater. The synthesized composite was analyzed in detail by the XRD, FTIR and SEM techniques. The FTIR investigation declared the successful synthesis of the rGO, MNPs/rGO, PMMA and MNPs/rGO/PMMA composite, while the SEM analysis confirmed the rougher surface morphology of the MNPs/rGO/PMMA composite. In batch adsorption, maximum adsorption of Cr(VI) (99.32 $\pm$ 2%) was attained under the optimum conditions of pH 2, sample volume of 10 mL, adsorbent amount of 150 mg and shacking time of 30 min. The kinetic investigations showed that the adsorption followed a pseudo-1st order kinetic model. The experimental data of adsorption followed the Langmuir model. Cr(VI) particles were effectively desorbed from the adsorbent using 30 mL of 2 M NaOH solution, while the adsorbent remained stable for five consecutive reuses. In the case of continuous column mode adsorption, the breakthrough time decreased with the increase in the feed flow rate as well as that of the mass of the composite adsorbent. According to the Thomas model, the Cr adsorption capacity of the composite with a column bed height of 2.7 cm and feed flow rate of 1 mL/min was found to be 135 mg/g. This study provides useful evidence for the remediation of hazardous Cr(VI) from tannery wastewater via a cost-effective and mechanically feasible strategy on a large scale.

Supplementary Materials: The following are available online at https://www.mdpi.com/1996-1944/14/22/6923/s1, Figure S1. Pseud 1st order of kinetic plot for the adsorption of Cr(VI) on MNPs/rGO/PMMA composite. Figure S2. Pseudo 2nd order of kinetic plot for the adsorption of Cr(VI) on MNPs/rGO/PMMA composite, Figure S3. Langmuir adsorption isotherm for the adsorption of Cr(VI) on MNPs/rGO/PMMA composite. Figure S4. Freundlich adsorption isotherm for the adsorption of Cr(VI) on MNPs/rGO/PMMA composite. Figure S5. Enthalpy and entropy changes for the adsorption of Cr(VI) on MNPs/rGO/PMMA composite. Figure S6. Thomas model for various flow rates i.e., 1, 2, and 3 mL/min for the adsorption of Cr(VI) on MNPs/rGO/PMMA composite. Figure S7. Yoon-Nelson model at different flow rates i.e., 1, 2, and 3 mL/min for the adsorption of Cr(VI) on MNPs/rGO/PMMA composite.

Author Contributions: R.U. and W.A., original concept and initial draft of the paper; R.I., M.Y., M.K. and M.I.K., materials synthesis, adsorption and analysis of experiments; W.A., M.Y., G.K. and R.U., processing, structural characterizations and data analysis; B.M.J. and W.A. supervised and coordinated the experimental work; G.K. and R.I., funding acquisition. All authors have read and agreed to the published version of the manuscript.

Funding: The authors would like to thank the Malaysia–Thailand Joint Authority under grant number IF062-2019 and Fundamental Research Grant Scheme FP050-2019A from the University of Malaya for providing funds during the course of this study. This work was also supported by the National Priorities Research Program Grant No. NPRP11S-1128-170042 from the Qatar National Research Fund (member of The Qatar Foundation) and co-financed by the European Union and Greek national funds through the Operational Program Competitiveness, Entrepreneurship and Innovation, under the call RESEARCH–CREATE–INNOVATE (2nd Cycle); acronym—SEMI-WEB; project code—T2EDK-02073. Finally, the support from the Higher Education Commission of Pakistan under NRPU program (Grant no. NRPU-9056) is highly acknowledged.

Institutional Review Board Statement: Not applicable.

Informed Consent Statement: Not applicable.

Data Availability Statement: Not applicable.

Acknowledgments: The authors are highly obliged for the support of the Centralized Resource Laboratory (CRL), University of Peshawar, KPK, Pakistan, for having provided analytical instruments during this work. In addition, the Pakistan Council of Scientific and Industrial Research (PCSIR) laboratories complex, Peshawar, KPK, Pakistan, is acknowledged for having provided their research facilities.

Conflicts of Interest: The authors declare that there are no conflict of interest regarding the publication of this manuscript.

References

1. Shi, X.; Wei, W.; Wu, L.; Ni, B.-J. Zero-valent iron mediated biological wastewater and sludge treatment. *Chem. Eng. J.* **2021**, *426*, 130821. [CrossRef]
2. Zhu, Y.; Wang, X.; Li, Z. Husbandry waste derived coralline-like composite biomass material for efficient heavy metal ions removal. *Bioresour. Technol.* **2021**, *337*, 125408. [CrossRef] [PubMed]
3. Nigam, M.; Rajoriya, S.; Singh, S.R.; Kumar, P. Adsorption of Cr (VI) ion from tannery wastewater on tea waste: Kinetics, equilibrium and thermodynamics studies. *J. Environ. Chem. Eng.* **2019**, *7*, 103188. [CrossRef]
4. Lofrano, G.; Meric, S.; Zengin, G.E.; Orhon, D. Chemical and biological treatment technologies for leather tannery chemicals and wastewaters: A review. *Sci. Total Environ.* **2013**, *461–462*, 265–281. [CrossRef]
5. Guo, Z.; Zhang, G.; Fang, J.; Dou, X. Enhanced chromium recovery from tanning wastewater. *J. Clean. Prod.* **2006**, *14*, 75–79. [CrossRef]
6. Mella, B.; Glanert, A.C.; Gutterres, M. Removal of chromium from tanning wastewater and its reuse. *Process. Saf. Environ. Prot.* **2015**, *95*, 195–201. [CrossRef]
7. Fernández, P.M.; Viñarta, S.C.; Bernal, A.R.; Cruz, E.L.; Figueroa, L.I. Bioremediation strategies for chromium removal: Current research, scale-up approach and future perspectives. *Chemosphere* **2018**, *208*, 139–148. [CrossRef]
8. Martínez-Huitle, C.A.; Ferro, S. Electrochemical oxidation of organic pollutants for the wastewater treatment: Direct and indirect processes. *Chem. Soc. Rev.* **2006**, *35*, 1324–1340. [CrossRef]
9. Wang, J.; Pan, K.; He, Q.; Cao, B. Polyacrylonitrile/polypyrrole core/shell nanofiber mat for the removal of hexavalent chromium from aqueous solution. *J. Hazard. Mater.* **2013**, *244–245*, 121–129. [CrossRef]
10. Pradhan, D.; Sukla, L.B.; Sawyer, M.; Rahman, P. Recent bioreduction of hexavalent chromium in wastewater treatment: A review. *J. Ind. Eng. Chem.* **2017**, *55*, 1–20. [CrossRef]
11. Zhang, H.; Li, P.; Wang, Z.; Cui, W.W.; Zhang, Y.; Zheng, S.; Zhang, Y. Sustainable Disposal of Cr(VI): Adsorption–Reduction Strategy for Treating Textile Wastewaters with Amino-Functionalized Boehmite Hazardous Solid Wastes. *ACS Sustain. Chem. Eng.* **2018**, *6*, 6811–6819. [CrossRef]
12. Zhitkovich, A. Chromium in Drinking Water: Sources, Metabolism, and Cancer Risks. *Chem. Res. Toxicol.* **2011**, *24*, 1617–1629. [CrossRef]
13. Hong, J.; Xie, J.; Mirshahghassemi, S.; Lead, J. Metal (Cd, Cr, Ni, Pb) removal from environmentally relevant waters using polyvinylpyrrolidone-coated magnetite nanoparticles. *RSC Adv.* **2020**, *10*, 3266–3276. [CrossRef]
14. Jin, W.; Du, H.; Zheng, S.; Zhang, Y. Electrochemical processes for the environmental remediation of toxic Cr(VI): A review. *Electrochim. Acta* **2016**, *191*, 1044–1055. [CrossRef]
15. Qasim, W.; Mane, A. Characterization and treatment of selected food industrial effluents by coagulation and adsorption techniques. *Water Resour. Ind.* **2013**, *4*, 1–12. [CrossRef]
16. Mubeena, K.; Muthuraman, G. Extraction and stripping of Cr(VI) from aqueous solution by solvent extraction. *Desalination Water Treat.* **2014**, *55*, 1–8. [CrossRef]
17. Alvarado, L.; Torres, I.R.; Chen, A. Integration of ion exchange and electrodeionization as a new approach for the continuous treatment of hexavalent chromium wastewater. *Sep. Purif. Technol.* **2013**, *105*, 55–62. [CrossRef]
18. Zhao, R.; Li, X.; Sun, B.; Li, Y.; Li, Y.; Yang, R.; Wang, C. Branched polyethylenimine grafted electrospun polyacrylonitrile fiber membrane: A novel and effective adsorbent for Cr(vi) remediation in wastewater. *J. Mater. Chem. A* **2017**, *5*, 1133–1144. [CrossRef]
19. Qi, W.; Zhao, Y.; Zheng, X.; Ji, M.; Zhang, Z. Adsorption behavior and mechanism of Cr(VI) using Sakura waste from aqueous solution. *Appl. Surf. Sci.* **2016**, *360*, 470–476. [CrossRef]
20. Deng, H.; Chen, G.Q.; Gras, S.L.; Kentish, S.E. The effect of restriction membranes on mass transfer in an electrodialysis with filtration membrane process. *J. Membr. Sci.* **2017**, *526*, 429–436. [CrossRef]
21. Ahmad, W.; Qaiser, S.; Ullah, R.; Mohamed Jan, B.; Karakassides, M.A.; Salmas, C.E.; Kenanakis, G.; Ikram, R. Utilization of Tires Waste-Derived Magnetic-Activated Carbon for the Removal of Hexavalent Chromium from Wastewater. *Materials* **2021**, *14*, 34. [CrossRef]
22. Bagri, A.; Mattevi, C.; Acik, M.; Chabal, Y.J.; Chhowalla, M.; Shenoy, V.B. Structural evolution during the reduction of chemically derived graphene oxide. *Nat. Chem.* **2010**, *2*, 581–587. [CrossRef]
23. Mukherjee, R.; Bhunia, P.; De, S. Impact of graphene oxide on removal of heavy metals using mixed matrix membrane. *Chem. Eng. J.* **2016**, *292*, 284–297. [CrossRef]
24. Lingamdinne, L.P.; Koduru, J.R.; Karri, R.R. A comprehensive review of applications of magnetic graphene oxide based nanocomposites for sustainable water purification. *J. Environ. Manag.* **2019**, *231*, 622–634. [CrossRef] [PubMed]
25. Shang, J.; Guo, Y.; He, D.; Qu, W.; Tang, Y.; Zhou, L.; Zhu, R. A novel graphene oxide-dicationic ionic liquid composite for Cr(VI) adsorption from aqueous solutions. *J. Hazard. Mater.* **2021**, *416*, 125706. [CrossRef]
26. Kong, D.; He, L.; Li, H.; Zhang, F.; Song, Z. Preparation and characterization of graphene oxide/chitosan composite aerogel with high adsorption performance for Cr(VI) by a new crosslinking route. *Colloids Surf. A Physicochem. Eng. Asp.* **2021**, *625*, 126832. [CrossRef]

27. Shan, H.; Zeng, C.; Zhao, C.; Zhan, H. Iron oxides decorated graphene oxide/chitosan composite beads for enhanced Cr(VI) removal from aqueous solution. *Int. J. Biol. Macromol.* **2021**, *172*, 197–209. [CrossRef] [PubMed]
28. Pipíška, M.; Zarodňanská, S.; Horník, M.; Ďuriška, L.; Holub, M.; Šafařík, I. Magnetically functionalized moss biomass as biosorbent for efficient Co2+ ions and thioflavin T removal. *Materials* **2020**, *13*, 3619. [CrossRef]
29. Liu, J.-F.; Zhao, Z.-S.; Jiang, G.-B. Coating Fe3O4 magnetic nanoparticles with humic acid for high efficient removal of heavy metals in water. *Environ. Sci. Technol.* **2008**, *42*, 6949–6954. [CrossRef] [PubMed]
30. Almomani, F.; Bhosale, R.; Khraisheh, M.; Kumar, A.; Almomani, T. Heavy metal ions removal from industrial wastewater using magnetic nanoparticles (MNP). *Appl. Surf. Sci.* **2020**, *506*, 144924. [CrossRef]
31. Wang, Z.; Yao, M.; Wang, X.; Li, S.; Liu, Y.; Yang, G. Influence of reaction media on synthesis of dialdehyde cellulose/GO composites and their adsorption performances on heavy metals. *Carbohydr. Polym.* **2020**, *232*, 115781. [CrossRef]
32. Lingamdinne, L.P.; Choi, J.-S.; Angaru, G.K.R.; Karri, R.R.; Yang, J.-K.; Chang, Y.-Y.; Koduru, J.R. Magnetic-watermelon rinds biochar for uranium-contaminated water treatment using an electromagnetic semi-batch column with removal mechanistic investigations. *Chemosphere* **2022**, *286*, 131776. [CrossRef]
33. Lee, S.; Lingamdinne, L.P.; Yang, J.-K.; Chang, Y.-Y.; Koduru, J.R. Potential electromagnetic column treatment of heavy metal contaminated water using porous Gd2O3-doped graphene oxide nanocomposite: Characterization and surface interaction mechanisms. *J. Water Process. Eng.* **2021**, *41*, 102083. [CrossRef]
34. Shahriary, L.; Athawale, A.A. Graphene oxide synthesized by using modified hummers approach. *Int. J. Renew. Energy Environ. Eng.* **2014**, *2*, 58–63.
35. Ren, P.-G.; Yan, D.-X.; Ji, X.; Chen, T.; Li, Z.-M. Temperature dependence of graphene oxide reduced by hydrazine hydrate. *Nanotechnology* **2010**, *22*, 055705. [CrossRef]
36. Gonçalves, G.; Marques, P.A.A.P.; Barros-Timmons, A.; Bdkin, I.; Singh, M.K.; Emami, N.; Grácio, J. Graphene oxide modified with PMMA via ATRP as a reinforcement filler. *J. Mater. Chem.* **2010**, *20*, 9927–9934. [CrossRef]
37. Hoan, V.; Thu, N.T.; Duc, H.V. Fe_3O_4/reduced graphene oxide nanocomposite: Synthesis and its application for toxic metal ion removal. *J. Chem.* **2016**, *2016*, 2418172.
38. Chang, Y.-P.; Ren, C.-L.; Qu, J.-C.; Chen, X.-G. Preparation and characterization of Fe_3O_4/graphene nanocomposite and investigation of its adsorption performance for aniline and p-chloroaniline. *Appl. Surf. Sci.* **2012**, *261*, 504–509. [CrossRef]
39. Zhang, M.; Jia, M. High rate capability and long cycle stability Fe3O4–graphene nanocomposite as anode material for lithium ion batteries. *J. Alloy Compd.* **2013**, *551*, 53–60. [CrossRef]
40. Chen, Y.; Wang, Y.; Zhang, H.-B.; Li, X.; Gui, C.-X.; Yu, Z.-Z. Enhanced electromagnetic interference shielding efficiency of polystyrene/graphene composites with magnetic Fe_3O_4 nanoparticles. *Carbon* **2015**, *82*, 67–76. [CrossRef]
41. Zhang, X.; Yi, G.; Zhang, Z.; Yu, J.; Fan, H.; Li, P.; Zeng, H.; Xing, B.; Chen, L.; Zhang, C. Magnetic graphene-based nanocomposites as highly efficient absorbents for Cr(VI) removal from wastewater. *Environ. Sci. Pollut. Res.* **2021**, *28*, 14671–14680. [CrossRef]
42. Amiri, M.; Shabani, A.M.H.; Dadfarnia, S.; Sadjadi, S. Simultaneous Functionalization and Reduction of Magnetic Graphene Oxide by l-Histidine and its Application for Magnetic Separation/Preconcentration of Antioxidant Trace Elements. *Biol. Trace Elem. Res.* **2018**, *190*, 262–272. [CrossRef] [PubMed]
43. Su, C.-Y.; Wang, J.-C.; Chen, C.-Y.; Chu, K.; Lin, C.-K. Spherical Composite Powder by Coupling Polymethyl Methacrylate and Boron Nitride via Spray Drying for Cosmetic Application. *Materials* **2019**, *12*, 706. [CrossRef] [PubMed]
44. Wang, E.R.; Shih, K.Y. Facile Microwave Hydrothermal Synthesis of ZnFe2O4/rGO Nanocomposites and Their Ultra-Fast Adsorption of Methylene Blue Dye. *Materials* **2021**, *14*, 5394. [CrossRef] [PubMed]
45. Cunha, G.d.C.; Silva, I.A.A.; Alves, J.R. Magnetic hybrids synthesized from agroindustrial byproducts for highly efficient removal of total chromium from tannery effluent and catalytic reduction of 4-nitrophenol. *Cellulose* **2018**, *25*, 7409–7422. [CrossRef]
46. Sharma, M.; Joshi, M.; Nigam, S.; Shree, S.; Avasthi, D.K.; Adelung, R.; Srivastava, S.K.; Mishra, Y. ZnO tetrapods and activated carbon based hybrid composite: Adsorbents for enhanced decontamination of hexavalent chromium from aqueous solution. *Chem. Eng. J.* **2019**, *358*, 540–551. [CrossRef]
47. Cai, W.; Zhu, F.; Liang, H.; Jiang, Y.; Tu, W.; Cai, Z.; Wu, J.; Zhou, J. Preparation of thiourea-modified magnetic chitosan composite with efficient removal efficiency for Cr(VI). *Chem. Eng. Res. Des.* **2019**, *144*, 150–158. [CrossRef]
48. Fazlzadeh, M.; Khosravi, R.; Zarei, A. Green synthesis of zinc oxide nanoparticles using Peganum harmala seed extract, and loaded on Peganum harmala seed powdered activated carbon as new adsorbent for removal of Cr(VI) from aqueous solution. *Ecol. Eng.* **2017**, *103*, 180–190. [CrossRef]
49. Yang, J.; Yu, M.; Chen, W. Adsorption of hexavalent chromium from aqueous solution by activated carbon prepared from longan seed: Kinetics, equilibrium and thermodynamics. *J. Ind. Eng. Chem.* **2015**, *21*, 414–422. [CrossRef]
50. Sarin, V.; Pant, K. Removal of chromium from industrial waste by using eucalyptus bark. *Bioresour. Technol.* **2006**, *97*, 15–20. [CrossRef]
51. Garg, U.K.; Kaur, M.; Garg, V.; Sud, D. Removal of hexavalent chromium from aqueous solution by agricultural waste biomass. *J. Hazard. Mater.* **2007**, *140*, 60–68. [CrossRef]
52. Muhammad, A.; Shah, A.-U.-H.A.; Bilal, S.; Rahman, G. Basic Blue Dye Adsorption from Water Using Polyaniline/Magnetite (Fe3O4) Composites: Kinetic and Thermodynamic Aspects. *Materials* **2019**, *12*, 1764. [CrossRef] [PubMed]
53. Rauf, M.; Bukallah, S.B.; Hamour, F.A.; Nasir, A.S. Adsorption of dyes from aqueous solutions onto sand and their kinetic behavior. *Chem. Eng. J.* **2008**, *137*, 238–243. [CrossRef]

54. Samir, L.; Samira, A.; Mekatel, E.H.; Djamel, N. Adsorption of Cr(VI) on *Stipa tenacissima* L (Alfa): Characteristics, kinetics and thermodynamic studies. *Sep. Sci. Technol.* **2018**, *54*, 876–887. [CrossRef]
55. Hashem, A.; Momen, A.; Hasan, M.; Nur-A-Tomal, S.; Sheikh, H.R. Chromium removal from tannery wastewater using Syzygium cumini bark adsorbent. *Int. J. Environ. Sci. Technol.* **2019**, *16*, 1395–1404. [CrossRef]
56. Nithya, K.; Sathish, A.; Kumar, P.S. Packed bed column optimization and modeling studies for removal of chromium ions using chemically modified Lantana camara adsorbent. *J. Water Process. Eng.* **2020**, *33*, 101069. [CrossRef]
57. Aranda-García, E.; Cristiani-Urbina, E. Hexavalent chromium removal and total chromium biosorption from aqueous solution by Quercus crassipes acorn shell in a continuous up-flow fixed-bed column: Influencing parameters, kinetics, and mechanism. *PLoS ONE* **2020**, *15*, e0227953. [CrossRef]
58. Agrawal, P.; Bajpai, A.K. Dynamic Column Adsorption Studies of Toxic Cr(VI) Ions onto Iron Oxide Loaded Gelatin Nanoparticles. *J. Dispers. Sci. Technol.* **2011**, *32*, 1353–1362. [CrossRef]
59. Ajmani, A.; Patra, C.; Subbiah, S.; Narayanasamy, S. Packed bed column studies of hexavalent chromium adsorption by zinc chloride activated carbon synthesized from Phanera vahlii fruit biomass. *J. Environ. Chem. Eng.* **2020**, *8*, 103825. [CrossRef]
60. Srivastava, S.; Agrawal, S.B.; Mondal, M.K. A fixed bed column study of natural and chemically modified Lagerstroemia speciosa bark for removal of synthetic Cr(VI) ions from aqueous solution. *Int. J. Phytoremediation* **2020**, *22*, 1233–1241. [CrossRef] [PubMed]
61. Noonpui, S.; Thiravetyan, P.; Nakbanpote, W.; Netpradit, S. Color removal from water-based ink wastewater by bagasse fly ash, sawdust fly ash and activated carbon. *Chem. Eng. J.* **2010**, *162*, 503–508. [CrossRef]
62. Schmuhl, R.; Krieg, H.; Keizer, K. Adsorption of Cu(II) and Cr(VI) ions by chitosan: Kinetics and equilibrium studies. *Water SA* **2004**, *27*, 1–8. [CrossRef]
63. Lee, M.-Y.; Hong, K.-J.; Kajiuchi, T.; Yang, J.-W. Synthesis of chitosan-based polymeric surfactants and their adsorption properties for heavy metals and fatty acids. *Int. J. Biol. Macromol.* **2005**, *36*, 152–158. [CrossRef] [PubMed]
64. Argun, M.E.; Dursun, S.; Ozdemir, C.; Karatas, M. Heavy metal adsorption by modified oak sawdust: Thermodynamics and kinetics. *J. Hazard. Mater.* **2007**, *141*, 77–85. [CrossRef] [PubMed]
65. Liu, Y.-X.; Yuan, D.-X.; Yan, J.-M.; Li, Q.-L.; Ouyang, T. Electrochemical removal of chromium from aqueous solutions using electrodes of stainless steel nets coated with single wall carbon nanotubes. *J. Hazard. Mater.* **2011**, *186*, 473–480. [CrossRef] [PubMed]

materials

MDPI

Review

Fabrication, Structure, Performance, and Application of Graphene-Based Composite Aerogel

Dequan Wei, Xiang Liu *, Shenghua Lv *, Leipeng Liu, Lei Wu, Zexiong Li and Yonggang Hou

College of Bioresources Chemical and Materials Engineering, Shaanxi University of Science and Technology, Xi'an 710021, China; Weidequan@hotmail.com (D.W.); 4233@sust.edu.cn (L.L.); wuleistarry@gmail.com (L.W.); lizexiong0622@sohu.com (Z.L.); houyonggang@sohu.com (Y.H.)

* Correspondence: LiuXiang@sust.edu.cn (X.L.); lvsh@sust.edu.cn (S.L.); Tel.: +86-298-616-8291 (X.L.)

Abstract: Graphene-based composite aerogel (GCA) refers to a solid porous substance formed by graphene or its derivatives, graphene oxide (GO) and reduced graphene oxide (rGO), with inorganic materials and polymers. Because GCA has super-high adsorption, separation, electrical properties, and sensitivity, it has great potential for application in super-strong adsorption and separation materials, long-life fast-charging batteries, and flexible sensing materials. GCA has become a research hotspot, and many research papers and achievements have emerged in recent years. Therefore, the fabrication, structure, performance, and application prospects of GCA are summarized and discussed in this review. Meanwhile, the existing problems and development trends of GCA are also introduced so that more will know about it and be interested in researching it.

Keywords: aerogel; applications; composite; graphene; graphene oxide; inorganic nanoparticle; natural polymer; reduced graphene oxide; structure; synthetic polymer

1. Introduction

1.1. Graphene-Based Composite Aerogel

Aerogel, a highly porous material with low density and high specific surface area, is obtained by replacing the liquid in wet gel with gas without significantly changing the structure and volume of the gel network. Graphene-based composite aerogel (GCA) is composed of graphene and its derivatives graphene oxide (GO) and reduced GO (rGO) with other matrix materials. Its functions are mainly derived from graphene and its derivatives (graphene-based materials), while its structure and volume stability are mainly determined by other matrix materials [1]. The research results indicate that GCA has lower density, higher porosity, smaller pore diameter, larger specific surface area, and more stable morphology compared to general aerogels, but more importantly it has some unique characteristics, such as higher heat resistance, better electrical conductivity, and higher absorbability of metal ions [2,3]. Therefore, GCA can be used not only as a thermal insulation, sound insulation, damping, and adsorptive material, but also as an electrode material for sensors and energy storage devices [4], which has become a research hotspot and attracted people's attention in recent years. Figure 1 shows the structure, properties, and application of GCA, and Figure 2 displays a structural schematic of graphene-based materials [5].

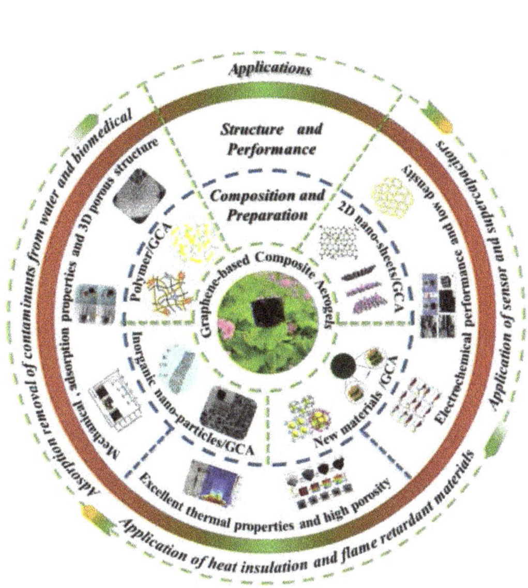

Figure 1. Composition, structure, properties, and applications of GCA.

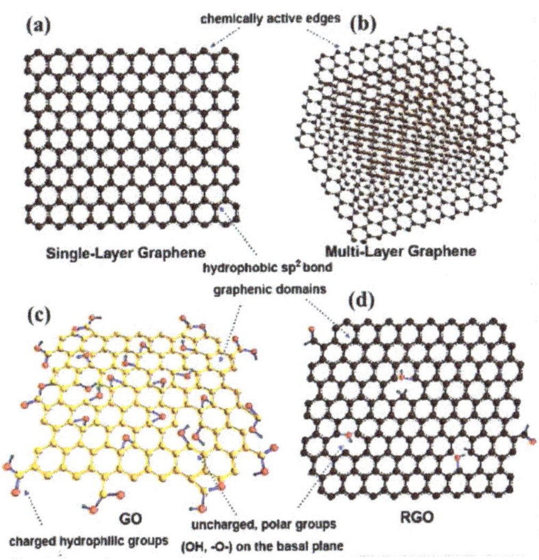

Figure 2. Structural schematic of graphene-based materials: (**a**) single-layer graphene, (**b**) multi-layer graphene, (**c**) GO, (**d**) rGO [6].

1.2. Preparation Principle of GCA

The preparation principle of GCA includes three key processes: sol, gel, and drying. In the sol-gel process, the reactants are uniformly mixed and reacted in the liquid phase to form clusters of different structures. The sol of graphene-based materials and matrix may be obtained by chemical vapor deposition (CVD), hydrothermal reaction, chemical reduction, and vacuum carbonization [7]. Then composite gel is constructed by the self-assembly, chemical cross-linking, template method and 3D printing. The gel contains a large amount of water or other solvents, up to more than 90%, with stable volume and no fluidity [8]. Then the solvent is removed from the gel by freeze-drying or supercritical drying to obtain GCA. The GCA always maintains higher porosity and larger specific surface area and has a similar network structure consisting of graphene-based and other matrix materials. Table 1 displays the preparation principle of GCA.

Table 1. Preparation methods and applications of common GCA materials.

GCA	Preparation Principle	Applications	Reference
2D nanosheets/GCA	3D printing	Energy storage materials	[9]
	Electrospinning	Wave-absorbing materials	[10]
	Self-assembly	Electrocatalysts	[11]
Inorganic nanoparticle materials/GCA	Self-assembly	Wearable piezoresistive sensors	[12]
	Sol-gel method	Energy storage material	[13]
	Sol-gel method	Gas sensing materials	[14]
	CVD	Adsorption materials	[15]
	Self-assembly	Catalytic material	[16]
Synthetic polymer/GCA	Sol-gel method	Pressure response sensor	[17]
	Template method	Thermal insulation materials	[18]
	Immersion method	Porous electrode	[19]
	Self-assembly	Energy storage material	[20]
Natural polymer/GCA	CVD	Adsorption materials	[21]
	Template method	Thermal insulation materials	[22]
	Self-assembly	Supercapacitor materials	[23]
	Sol-gel method	Adsorption materials	[24]
	Chemical cross-linking	Adsorption materials	[25]

1.3. Current Research Situation

Figure 3 shows the number of papers published on GCA from 2010 to 2021, indicating that the research began in 2010 and the number of papers grew exponentially in the last 10 years. According to the literature reports, GCA is usually composed of graphene and its derivatives and matrix materials. The properties and functions of GCA are mainly determined by graphene and its derivatives, while the porous structure and stability are mainly determined by matrix materials. The matrix materials include inorganic nanomaterials, synthetic polymers and natural polymers. Therefore, the research process on the fabrication method, material selection, structure construction, performance and application design of GCA is very complex, and it is necessary to summarize and guide based on these research results. Currently there is also a lack of targeted summary articles. In this review, we focus primarily on reviewing the fabrication of GCA with inorganic nanomaterials, synthetic polymer, and natural polymers along with its structure, performance, and applications.

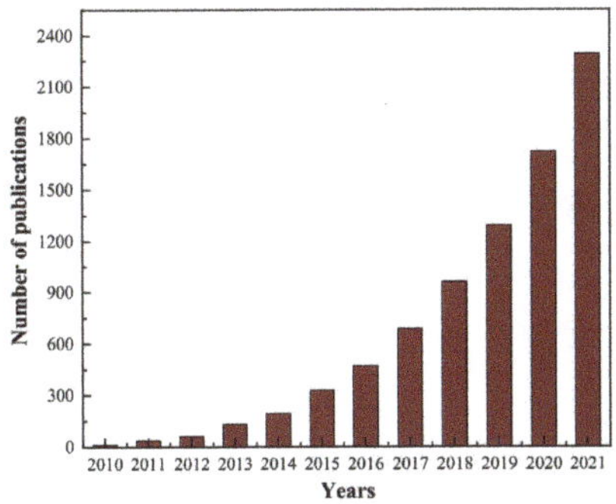

Figure 3. Number of papers on GCA in the last 12 years (as of 20 September 2021 by Science Direct record).

2. Composition of GCA

GCA usually consists of graphene-based materials and inorganic 2D nanosheets materials, inorganic nanoparticle materials, synthetic polymers, and natural polymers.

2.1. 2D Nanosheets/GCA

Inorganic 2D nanosheet materials can be compounded with graphene-based materials to prepare aerogels with better properties, including $Ti_3C_2T_x$ (MXenes), transition metal sulfide (TMD), metal organic framework (MOF), hexagonal boron nitride (h-BN), layered double hydroxide (LDH), perovskite and black phosphorus (BP), and other hot materials [26].

Mxenes is a kind of flexible 2D nanosheet with light weight, high electrical conductivity, high surface activity, and excellent electrochemical properties. Li et al. [10] prepared GCA by forming MXenes@GO composites by electrospinning, and the GCA as a wave-absorbing material had the advantages of light weight and elongated attenuating paths. The preparation process and mechanism are shown in Figure 4. 2D MoS_2 nanowires were fixed on the 2D rGO nanowires by 3D printing technology, which could solve the disadvantage of low conductivity of MoS_2 and prepare GCA for sodium ion batteries. [9] The introduction of 2D nanomaterials in GCA has aroused widespread interest, but reasonable design steps are needed to achieve superior properties of GCA, including the need to consider the problem of easy agglomeration in the composite process.

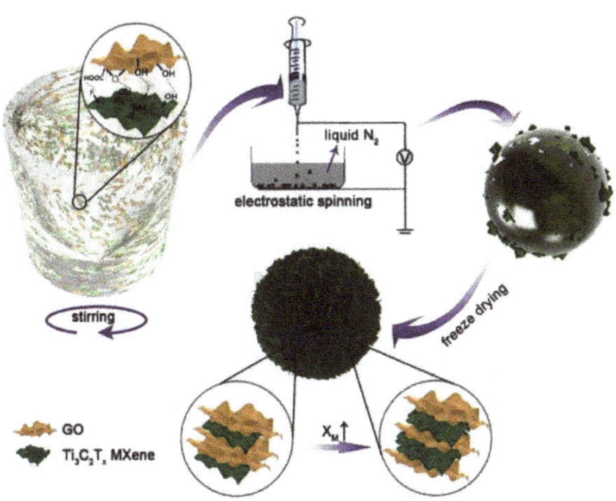

Figure 4. Assembly mechanism of MXenes@GO composite aerogel [10].

2.2. Inorganic Nanoparticle Materials/GCA

Inorganic nanoparticles, such as SnO_2, SiO_2, TiO_2, IO, etc. [27], have been widely used to prepare GCA. Nanoparticle GCA can be prepared by surface modification of inorganic nanoparticles or as a composite with graphene and hydrothermal assembly [28]. Cheng et al. [13] reported that TiO_2/GCA prepared by template-free method had excellent electrochemical properties and could be used as lithium battery anodes and high-performance energy storage. In addition, the solvothermal method of depositing ZnO nanoparticles on graphene nanosheets to prepare ZnO/GCA (Figure 5) showed good thermal conductivity performance, a porous network structure, and high specific surface area, and had greatly increased contact with gas, and these characteristics make ZnO/GCA suitable for preparing gas-sensing materials [14]. IO/GCA was prepared by the in situ growth method in a high-gravity reactor, with good catalytic efficiency for catalyzing the photo-Fenton reaction [16]. Graphene, combined with the good electronic, optical, and thermal conductivity of inorganic nanoparticles for preparing aerogels, broadens its potential applications in batteries, catalysis, sensors, and so on.

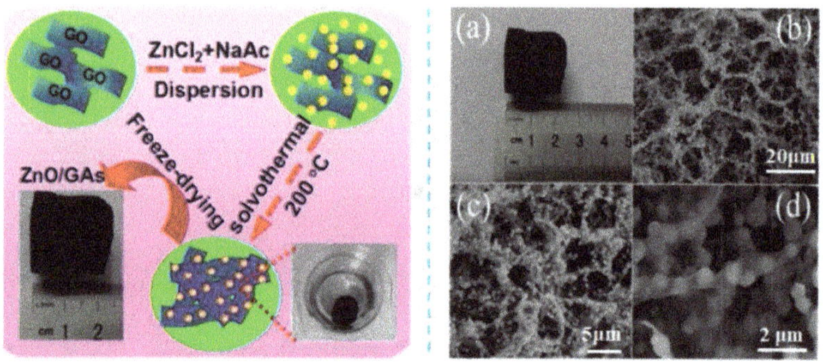

Figure 5. Preparation process of ZnO/GCA: (**a**) Macro-morphology and (**b–d**) Micro-morphology by SEM [14].

2.3. Synthetic Polymer/GCA

The weak strength of pure graphene aerogels greatly limits their practical application. However, this situation can be changed by forming composite GCA with some synthetic polymers, such as PE, PVA, PDMS, PANI, PAM, PM. Synthetic polymer/GCA has higher strength, lower density, larger specific surface area, and good strength and electrochemical properties, and can make sensors, adsorbents, catalysts, and other items [29]. As shown in Figure 6, Xiang et al. [17] prepared PVA/GCA by using γ-oxo-1-pyranobutyric acid (OPBA) as adhesive to hold the PVA coating and graphene skeleton together. The GCA can be used as a pressure response sensor due to its compressibility and deformation recovery. Zhang et al. [18] produced PDMS/GCA aerogel by permeation of PDMS into the interior of graphene aerogel to obtain aerogel with ultra-high electrical conductivity (1 $S{\cdot}cm^{-1}$), thermal conductivity (0.58 $W{\cdot}m^{-1}{\cdot}K^{-1}$), high hydrophobicity (contact angle 135°), and excellent strength and thermal stability. An et al. [19] deposited polyaniline (PANI) into porous graphene microspheres to make conductive spherical PANI/GCA, and the inclusion of PANI enhanced the graphene network and made the microspheres more resistant to deformation (Figure 7), which had the characteristics of shrinkage after water loss, recovery after dissolution, high specific capacitance, and good cycle stability, so they can be used as porous electrodes for energy conversion.

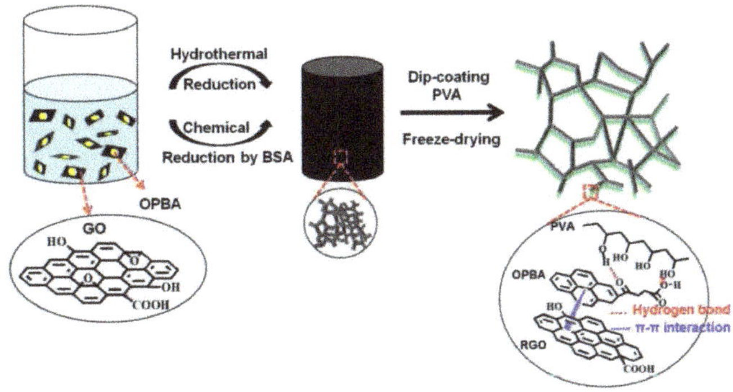

Figure 6. Schematic diagram of preparation process of PVA/GCA [17].

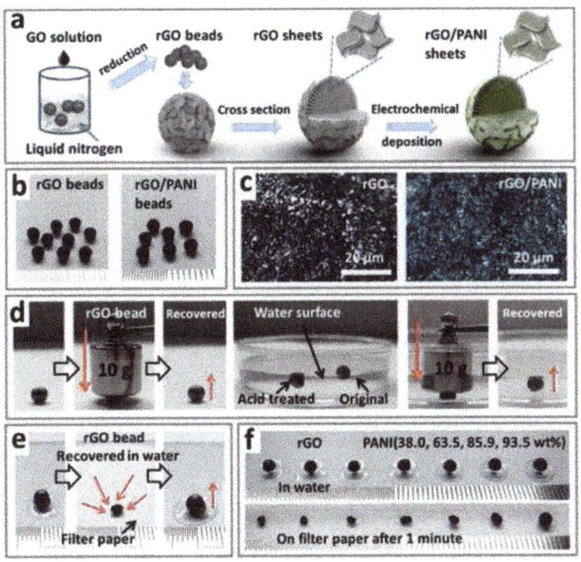

Figure 7. Schematic diagram of preparation of PANI/GCA: (**a**) Freeze-casting to prepare PANI/rGO aerogels; (**b**) PANI/rGO aerogel samples; (**c**) Microstructure of PANI/rGO aerogels; (**d**) Preparing rGO aerogels; (**e**) Recovery of shrunken rGO aerogels in water; (**f**) Shrinkage degree of GO and PANI/rGO aerogel by loss of water [19].

In short, the disadvantage of a weak structure in the reapplication of 3D graphene can be solved by preparing synthetic polymer modified GCA in various shapes, such as sol-gel, immersion, hydrothermal reaction, or chemical weather deposition. With better strength, adsorption, and electrical properties, the new synthetic polymer/GCA has great development prospects. Therefore, in the future, new forms of synthetic polymer/GCA should be developed by seeking new synthetic polymers and preparation methods for better performance.

2.4. Natural Polymer/GCA

The natural polymers mainly include cellulose, starch, chitosan, sodium alginate, carrageenan, and pectin. Aerogels of natural polymers have abundant resource advantages, good biocompatibility and biodegradability, and can be exploited in medicine, environmental engineering, and food packaging [30]. The introduction of graphene-based materials not only improves the shortcomings of low mechanical strength, poor brittleness, and poor stability of natural polymeric aerogels, but affects their structure and properties.

2.4.1. Cellulose/GCA

Cellulose is the most abundant natural polymer in nature [31]; it is divided into cellulose nanocrystal (CNC), [32] cellulose nanofiber (CNF), and bacterial cellulose (BC) according to the formation conditions and sources [33].

Cellulose/GCA can be prepared by a self-assembly process [34], and the formation mechanism is shown in Figure 8. Mi et al. [21] obtained cellulose/GCA by bi-directional freezing and CVD (Figure 8). This composite aerogel presented a porous structure with ultra-low density (0.0059 $mg \cdot cm^{-3}$) and high surface area (47.3 $m^2 \cdot g^{-1}$), and had good selective adsorption effect for oil. Using boric acid (BA) as cross-linker of GO and carboxymethyl cellulose (CMC), Ge et al. [22] synthesized CMC/GCA by the ice template method (Figure 9). The results showed that the GO content had a significant influence on the morphology and strength of the aerogel. When the content of GO reached 5 wt%, the strength was excellent, and thermal conductivity (0.0417 $W \cdot m^{-1} \cdot K^{-1}$) was similar to that of polystyrene foam (0.03–0.04 $W \cdot m^{-1} \cdot K^{-1}$). There is a good interaction between cellulose and GCA with the 3D structure, which provides a way to improve the mechanical properties of aerogels. The self-assembly process, CVD method, and template method can introduce cellulose into graphene to produce aerogels with better mechanical properties, thermal insulation, and energy storage.

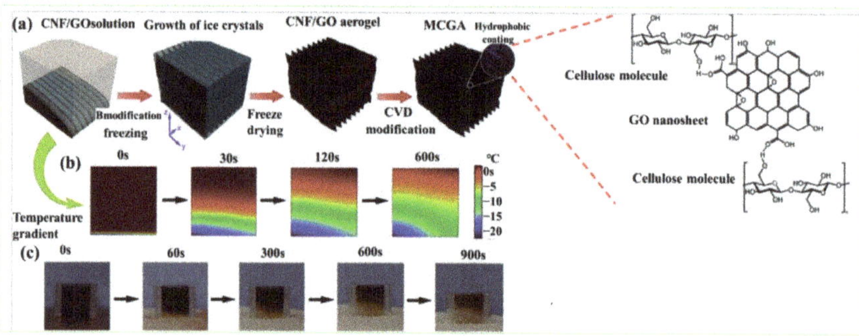

Figure 8. Schematic of hydrogen bond interactions between GCA and cellulose [34]: (**a**) Bi-directional freeze-drying preparation of cellulose/GCA; (**b**) Temperature gradient simulation in the freezing process; (**c**) Digital image of solution freezing process of two-way freezing [21].

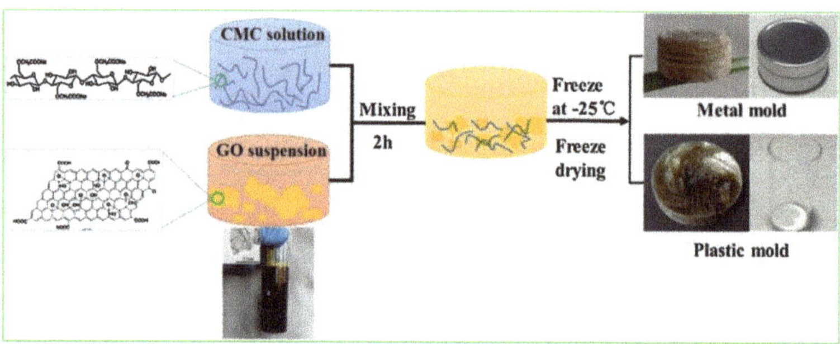

Figure 9. Schematic demonstration of CMC/GCA composite aerogel preparation [22].

2.4.2. Starch/GCA

Starch is one of the most common biopolymers and usually divided into amylose and amylopectin according to the chemical structure. Starch in its natural form can easily form a gel without a cross-linking agent [35]. Chen et al. [23] developed a simple and rapid method for preparing porous starch/GCA. The aerogels can be used as electrode materials to manufacture flexible and gel-type symmetrical supercapacitors with excellent capacitance performance and high energy density (Figure 10). Starch/GCA was prepared by chemical reduction and self-assembly of nanocrystalline starch, which has higher mechanical properties, capacitance performance, and adsorption capacity.

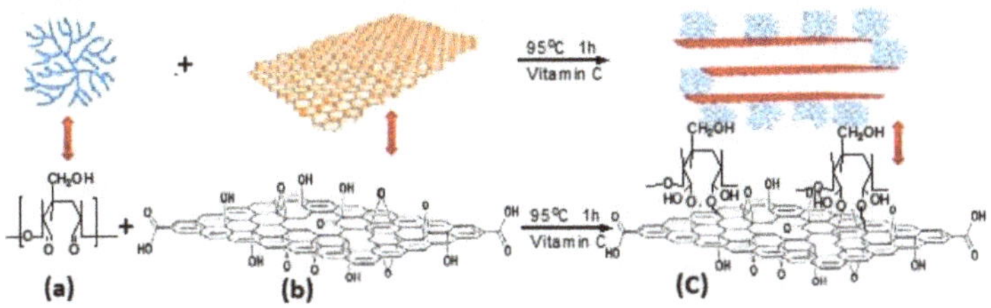

Figure 10. Fabrication and structure of starch/GCA [23]. (**a**) Starch; (**b**) Graphene oxide; (**c**) Starch/GCA.

2.4.3. Chitosan/GCA

Chitosan (CS) is extracted and processed from the shells of common arthropods [36]. The aerogel prepared by CS can be exploited as an adsorbent for sewage purification, but its strength is poor. Therefore, it must be modified to increase its strength. Interestingly, the introduction of graphene-based materials can improve not only the strength and stability of the CS aerogel, but also the purification efficiency and degree. Using solution mixing and injection methods, CS/GCA of various shapes can be prepared and applied in many fields [37], such as CS/GCA microspheres and membranes (Figure 11) [38]. In addition, both graphene and CS are used as the basic skeleton or filling material of GCA, which significantly improve the aerogel's properties. When CS was grafted onto a GO skeleton, the GCA with more ordered mesoporous and the adsorption capacity was significantly superior to that of pure CS aerogel [24]. CS/GCA has enough strength and stability and excellent absorbability for heavy metals, dyes, and organic solvents. Therefore, CS/GCA will still be a hot spot of research and application in the future.

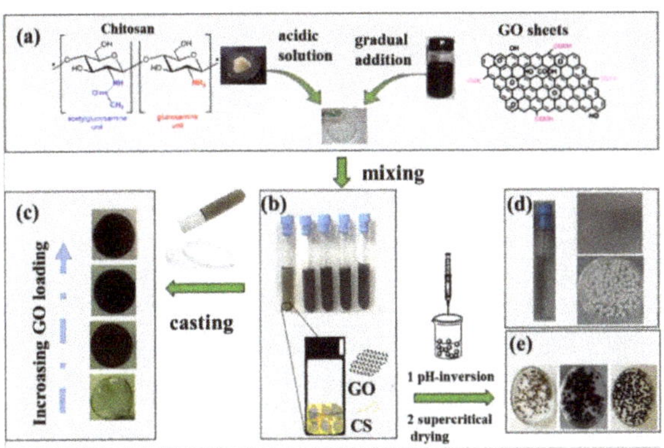

Figure 11. Schematic diagram of preparation process of CS/GCA [38]. (**a**) Chemical structure of chitosan and graphene oxide as two mixed dissimilar phases; (**b**) Digital photo of CS-GO acidic-aqueous solution with increasing GO weight percentage; (**c**) The as-prepared hybrid films (denoted as CS-GO) with increasing GO amount; (**d**) Self-standing chitosan-graphene oxide aerogel microspheres with increasing amount of GO in the mixture; (**e**) Blank materials prepared for comparison: chitosan aqueous acidic solution, chitosan film and chitosan microspheres.

2.4.4. Sodium Alginate/GCA

Sodium alginate (SA), an anionic polysaccharide with hydroxyl and carboxyl groups, comes from algae. It is widely used in medical stents, controlled-release drug carriers, and food thickeners, and also as a flocculant for treating wastewater. SA has different solubility affected by pH value. When the pH is less than 4, SA is insoluble, and when the pH is between 6 and 9, SA is a viscous solution. Although it is easy to form film and gel, SA-polysaccharide aerogel has weak strength and stability. Therefore, many studies have focused on functional modification and composite hybridization to overcome the limitations of the aerogel.

SA can be physically blended with graphene and chemically modified to improve its strength compared to the brittleness and easy collapse of pure SA aerogel (Figure 12) [31]. Shan et al. [25] applied in situ cross-linking to prepare SA/GCA, and the introduction of GO improved the uniformity of the spherical morphology and the efficiency and capacity of phosphorus ion adsorption in sewage, and was applied to remove phosphine pollution in water. In particular, the introduction of graphene can increase electrical conductivity and broaden the application of SA aerogel to other fields, such as its use as electrode material for supercapacitors and as a biomass aerogel catalyst with good catalytic activity.

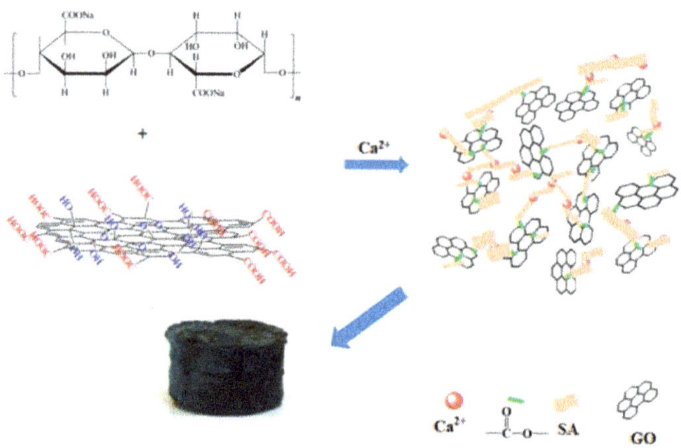

Figure 12. Synthesis route of SA/GO composite aerogel [39].

3. Structure of GCA

3.1. Porous Structure of GCA

GCA has a porous network structure and low density. Pore structure is closely related to strength, thermal performance, and electrochemical performance. In addition, GCA can uniformly disperse the graphene nanolayers in the matrix in the form of a single layer or a few layers to reduce agglomeration. Different pore diameters of aerogels also affect their density, pore volume, and specific surface area. For example, pore volume is the volume of pores per unit mass of aerogel and is related to the inner and outer diameter. With the same inner diameter, the pore volume decreases along with the increase of the outer diameter, and with the same outer diameter, the larger the inner diameter, the larger the pore volume. Like specific surface area, pore volume is another important factor affecting the surface adsorption and loading properties of aerogels. The structural parameters, including specific surface area, density, and pore size of GCA, are listed in Table 2.

Table 2. Fabrication methods and morphology of GCA materials.

Composite Aerogel Material	Fabrication Method	Specific Surface Area ($m^2 \cdot g^{-1}$)	Density ($mg \cdot cm^{-3}$)	Pore Volume ($cm^3 \cdot g^{-1}$)	Diameter (nm)	Reference
GNPA	Self-assembly	/	59.3	/	5	[12]
GO/SiO_2	Sol-gel method	889	/	3.72	16.75	[40]
VO_2/GA @NiS_2	Sol-gel method	141.1	/	/	17.3	[41]
Co-N-GA	Self-assembly	485	0.29	0.71	/	[11]
PPy@GA	Self-assembly	686	7.8	/	/	[20]
MGGNA	Freeze-drying	45.1	2.32	/	/	[42]
GNR	Self-assembly	113.1	9.33	/	/	[43]
MoS_2-RGO	3D printing	/	/	/	100–200	[9]
3D-GMOs	CVD	560	22	/	/	[15]
N-CMS/GA	Mixed	450	/	/	3.4–36.9	[44]

GNPA, graphene–PVA–co–PE nanofiber–PVA aerogel; Co-N-GA, hierarchically porous Co-N functionalized graphene aerogel; PPy@GA, poly-pyrrole nanosphere graphene aerogel; MGGNA, capillary-like hydrophobic GO/GONR-APTES composite aerogel; GNR, graphene nanoribbon aerogel; 3D-GMOs, high-density three-dimensional GA macroscopic objects; N-CMS/GA, nitrogen-doped carbon microsphere/graphene aerogel.

In terms of the morphology of aerogels, GCA can be roughly divided into two categories: porous structure and mesoporous core–shell structure. The common porous structures include honeycomb, microsphere, etc. The local compressibility and elasticity of aerogel honeycomb structure were well studied by density functional theory (DFT) [45]. The honeycomb structure [46] has the characteristics of ultra-low density, super-elasticity, good electrical conductivity, and high energy absorption efficiency (Figure 13), which is related to its inherent mechanical properties and elasticity [47]. By adjusting the method and process parameters, optimizing a complete honeycomb structure can control the specific surface area, density, pore size, appearance, and other morphological characteristics of GCA, and further change its mechanical and thermal properties and electrical conductivity.

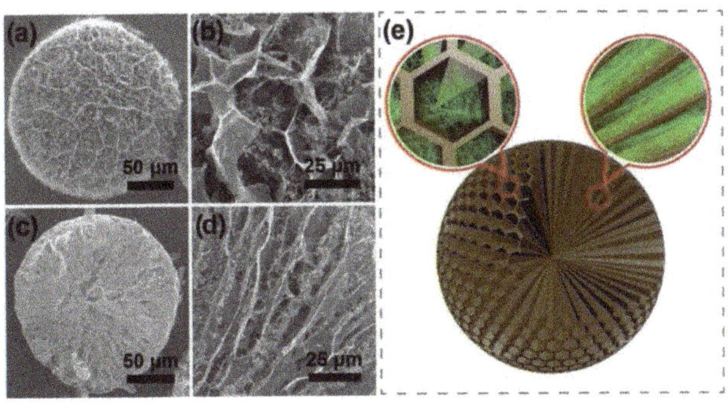

Figure 13. SEM images of: (**a**,**b**) Graphene aerogel, (**c**,**d**) GCA, and (**e**) Microchannel structures of GCA [47].

3.2. Mesoporous Core–Shell Structure of GCA

Microspheres with mesoporous and core–shell structure produced by GCA exhibit greater strength and have wider application. These aerogels can usually be obtained using the sol-gel, blending, or other method to obtain hydrogels, which are then transformed into aerogel by freeze-drying or super-zero boundary CO_2 drying. Liu et al. [48] prepared GCA with well-shaped microspheres and with internal honeycomb, which could be exploited as a photocatalyst and adsorbent. As shown in Figure 14, the shape of the honeycomb and the mesoporous microspheres with an internal radial microchannel structure help in shortening the diffusion path of pollutants and promoting a rapid adsorption equilibrium in the treatment of sewage. Therefore, the relationship between the design of the honeycomb structure and the performance and application of GCA needs further study.

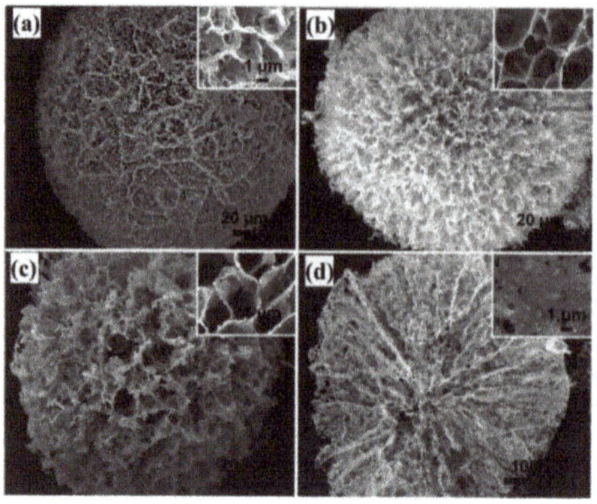

Figure 14. SEM images of: (**a**) GCA, (**b**) GCA1 (50% Ag_3PO_4); (**c**) GCA3 (75% Ag_3PO_4); (**d**) Cross-section of GCA microspheres [48].

Compared with the mesoporous microspheres structure, the core–shell structure is more complicated. Researchers have adjusted the pore structure of GCA aerogels by controlling the freezing temperature and direction of ice crystals formed between graphene layers. In addition, in order to form dense nuclei and sparse filling shells, the freezing temperature gradient is needed to control the nucleation and growth of ice crystals to obtain GCA microspheres with core–shell structure (Figure 15). This unique structure achieves high compressive strength by continuously distributing mechanical loads between core and shell, resulting in improved mechanical, electrical, and thermodynamic properties.

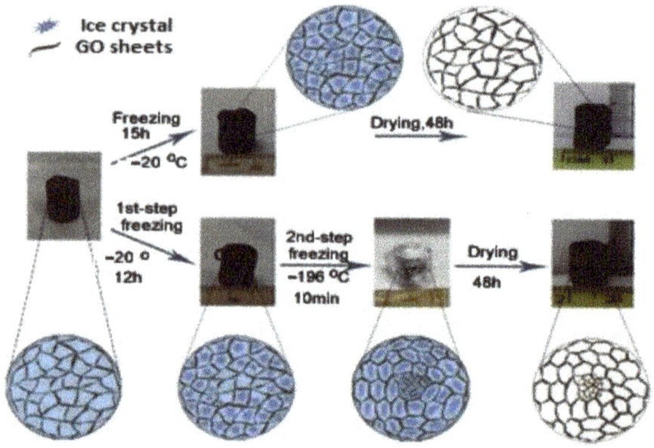

Figure 15. Schematic illustration of fabrication of GCA microspheres with core–shell structure [49].

3.3. Methods for Adjusting the Porous Structure of GCA

The pore structure of GCA can be adjusted by controlling the concentration of polymers, inorganic nanomaterials, and graphene nanosheets. The template effects and self-assembly process of graphene-based materials can be used to change the pore structure of aerogels and other materials [50]. In addition, the newly developed impregnation method and 3D printing technology have better effects on regulating the pore structure. Compared with traditional strategies, they greatly broaden the unique functional structure with controllable structural parameters [51].

4. Properties of GCA

4.1. Adsorptive Properties

Adsorption properties include physical and chemical adsorption. GCA aerogel's abilities are connected with the structural design and performance of the material. In terms of structural design, 3D GCA usually has a larger specific surface area, higher porosity, and an interconnected porous structure, which can increase its combination with other functional materials and promote the diffusion of ions and molecules [52]. In addition, magnetic inorganic nano-GCA (Co_3O_4, Fe_3O_4) is easy to recycle [53]. Introducing different functional groups, such as amino and phosphate groups, into 3D GCA leads to different adsorption mechanisms, such as ion exchange, complexation, chelation, electrostatic interaction, hydrogen bond, π–π, hydrophobic interaction, etc. [54].

The adsorption of metal ions requires a large number of oxygen-containing functional groups (–OH, –COOH, –O–) on the GO surface combined with positively charged heavy metal ions through electrostatic interactions, or surface complexes to form metal complexes [55]. At the same time, a porous network structure effectively prevents the accumulation of GO, promotes the free diffusion of heavy metal ions, and broadens the contact probability of heavy metal ions with an active center. Additionally, π–π interactions and hydrogen bonds were observed between the edges of graphene [56], which efficiently bound and adsorbed dye molecules and metal ions (Figure 16) [47].

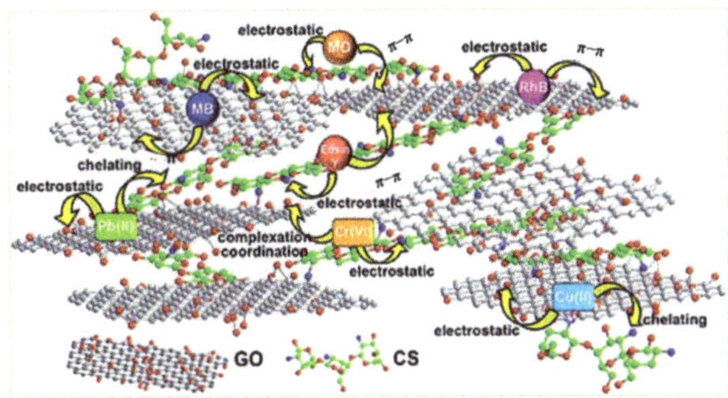

Figure 16. Schematic diagram of purification mechanisms of CS/GO aerogels for sewage [47].

Furthermore, after the design and construction of roughness, the introduction of hydrophobic groups, and the thermal annealing reduction process, GCA was shown to have hydrophobic and lipophilic adsorption properties, which benefited oil–water separation [57]. Using the unique 3D porous structure, hydrogen bonds, and π-π interaction characteristics also adsorbed antibiotics and other drugs. These adsorption characteristics make it valuable for application in sewage purification.

4.2. Mechanical Properties

The key mechanical properties of GCA include compressive strength and deformation resistance. However, pure graphene aerogel is too weak for practical use. For this reason, various cross-linking agents or structural enhancers are introduced to prepare GCA with high elasticity and mechanical properties [58]. For example, some 2D materials were used as structural reinforcement materials for GCA, such as MXenes, which helped the aerogel obtain a super-elastic structure [59]. Organic nanomaterials, synthetic polymers, and natural polymers are commonly used as mechanical reinforcements for GCA to maximize its mechanical performance [60].

As shown in Figure 17, cross-linking agents or structural strengthening materials improved the mechanical stability and the relationship between microstructure anisotropy and mechanical strength of the transverse (TD) and longitudinal (LD) direction of composite aerogels; compared with pure graphene, the maximum compressive strength of LD and

TD was increased by more than 10 times (10–50 kPa) [61]. Therefore, reasonable design of the preparation process is very important in order to improve the mechanical properties of composite aerogels without adversely affecting other functional performance.

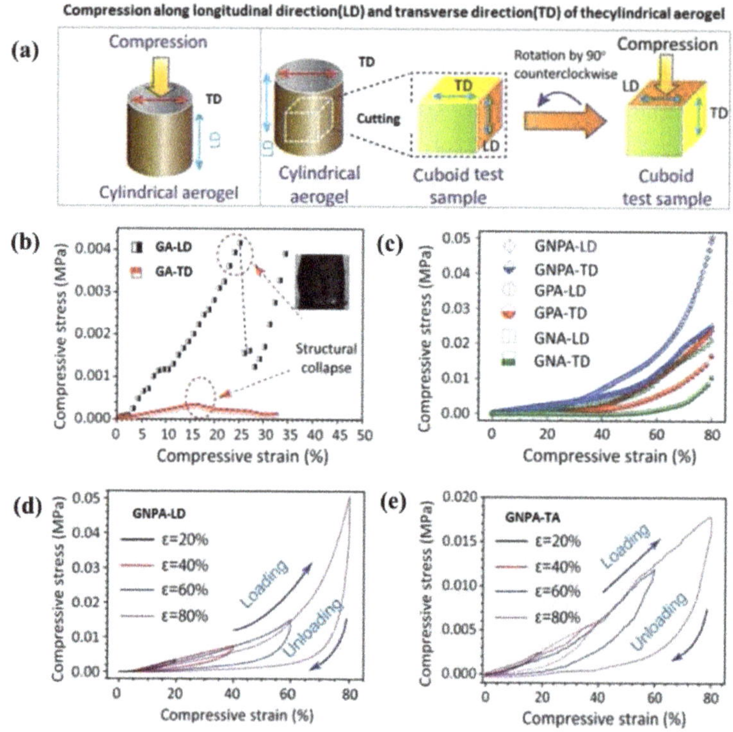

Figure 17. (**a**) Schematic of sampling instructions for expression test along TD and LD; (**b**,**c**) Stress–strain surveys of all testing samples in LD and TD; (**d**,**e**) Loading–unloading stress–strain surveys for GCA-LD and GCA-TD under different strains [61].

4.3. Thermal Properties

The thermal conductivity of graphene is 5000 $W \cdot m^{-1} \cdot K^{-1}$, which makes it a suitable filler for the preparation of composites, with excellent thermal properties [62]. The inherent characteristics of aerogels, such as light weight and very low thermal conductivity, provide new ideas for the development of heat insulation materials. However, the distribution of graphene nanosheets, the density of aerogel, and the pore structure of GCA affect its thermal properties; for example, uniformly distributed graphene nanosheets can greatly reduce or eliminate the contact thermal resistance and thus improve the thermal conductivity, whereas the low bulk density of the aerogel decreases thermal conductivity [63]. Therefore, in order to design materials with better thermal properties, the advantages of both need to be taken into account. For example, Wang et al. [64] prepared layered porous and continuous silk fibroin SF/GCA fibers by wet-spinning and freeze-drying, showed that the introduction of GO not only improved the mechanical properties, but also significantly raised the thermal properties under infrared radiation. Compared with pure SF aerogel fibers, the surface temperature of the SF/GCA was increased by 2.6 °C after infrared radiation for 30 s. At the same time, layered porous and hollow fiber structures reduced heat conduction and inhibited thermal radiation, providing good thermal insulation of SF/GCA fibers (Figure 18).

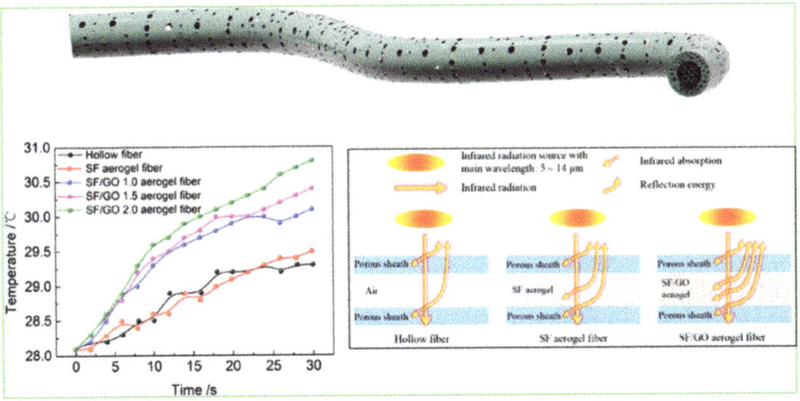

Figure 18. Radiative heating properties of SF/GCA fibers [64].

In short, the introduction of graphene-based materials improves the thermal performance of aerogel and provides a scheme to overcome the problem of energy consumption [65], which gives it great application potential in thermal insulation materials, flame retardant materials, and electronic devices.

4.4. Electrochemical Properties

The ultra-high electrical conductivity of graphene is one of its most attractive characteristics, which can reach 10^7–10^8 S/cm, and makes it possible to prepare GCA with excellent electrical performance [66]. However, due to the contact resistance between graphene nanosheets, the electrical conductivity of GCA usually decreases significantly. There are two common methods to further increase the electrical properties of graphene aerogels. The first method is to introduce conductive polymers or doping metal oxides into the aerogel, and the synergistic effect between components can improve the electrical conductivity and structural stability of the polymer/GCA. The second method is to uniformly coat conductive polymer/GA materials on the substrate by spraying or spinning [67], which will significantly improve the conductivity of the aerogel as a supercapacitor and energy storage material.

Common metal oxides such as Fe_2O_3, Co(OH), Co_3O_4, MnO_2, and MoS_2 increase the electrical conductivity of GCA. For example, with the hydrothermal method, GCA was obtained by doping nano-Fe_2O_3 into graphene aerogel, and its specific capacitance was 81.3 $F{\cdot}g^{-1}$ at a constant current density of 1 $A{\cdot}g^{-1}$ and working potential of −0.8–0.8 V. GCA can also be functionalized by O, N, S, B, etc. [68]. Yun et al. [69] doped N into carbon quantum dots (CQDs) and then combined them with rGO and different ratios of Fe_2O_3 to produce GCA by form N-CQDs/rGO/Fe_2O_3 composite, and the preparation process is shown in Figure 19. The composite aerogel had excellent electrochemical performance and ultra-high specific capacity due to its high surface area and porous layered structure, as well as the synergistic effect of high-capacity Fe_2O_3 and stable high-conductivity N-CQD/rGO. The proper doping proportion can further improve electrical conductivity [70]. Yang et al. [71] successfully prepared 3D MXene/rGO composite aerogel by the ice template method and coating with carboxylated polyurethane (PU). This not only had excellent electrochemical performance, but also good self-healing ability, and the capacitance retention rate reached 91% in 15,000 cycles, providing a method for use in long-life multi-function electronic devices. Therefore, the preparation of GCA with excellent electrochemical performance, whether coated with polymer conductive materials or doped with inorganic nanoparticles or multicomponent composites, mainly depends on the filler ratio, the inherent electrical conductivity of the graphene-based nanosheets and materials used, and the control of the micro-morphology.

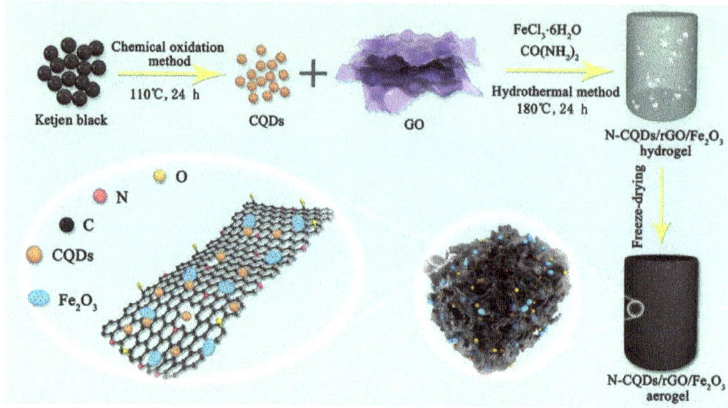

Figure 19. Schematic demonstration of GCA aerogel synthesis [67].

5. Application of GCA

5.1. Adsorption Removal of Contaminants from Water

Adsorption is a popular sewage treatment because of low cost, simple operation, large adsorption capacity, and high removal rate [72]. GCA is an ideal adsorbent for sewage treatment owing to its higher adsorption capacity and easy reuse [73]. Zang et al. found that more porous CS/GCA had good adsorption and removal effects for Pb^{2+}. When the GO content in the aerogel was 5 wt%, the adsorption capacity of Pb^{2+} increased from 68.5 to 100 $mg \cdot g^{-1}$. The adsorption mechanism of common metal ions Pb^{2+} and Cu^{2+} may consist of intergroup coordination and complexation (Figure 20) [74].

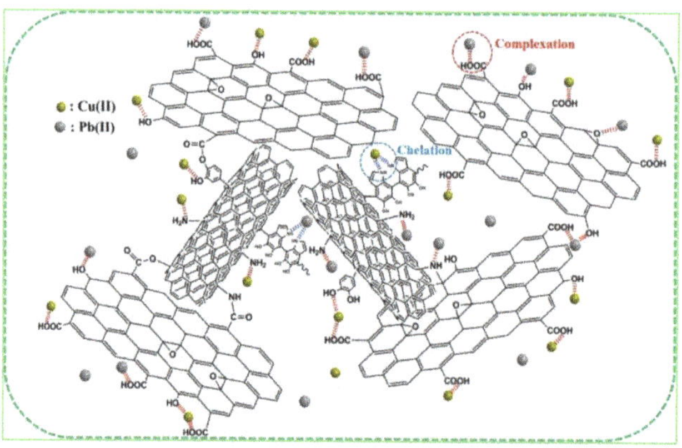

Figure 20. Complexation (red) and communication (blue) interactions of Pb (II) and Cu (II) adsorption on GCA [74].

The surface properties of the adsorbent and the chemical properties (structure, hydrophobicity, polarity) of the aerogels [75] determine the types of pollutants that can be adsorbed [76]. Preparation of GCA with high hydrophobicity and high specific surface area provides a feasible solution for the removal of organic oil pollution. Yang et al. [77] introduced fluoroalkyl-silane into GCA by gas–liquid deposition to obtain superhydrophobic graphene aerogels (SGAs) with super hydrophobicity, super lipophilicity, ultra-low density, large specific surface area, excellent adsorption capacity, and adsorption recycling (Figure 21), which has great potential in the field of oil–water separation.

Table 3 lists the studies on the adsorption properties of GCA for heavy metals, organic compounds, dyes, and other pollutants. In terms of recyclability, most studies have shown that recovery and desorption efficiency were improved by selecting suitable desorbents (acids, alkalis, and organic solvents), and the 3D structure of GA had a unique effect in this regard [78]. This shows that GCA has broad prospects in adsorption treatment of water pollution.

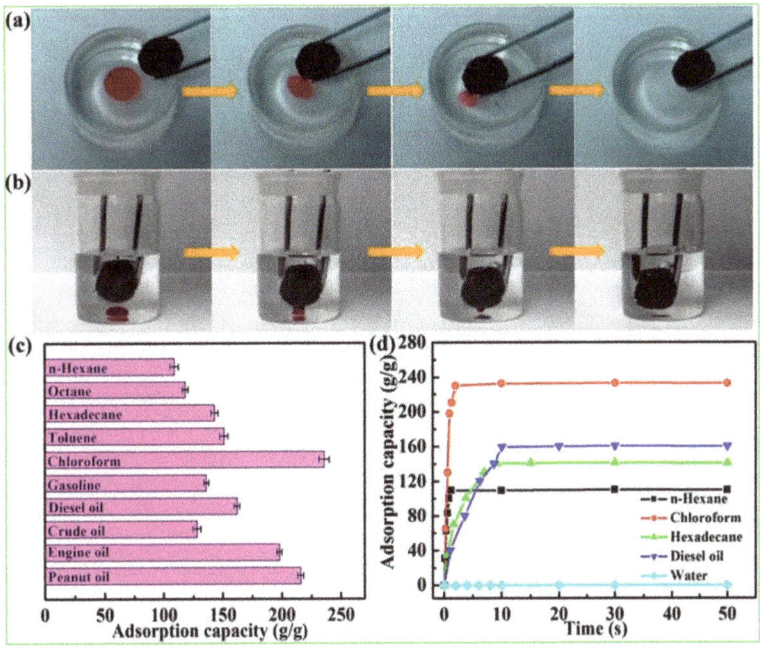

Figure 21. Adsorption performance of SGAs for oil. (**a**) Snapshots showing the SGAs adsorbs hexadecane (dyed by oil red) floating on the water; (**b**) Snapshots showing the SGAs adsorbs chloroform (dyed by oil red) from the bottom of the water; (**c**) The adsorption capacity of the SGAs for various kinds of organic solvents and oils; (**d**) The time-dependent sorption behaviour of various oily compounds and water by the SGAs [77].

Table 3. Adsorption properties of different GCAs.

Adsorbent	Contaminants	Adsorption Conditions	Isotherm Model	q_m ($mg \cdot g^{-1}$)	Reference
RGO/ZIF-67	CV MO	pH 3–9, T = 298 K, t = 0–16 h	Langmuir	1714 426	[79]
MWCNT-PDA	Cu^{2+} Pb^{2+}	pH 2–7, T = 298 K, t = 0–10 h	Langmuir	318.47 350.87	[74]
3D-Fe_3O_4/GA	As^{5+}	pH 7, T = 298 K, t = 0–16 h	Langmuir	40.05	[80]
3D-SRGO	Cd^{2+}	pH 2–9, T = 298 K, t = 0–24 h	Langmuir	234.8	[81]
GA/SiO_2	Hg^{2+}	pH 2–10, T = 296 K, t = 0–1.5 h	Langmuir	500.0	[82]
3D δ-MnO_2	Pb^{2+} Cd^{2+} Cu^{2+}	pH 2–6, T = 298 K, t = 0–3 h	Langmuir	643.62 250.31 228.46	[83]
3D GO/SA	MB	pH 4–9, T = 293 K, t = 0–24 h	Langmuir	4.25	[84]
PAM/GO	Magenta	pH 2.6–8.9, T = 303 K, t = 0–55 h	Langmuir	1034.3	[85]
GO-AL	MB	pH 3.0, T = 303 K, t = 0–4 h	Langmuir	1185.98	[86]
PPGA	N-hexane, MO	pH 3.0, T = 303 K, t = 0–4 h	Langmuir	63.5 202.8	[87]

Table 3. *Cont.*

Adsorbent	Contaminants	Adsorption Conditions	Isotherm Model	q_m ($mg{\cdot}g^{-1}$)	Reference
RGO/REMO	RhB,	t = 0–24 h	Langmuir	243.4	[88]
GAS-MS	phenol, catechol, resorcinol, hydroquinone	pH 3.0, T = 298 K, t = 24 h	Langmuir&Freundlich	90 66 22 67	[28]
CNF/GO	Tetracycline	pH 2–12, T = 298 K, t = 24 h	Langmuir	454.6	[89]
3DG	Methyl bromide	pH 7.5, T = 298 K, t = 0–5 h	Langmuir	685	[90]

RGO/ZIF-67, three-dimensional rGO/zeolitic imidazolate framework-67 aerogel; MWCNT-PDA, graphene/polydopamine modified multiwalled carbon nanotube hybrid aerogel; 3D-SRGO, 3D sulfonated reduced GO aerogel; 3D δ-MnO_2, 3D graphene/delta-MnO_2 aerogel; GO-AL, GO/alkali lignin aerogel composite; PPGA, polydopamine and poly-ethylenimine co-functionalized GO aerogel; RGO/REMO, RGO/rare earth–metal oxide aerogel; GAS-MS, 3D graphene aerogel–mesoporous silica; CNF/GO, cellulose nanofibril/graphene oxide hybrid aerogel; 3DG, three-dimensional graphene aerogel.

5.2. Application of Sensors and Supercapacitors

The unique porous structure of GCA gives it good flexibility and elasticity, and it has become the preferred material for piezoelectric resistance sensors. In addition, GCA has great application potential in the field of energy storage and sensing, including supercapacitors, lithium batteries, solar cells, and fuel cells, because of its high conductivity, high electrochemical stability, and good mechanical stability [91].

Pressure sensitivity plays a key role in piezoelectric resistance materials, which determines the sensitivity of resistance materials. Generally, a highly sensitive aerogel can be obtained by controlling the composition of the GCA. Wei et al. [92] used borate as a reinforcing material in a 3D graphene aerogel structure to obtain nitrogen and boron co-modification (NBGC) aerogel with excellent elasticity and compressibility and good electrochemical properties. This kind of aerogel provided a fast external stress change current response with specific capacitance up to 336 $F{\cdot}g^{-1}$; when the stress increased from 0.05 N to 10 N, the response current varied from 0.44–2.89 mA to 10 N, so it could be applied in high-performance stress sensors (Figure 22).

The application of GCA in energy storage is shown in Table 4. The introduction of polymer or inorganic nano-active materials into the aerogel structure improves the electrochemical energy storage of super capacitors [93]. For example, using high-conductivity Cu/Cu_xO to modify the rGO network structure, C aerogel with high apparent conductivity was obtained [94], which was two to three orders of magnitude higher than pure graphene aerogel (0.1 S/m). Li et al. [88] designed a new type of PPy layer coated sulfur/GCA by vapor deposition, which was used as the cathode of a lithium–sulfur battery. PPy as a uniform coating ensures long-time, stable cycle performance of lithium–sulfur batteries, and it also has excellent electrochemical properties, such as high specific capacity. The discharge capacity at 0.2 C after 500 cycles reached 1167 $mAh{\cdot}g^{-1}$ after 500 cycles. In addition, Table 4 also shows that the electrochemical energy storage and conductivity of GCA doped with heteroatoms N or B are greatly improved, and the active specific surface area is increased. The synergistic effect of N and B co-doping promotes the charge transfer between adjacent carbon atoms, improves the electrochemical performance of carbon-based materials, and gives them excellent power density and charge–discharge rate, which makes this a very promising super energy storage capacitor material [95]. This shows that 3D GCA has great research value with regard to energy storage materials. In the future, designing GCA with a porous structure and a larger specific surface area, while maintaining a good conductive path for effective charge transfer, is a problem that researchers will need to pay attention to.

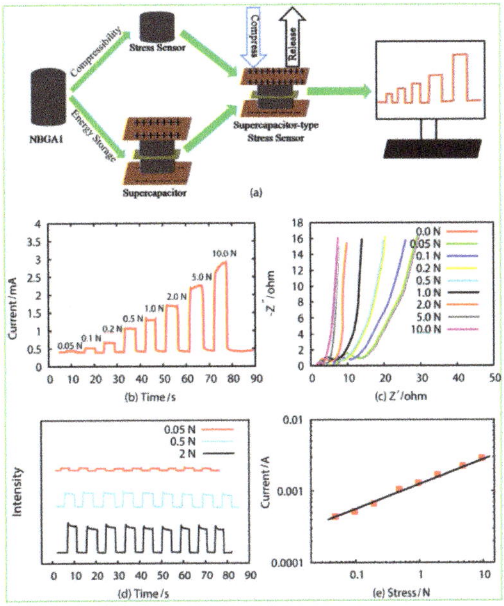

Figure 22. Electrochemical and pressure response performance of NBGC aerogel as supercapacitor and stress sensor. (**a**) A schematic illustration of assembled NBGC aerogel as a supercapacitor stress sensor; (**b**) The response currentetime (Iet) characteristic curves of the NBGC aerogel stress sensor at different stress (T) of 0.05e10 N; (**c**) The EIS of the cell under different stress; (**d**) The stress sensor cycle stability of device at different stress of 0.05 N, 0.5 N and 2 N, respectively; (**e**) The relationship between the response current and the stress. [92].

Table 4. Energy storage performance of different GCA materials.

Composite Aerogel Materials	Specific Capacitance ($F{\cdot}g^{-1}$)	Energy Density ($Wh{\cdot}kg^{-1}$)	Cyclic Behavior	Electrolyte	Reference
F-Fe_2O_3@MGA	1119@1 Ag^{-1}	800	98.9% after 2000 cycles	3M KOH	[96]
LMP/rGO	4645@0.5Ag^{-1}	11.79	91.2% after 10,000 cycles	6M KOH	[97]
PANI/CRGO/Co_3O_4	1247@1 Ag^{-1}	190	92% after 3500 cycles	6M KOH	[90]
PPy/CRGO	304@0.5 Ag^{-1}	/	58% after 50 cycles	6M KOH	[98]
GR-CNT	375@1 Ag^{-1}	/	94.8% after 5000 cycles	6M KOH	[99]
CA	467@20 Ag^{-1}	22.75	90.9% after 10,000 cycles	1M KOH	[100]
N,S-MGA	4929@2 Ag^{-1}	686.7	98.7% after 5000 cycles	1M KOH	[101]
SnO_2-GA	541@5 Ag^{-1}	160	97.3% after 10,000 cycles	3M KOH	[102]
MnO_2/P-RGO	645@1 Ag^{-1}	59.2	94.6% after 10,000 cycles	1M Na_2SO_4	[103]

F-Fe_2O_3@MGA, flower-like Fe_2O_3@ multiple graphene aerogel; LMP/rGO, $LiMnPO_4$/reduced GO aerogel; PANI/CRGO/Co_3O_4, self-assembled graphene/polyaniline/Co_3O_4 ternary hybrid aerogel; PPy/CRGO, conductive graphene/poly-pyrrole hybrid aerogel; GR-CNT, nitrogen-doped carbon aerogel; CA, nitrogen-doped carbon aerogel; N,S-MGA, nitrogen and sulfur-functionalized multiple graphene; MnO_2/P-RGO, phytic acid modified manganese dioxide/GCA.

5.3. Application of Heat-Insulation and Flame-Retardant Materials

Aerogels have unparalleled advantages as thermal insulation materials. The reason is that the ultra-high porosity of aerogel reduces heat conduction, and the pore walls in the aerogel network can effectively restrain thermal radiation. When the aerogel has smaller pore size, thermal convection will be reduced [104]. Some polymer aerogels, such as PVA [105], cellulose [106], and pectin [107] aerogels, are excellent thermal insulation materials, but their application is limited because of poor thermal stability and flame retardancy. Fortunately, carbon nanomaterials such as graphene materials can improve the physical and thermal properties of polymer aerogels. Because the density, porosity, and complex skeleton structure have an influence on the porous thermal insulation materials, composite aerogel with low density, low thermal conductivity, and high strength prepared by quartz fiber and nitrogen-doped graphene has great application potential in aviation for thermal insulation.

In terms of flame-retardant materials, highly thermal stable graphene aerogel with large porosity eliminates heat rapidly during combustion [108]. Hence, taking advantage of the flame retardancy of graphene aerogel and the low thermal conductivity of composite, aerogels prepared with phenolic resin have ultra-low thermal conductivity, high thermal stability, and good flame retardancy [109]. Ceramic fillers with Al_2O_3 ceramic and graphene have been designed with a layered honeycomb microstructure, showing a coupling strengthening effect between the graphene skeleton and the Al_2O_3 ceramic nanolayer. The composite aerogel not only has ultra-light density, reversible compressibility, and high electrical conductivity, but also great application prospects in flame retardancy, thermal insulation, and so on [110].

5.4. Biomedical Applications

In the biomedical field, some degrees of biocompatibility, biodegradability and antibacterial properties of graphene-based materials are beneficial [111]. The toxicity of graphene-based materials is related to many aspects such as content, size, surface chemistry, cell lines, morphologies, administration route, etc., and this is also applicable to similar GCA studies [112]. For example, in the study of GCA used in in situ bone regeneration, the content of nano graphene-based materials can affect the biocompatibility and biodegradability of GCA, and the content of GO is 0.10% aerogel is more conducive to cell increment [113]. Nanoscale small-sized GO has a lower level of cytotoxicity so that it can be used as a drug delivery carrier; the graphene aerogel nanoparticles (GANPs) were prepared as drug delivery carriers with high pH sensitivity, and released after 5 days in vivo, and are expected to be used in nanomedicine in the future [114]. The toxicity of graphene-based materials is different in vivo and in vitro, which is affected by their physical and chemical properties, such as functional groups, charges, sizes, stiffness, hydrophobicity and structural defects [112]. Luo et al. synthesized three-dimensional multi-functional GCA materials using tannic acid as raw material, which showed high porosity, low density, good hydrophobicity, good mechanical properties, high thermal stability, strong antibacterial properties, and sterilization rates of 58.12 and 99.99% for *Escherichia coli* and *Staphylococci*, respectively [115]. Therefore, a large number of studies in the biomedical field have focused on its safety and reducing its cytotoxicity, such as through the introduction of some biocompatible materials such as chitosan, collagen, gelatin, serum albumin and so on [116]. In addition, graphene-based nanosheets are used as structural reinforcement materials to prevent cells from collapsing during growth, thus stabilizing the three-dimensional structure, which is more obvious in the case of good dispersion of nanoparticles and good affinity of polymer fillers. For example, a thin biocompatible coating can be formed on the surface of three-dimensional graphene, which can be used as a high-strength biocompatible scaffold material for nervous system regeneration and musculoskeletal tissue engineering. This presented good tissue integrity renewal ability and inhibition of lesion expansion after spinal cord injury [117].

Furthermore, these materials could also be applied as biosensors to provide a wealth of information for early diagnosis of diseases and prevention of their evolution [118]. Composite material modified with 3D GCA and nano-CuO, with high sensitivity, was exploited as a biosensor for the detection of ascorbic acid [119]. Graphene aerogel was shown to detect dopamine with high sensitivity (619.6 $\mu A \cdot \mu M^{-1} \cdot cm^{-2}$), which was attributed to the highly conductive 3D multichannel, high charge transfer rate, and efficient transport guaranteed by the close interaction between dopamine and graphene [120]. The multifunctional silica/GCA material can be used as a biosensor for the detection of insulin (INS) with high selectivity and sensitivity [121]. The extensive research on GCA in the biomedical field is focused on its convenience as a drug carrier, antibacterial, biological scaffold, sensor, and so on. Designing GCA material with flexible biological and intelligent properties in the future will broaden its application in the biomedical field. Research on GCA in biomedicine has great potential, and is worth promoting further.

6. Conclusions and Outlook

This paper reviews the material composition, preparation methods, structural characteristics, properties, and applications of GCA. Here we focus on several types of GCA, including graphene-based/2D nanomaterial (MXenes) aerogel for sensors and supercapacitors, graphene-based/inorganic nanomaterial (SiO_2, SnO_2, SnO_2, TiO_2) aerogel and heat-insulating flame-retardant materials, and graphene-based/synthetic polymer aerogel and graphene-/natural sugar-based polymer aerogel used as adsorbents to remove metal ions and dye contaminants in water.

In addition, this paper also describes the influence of differences in the morphological structure (specific surface area, density, pore size, etc.) of GCA on its mechanical, thermal, and electrochemical properties. At the same time, the commonly used methods for preparing GCA are briefly described, including template method, self-assembly, chemical vapor deposition, 3D printing, and performance improvement strategies (doping, coating, cross-linking). GCA not only makes up for the deficiencies in the mechanical properties of graphene aerogel, but also retains excellent electrical conductivity, good mechanical flexibility, low thermal expansion coefficient, and other physical properties. It meets the special requirements for material properties in new fields, such as environmental purification, sensing, energy storage, biomedicine, heat insulation, and fire protection, and has become a hot spot in the field of graphene research in recent years. However, at present, the preparation method of graphene aerogel is relatively cumbersome, and it is necessary to find a quick preparation method to realize large-scale industrial production, although the chemical exfoliation method described in previous studies can be used to easily prepare graphene and its derivatives on a large scale. This still needs to be optimized to control the structure and size of materials. In terms of well-designed GCA, it is necessary to further explore the relationship between its properties and microstructure and optimize the preparation process parameters. For example, in the previous introduction, the pore structure of the aerogel was adjusted by controlling the formation of ice crystals between graphene sheets, but it is difficult to obtain uniform pore structure due to the existence of a temperature gradient, so it is still a difficult problem to skillfully control the morphology and size distribution of the product. In addition, most of the research has remained in the laboratory stage. Researchers should combine studies with production demand and broaden the scope of application.

In summary, the future research direction of GCA will move toward the design and preparation of new graphene derivatives and their composite materials, with multi-dimensionality and better performance. Compared with GCA, it is difficult to use other materials in so many areas, especially energy storage, sensing, and adsorption, which are based on the characteristics of graphene-based and aerogel materials. With the continuous advancement of science and technology, GCA that is more environmentally friendly and has excellent application performance and better structural properties will be prepared in the future, which is both an opportunity and a challenge.

Author Contributions: Conceptualization, S.L. and X.L.; methodology, X.L. and S.L.; data curation, X.L., L.L. and D.W.; writing—original draft preparation, X.L. and S.L.; writing—review and editing, S.L., X.L., D.W. and L.L.; investigation, X.L., S.L., Y.H., L.W. and Z.L.; project administration, S.L.; funding acquisition, S.L. All authors have read and agreed to the published version of the manuscript.

Funding: This research was funded by the Innovational Industrialization Foundation of Shaanxi Province of China (grant number 2021QFY04-04) and the National Natural Science Foundation of China (grant number 21276152).

Institutional Review Board Statement: Not applicable.

Informed Consent Statement: Not applicable.

Data Availability Statement: Publicly available datasets were analyzed in this study. This data can be found here: https://www.webofscience.com/wos/alldb/basic-search (accessed on 25 November 2021).

Conflicts of Interest: We declare that we have no financial and personal relationships with other people or organizations that could inappropriately influence our work.

References

1. ElKhatat, A.M.; Al-Muhtaseb, S.A. Advances in tailoring resorcinol-formaldehyde organic and carbon gels. *Adv. Mater.* **2011**, *23*, 2887–2903. [CrossRef] [PubMed]
2. Guo, F.; Jiang, Y.; Xu, Z.; Xiao, Y.; Fang, B.; Liu, Y.; Gao, W.; Zhao, P.; Wang, H.; Gao, C. Highly stretchable carbon aerogels. *Nat. Commun.* **2018**, *9*, 881. [CrossRef]
3. Khetib, Y.; Alahmadi, A.A.; Alzaed, A.; Sajadi, S.M.; Cheraghian, G.; Sharifpur, M. Numerical study of the effect of graphene nanoparticles in calcium chloride hexahydrate-based phase change material on melting and freezing time in a circular cavity with a triangular obstacle. *J. Energy Storage* **2021**, *43*, 103243. [CrossRef]
4. Abidi, A.; Rawa, M.; Khetib, Y.; Sindi, H.F.A.; Sharifpur, M.; Cheraghian, G. Simulation of melting and solidification of graphene nanoparticles-PCM inside a dual tube heat exchanger with extended surface. *J. Energy Storage* **2021**, *44*, 103265. [CrossRef]
5. Bi, H.; Yin, K.; Xie, X.; Zhou, Y.; Wan, N.; Xu, F.; Banhart, F.; Sun, L.; Ruoff, R.S. Low temperature casting of graphene with high compressive strength. *Adv. Mater.* **2012**, *24*, 5124–5129. [CrossRef]
6. Cheng, C.; Li, S.; Thomas, A.; Kotov, N.A.; Haag, R. Functional graphene nanomaterials based architectures: Biointeractions, fabrications, and emerging biological applications. *Chem. Rev.* **2017**, *117*, 1826–1914. [CrossRef]
7. Olszowska, K.; Pang, J.; Wrobel, P.S.; Zhao, L.; Ta, H.Q.; Liu, Z.; Trzebicka, B.; Bachmatiuk, A.; Rummeli, M.H. Three-dimensional nanostructured graphene: Synthesis and energy, environmental and biomedical applications. *Synth. Met.* **2017**, *234*, 53–85. [CrossRef]
8. Kistler, S.S. Coherent expanded aerogels and jellies. *Nature* **1931**, *127*, 741. [CrossRef]
9. Brown, E.; Yan, P.; Tekik, H.; Elangovan, A.; Wang, J.; Lin, D.; Li, J. 3D printing of hybrid MoS_2-graphene aerogels as highly porous electrode materials for sodium ion battery anodes. *Mater. Des.* **2019**, *170*, 107689. [CrossRef]
10. Li, Y.; Meng, F.; Mei, Y.; Wang, H.; Guo, Y.; Wang, Y.; Peng, F.; Huang, F.; Zhou, Z. Electrospun generation of Ti_3C_2Tx MXene@ graphene oxide hybrid aerogel microspheres for tunable high-performance microwave absorption. *Chem. Eng. J.* **2020**, *391*, 123512. [CrossRef]
11. Fu, X.; Choi, J.-Y.; Zamani, P.; Jiang, G.; Hoque, M.A.; Hassan, F.M.; Chen, Z. Co–N decorated hierarchically porous graphene aerogel for efficient oxygen reduction reaction in acid. *ACS Appl. Mater. Interfaces* **2016**, *8*, 6488–6495. [CrossRef]
12. Wang, Y.; Wang, L.; Zhu, H.W. Wearable and highly sensitive graphene strain sensors for human motion monitoring. *Adv. Funct. Mater.* **2014**, *24*, 4666–4670.
13. Cheng, L.; Qiao, D.; Zhao, P.; He, Y.; Sun, W.; Yu, H.; Jiao, Z. Template-free synthesis of mesoporous succulents-like TiO2/graphene aerogel composites for lithium-ion batteries. *Electrochim. Acta* **2019**, *300*, 417–425. [CrossRef]
14. Liu, X.; Sun, J.; Zhang, X. Novel 3D graphene aerogel–ZnO composites as efficient detection for NO_2 at room temperature. *Sens. Actuators B Chem.* **2015**, *211*, 220–226. [CrossRef]
15. Li, W.; Gao, S.; Wu, L.; Qiu, S.; Guo, Y.; Geng, X.; Chen, M.; Liao, S.; Zhu, C.; Gong, Y. High-density three-dimension graphene macroscopic objects for high-capacity removal of heavy metal ions. *Sci. Rep.* **2013**, *3*, 2125. [CrossRef]
16. Zhao, Z.; Wang, Z.; Wang, D.; Wang, J.-X.; Foster, N.R.; Pu, Y.; Chen, J.-F. Preparation of 3D graphene/iron oxides aerogels based on high-gravity intensified reactive precipitation and their applications for photo-Fenton reaction. *Chem. Eng. Process.* **2018**, *129*, 77–83. [CrossRef]
17. Xiang, Y.; Liu, L.; Li, T.; Dang, Z. Compressible, amphiphilic graphene-based aerogel using a molecular glue to link graphene sheets and coated-polymer layers. *Mater. Des.* **2016**, *110*, 839–848. [CrossRef]
18. Zhang, Q.; Xu, X.; Li, H.; Xiong, G.; Hu, H.; Fisher, T.S. Mechanically robust honeycomb graphene aerogel multifunctional polymer composites. *Carbon* **2015**, *93*, 659–670. [CrossRef]
19. Ouyang, A.; Cao, A.; Hu, S.; Li, Y.; Xu, R.; Wei, J.; Zhu, H.; Wu, D. Polymer-coated graphene aerogel beads and supercapacitor application. *ACS Appl. Mater. Interfaces* **2016**, *8*, 11179–11187. [CrossRef] [PubMed]
20. Zhang, E.; Liu, W.; Liang, Q.; Liu, X.; Zhao, Z.; Yang, Y. Polypyrrole nanospheres@ graphene aerogel with high specific surface area, compressibility, and proper water wettability prepared in dimethylformamide-dependent environment. *Polymer* **2019**, *185*, 121974. [CrossRef]
21. Mi, H.-Y.; Jing, X.; Politowicz, A.L.; Chen, E.; Huang, H.-X.; Turng, L.-S. Highly compressible ultra-light anisotropic cellulose/graphene aerogel fabricated by bidirectional freeze drying for selective oil absorption. *Carbon* **2018**, *132*, 199–209. [CrossRef]

22. Ge, X.; Shan, Y.; Wu, L.; Mu, X.; Peng, H.; Jiang, Y. High-strength and morphology-controlled aerogel based on carboxymethyl cellulose and graphene oxide. *Carbohydr. Polym.* **2018**, *197*, 277–283. [CrossRef] [PubMed]
23. Wang, C.-C.; Liang, J.; Liao, Y.-H.; Lu, S.-Y. 3D porous graphene nanostructure from a simple, fast, scalable process for high performance flexible gel-type supercapacitors. *ACS Sustain. Chem. Eng.* **2017**, *5*, 4457–4467. [CrossRef]
24. Hsan, N.; Dutta, P.; Kumar, S.; Bera, R.; Das, N. Chitosan grafted graphene oxide aerogel: Synthesis, characterization and carbon dioxide capture study. *Int. J. Biol. Macromol.* **2019**, *125*, 300–306. [CrossRef] [PubMed]
25. Shan, S.; Tang, H.; Zhao, Y.; Wang, W.; Cui, F. Highly porous zirconium-crosslinked graphene oxide/alginate aerogel beads for enhanced phosphate removal. *Chem. Eng. J.* **2019**, *359*, 779–789. [CrossRef]
26. Zhang, H. Ultrathin two-dimensional nanomaterials. *ACS Nano* **2015**, *9*, 9451–9469. [CrossRef]
27. Hanemann, T.; Szabó, D.V. Polymer-nanoparticle composites: From synthesis to modern applications. *Materials* **2010**, *3*, 3468–3517. [CrossRef]
28. Karamikamkar, S.; Behzadfar, E.; Naguib, H.E.; Park, C.B. Insights into in-situ sol-gel conversion in graphene modified polymer-based silica gels for multifunctional aerogels. *Chem. Eng. J.* **2020**, *392*, 123813. [CrossRef]
29. Parameswaranpillai, J.; Joseph, G.; Shinu, K.; Sreejesh, P.; Jose, S.; Salim, N.V.; Hameed, N. The role of SEBS in tailoring the interface between the polymer matrix and exfoliated graphene nanoplatelets in hybrid composites. *Mater. Chem. Phys.* **2015**, *163*, 182–189. [CrossRef]
30. Wang, Y.; Su, Y.; Wang, W.; Fang, Y.; Riffat, S.B.; Jiang, F. The advances of polysaccharide-based aerogels: Preparation and potential application. *Carbohydr. Polym.* **2019**, *226*, 115242. [CrossRef]
31. Dai, J.; Xie, A.; Zhang, R.; Ge, W.; Chang, Z.; Tian, S.; Li, C.; Yan, Y. Scalable preparation of hierarchical porous carbon from lignin for highly efficient adsorptive removal of sulfamethazine antibiotic. *J. Mol. Liq.* **2018**, *256*, 203–212. [CrossRef]
32. Li, Y.; Grishkewich, N.; Liu, L.; Wang, C.; Tam, K.C.; Liu, S.; Mao, Z.; Sui, X. Construction of functional cellulose aerogels via atmospheric drying chemically cross-linked and solvent exchanged cellulose nanofibrils. *Chem. Eng. J.* **2019**, *366*, 531–538. [CrossRef]
33. Kontturi, E.; Laaksonen, P.; Linder, M.B.; Gröschel, A.H.; Rojas, O.J.; Ikkala, O. Advanced materials through assembly of nanocelluloses. *Adv. Mater.* **2018**, *30*, 1703779. [CrossRef] [PubMed]
34. Wan, C.; Li, J. Graphene oxide/cellulose aerogels nanocomposite: Preparation, pyrolysis, and application for electromagnetic interference shielding. *Carbohydr. Polym.* **2016**, *150*, 172–179. [CrossRef]
35. Dogenski, M.; Navarro-Díaz, H.J.; de Oliveira, J.V.; Ferreira, S.R.S. Properties of starch-based aerogels incorporated with agar or microcrystalline cellulose. *Food Hydrocoll.* **2020**, *108*, 106033. [CrossRef]
36. Li, J.; Zhang, K.; Zhang, H. Adsorption of antibiotics on microplastics. *Environ. Pollut.* **2018**, *237*, 460–467. [CrossRef]
37. Huang, T.; Shao, Y.-W.; Zhang, Q.; Deng, Y.-F.; Liang, Z.-X.; Guo, F.-Z.; Li, P.-C.; Wang, Y. Chitosan-cross-linked graphene oxide/carboxymethyl cellulose aerogel globules with high structure stability in liquid and extremely high adsorption ability. *ACS Sustain. Chem. Eng.* **2019**, *7*, 8775–8788. [CrossRef]
38. Frindy, S.; Primo, A.; Ennajih, H.; el Kacem Qaiss, A.; Bouhfid, R.; Lahcini, M.; Essassi, E.M.; Garcia, H.; El Kadib, A. Chitosan–graphene oxide films and CO_2-dried porous aerogel microspheres: Interfacial interplay and stability. *Carbohydr. Polym.* **2017**, *167*, 297–305. [CrossRef]
39. Yu, M.; Zhang, H.; Yang, F. Hydrophilic and compressible aerogel: A novel draw agent in forward osmosis. *ACS Appl. Mater. Interfaces* **2017**, *9*, 33948–33955. [CrossRef] [PubMed]
40. Zhang, M.; Xiao, Q.; Chen, C.; Li, L.; Yuan, W. Developing a heat-insulating composite phase change material with light-to-thermal conversion performance from graphene oxide/silica hybrid aerogel. *Appl. Therm. Eng.* **2020**, *174*, 115303. [CrossRef]
41. Chen, H.-C.; Lin, Y.-C.; Chen, Y.-L.; Chen, C.-J. Facile fabrication of three-dimensional hierarchical nanoarchitectures of VO_2/graphene@ NiS_2 hybrid aerogel for high-performance all-solid-state asymmetric supercapacitors with ultrahigh energy density. *ACS Appl. Energy Mater.* **2018**, *2*, 459–467. [CrossRef]
42. Hou, S.; Wu, X.; Lv, Y.; Jia, W.; Guo, J.; Wang, L.; Tong, F.; Jia, D. Ultralight, highly elastic and bioinspired capillary-driven graphene aerogels for highly efficient organic pollutants absorption. *Appl. Surf. Sci.* **2020**, *509*, 144818. [CrossRef]
43. Peng, Q.; Li, Y.; He, X.; Gui, X.; Shang, Y.; Wang, C.; Wang, C.; Zhao, W.; Du, S.; Shi, E. Graphene nanoribbon aerogels unzipped from carbon nanotube sponges. *Adv. Mater.* **2014**, *26*, 3241–3247. [CrossRef] [PubMed]
44. Zhang, E.; Liu, W.; Liu, X.; Zhao, Z.; Yang, Y. A facile route to prepare nitrogen-doped carbon microspheres/graphene aerogel with high compressibility and superior capacitive property. *Mater. Today Commun.* **2020**, *24*, 101125. [CrossRef]
45. Zhang, Z.; Kutana, A.; Yang, Y.; Krainyukova, N.V.; Penev, E.S.; Yakobson, B.I. Nanomechanics of carbon honeycomb cellular structures. *Carbon* **2017**, *113*, 26–32. [CrossRef]
46. Jiao, C.; Xiong, J.; Tao, J.; Xu, S.; Zhang, D.; Lin, H.; Chen, Y. Sodium alginate/graphene oxide aerogel with enhanced strength–toughness and its heavy metal adsorption study. *Int. J. Biol. Macromol.* **2016**, *83*, 133–141. [CrossRef] [PubMed]
47. Yu, R.; Shi, Y.; Yang, D.; Liu, Y.; Qu, J.; Yu, Z.-Z. Graphene oxide/chitosan aerogel microspheres with honeycomb-cobweb and radially oriented microchannel structures for broad-spectrum and rapid adsorption of water contaminants. *ACS Appl. Mater. Interfaces* **2017**, *9*, 21809–21819. [CrossRef] [PubMed]
48. Liu, Y.; Yang, D.; Shi, Y.; Song, L.; Yu, R.; Qu, J.; Yu, Z.-Z. Silver Phosphate/Graphene Oxide Aerogel Microspheres with Radially Oriented Microchannels for Highly Efficient and Continuous Removal of Pollutants from Wastewaters. *ACS Sustain. Chem. Eng.* **2019**, *7*, 11228–11240. [CrossRef]

49. Afroze, J.D.; Abden, M.J.; Yuan, Z.; Wang, C.; Wei, L.; Chen, Y.; Tong, L. Core-shell structured graphene aerogels with multifunctional mechanical, thermal and electromechanical properties. *Carbon* **2020**, *162*, 365–374. [CrossRef]
50. Lv, S.; Ma, Y.; Qiu, C.; Sun, T.; Liu, J.; Zhou, Q. Effect of graphene oxide nanosheets of microstructure and mechanical properties of cement composites. *Constr. Build. Mater.* **2013**, *49*, 121–127. [CrossRef]
51. Zhu, C.; Liu, T.; Qian, F.; Han, T.Y.-J.; Duoss, E.B.; Kuntz, J.D.; Spadaccini, C.M.; Worsley, M.A.; Li, Y. Supercapacitors based on three-dimensional hierarchical graphene aerogels with periodic macropores. *Nano Lett.* **2016**, *16*, 3448–3456. [CrossRef] [PubMed]
52. Li, P.; Jin, Z.; Peng, L.; Zhao, F.; Xiao, D.; Jin, Y.; Yu, G. Stretchable all-gel-state fiber-shaped supercapacitors enabled by macromolecularly interconnected 3D graphene/nanostructured conductive polymer hydrogels. *Adv. Mater.* **2018**, *30*, 1800124. [CrossRef]
53. Lv, S.H.; Zhu, L.L.; Li, Y.; Jia, C.M.; Sun, S.Y. The Adsorption capacity of GONs/CMC/Fe_3O_4 magnetic composite microspheres and applications for purifying dye wastewater. *Materials* **2017**, *10*, 58. [CrossRef] [PubMed]
54. Yagub, M.T.; Sen, T.K.; Afroze, S.; Ang, H.M. Dye and its removal from aqueous solution by adsorption: A review. *Adv. Colloid Interface Sci.* **2014**, *209*, 172–184. [CrossRef]
55. Gao, Y.; Chen, K.; Ren, X.; Alsaedi, A.; Hayat, T.; Chen, C. Exploring the aggregation mechanism of graphene oxide in the presence of radioactive elements: Experimental and theoretical studies. *Environ. Sci. Technol.* **2018**, *52*, 12208–12215. [CrossRef]
56. Labiadh, L.; Kamali, A.R. 3D graphene nanoedges as efficient dye adsorbents with ultra-high thermal regeneration performance. *Appl. Surf. Sci.* **2019**, *490*, 383–394. [CrossRef]
57. Liu, H.; Qiu, H. Recent advances of 3D graphene-based adsorbents for sample preparation of water pollutants: A review. *Chem. Eng. J.* **2020**, *393*, 124691. [CrossRef]
58. Hong, J.Y.; Bak, B.M.; Wie, J.J.; Kong, J.; Park, H.S. Reversibly compressible, highly elastic, and durable graphene aerogels for energy storage devices under limiting conditions. *Adv. Funct. Mater.* **2015**, *25*, 1053–1062. [CrossRef]
59. Ma, Y.; Yue, Y.; Zhang, H.; Cheng, F.; Zhao, W.; Rao, J.; Luo, S.; Wang, J.; Jiang, X.; Liu, Z. 3D synergistical MXene/reduced graphene oxide aerogel for a piezoresistive sensor. *ACS Nano* **2018**, *12*, 3209–3216. [CrossRef] [PubMed]
60. Ha, H.; Shanmuganathan, K.; Ellison, C.J. Mechanically stable thermally crosslinked poly (acrylic acid)/reduced graphene oxide aerogels. *ACS Appl. Mater. Interfaces* **2015**, *7*, 6220–6229. [CrossRef] [PubMed]
61. He, X.; Liu, Q.; Zhong, W.; Chen, J.; Sun, D.; Jiang, H.; Liu, K.; Wang, W.; Wang, Y.; Lu, Z.; et al. Strategy of Constructing Light-Weight and Highly Compressible Graphene-Based Aerogels with an ordered unique configuration for wearable piezoresistive sensors. *ACS Appl. Mater. Interfaces* **2019**, *11*, 19350–19362. [CrossRef]
62. Balandin, A.A.; Ghosh, S.; Bao, W.; Calizo, I.; Teweldebrhan, D.; Miao, F.; Lau, C.N. Superior thermal conductivity of single-layer graphene. *Nano Lett.* **2008**, *8*, 902–907. [CrossRef]
63. Zhong, Y.; Zhou, M.; Huang, F.; Lin, T.; Wan, D. Effect of graphene aerogel on thermal behavior of phase change materials for thermal management. *Sol. Energy Mater. Sol. Cells* **2013**, *113*, 195–200. [CrossRef]
64. Wang, Z.; Yang, H.; Li, Y.; Zheng, X. Robust Silk Fibroin/Graphene Oxide Aerogel Fiber for Radiative Heating Textiles. *ACS Appl. Mater. Interfaces* **2020**, *12*, 15726–15736. [CrossRef] [PubMed]
65. Noroozi, M.; Panahi-Sarmad, M.; Abrisham, M.; Amirkiai, A.; Asghari, N.; Golbaten-Mofrad, H.; Karimpour-Motlagh, N.; Goodarzi, V.; Bahramian, A.R.; Zahiri, B. Nanostructure of Aerogels and Their Applications in Thermal Energy Insulation. *ACS Appl. Energy Mater.* **2019**, *2*, 5319–5349. [CrossRef]
66. Chen, J.-H.; Jang, C.; Xiao, S.; Ishigami, M.; Fuhrer, M.S. Intrinsic and extrinsic performance limits of graphene devices on SiO_2. *Nat. Nanotechnol.* **2008**, *3*, 206. [CrossRef]
67. Song, L.-T.; Wu, Z.-Y.; Liang, H.-W.; Zhou, F.; Yu, Z.-Y.; Xu, L.; Pan, Z.; Yu, S.-H. Macroscopic-scale synthesis of nitrogen-doped carbon nanofiber aerogels by template-directed hydrothermal carbonization of nitrogen-containing carbohydrates. *Nano Energy* **2016**, *19*, 117–127. [CrossRef]
68. Song, Z.; Liu, W.; Xiao, P.; Zhao, Z.; Liu, G.; Qiu, J. Nano-iron oxide (Fe_2O_3)/three-dimensional graphene aerogel composite as supercapacitor electrode materials with extremely wide working potential window. *Mater. Lett.* **2015**, *145*, 44–47. [CrossRef]
69. Yun, X.; Li, J.; Chen, X.; Chen, H.; Xiao, L.; Xiang, K.; Chen, W.; Liao, H.; Zhu, Y. Porous Fe_2O_3 modified by nitrogen-doped carbon quantum dots/reduced graphene oxide composite aerogel as a high-capacity and high-rate anode material for alkaline aqueous batteries. *ACS Appl. Mater. Interfaces* **2019**, *11*, 36970–36984. [CrossRef] [PubMed]
70. Chen, M.; Duan, S.; Zhang, L.; Wang, Z.; Li, C. Three-dimensional porous stretchable and conductive polymer composites based on graphene networks grown by chemical vapour deposition and PEDOT: PSS coating. *Chem. Commun.* **2015**, *51*, 3169–3172. [CrossRef] [PubMed]
71. Yue, Y.; Liu, N.; Ma, Y.; Wang, S.; Liu, W.; Luo, C.; Zhang, H.; Cheng, F.; Rao, J.; Hu, X. Highly self-healable 3D microsupercapacitor with MXene–graphene composite aerogel. *ACS Nano* **2018**, *12*, 4224–4232. [CrossRef]
72. Shannon, M.A.; Bohn, P.W.; Elimelech, M.; Georgiadis, J.G.; Marinas, B.J.; Mayes, A.M. Science and technology for water purification in the coming decades. *Nature* **2008**, *452*, 301–310. [CrossRef]
73. Kim, H.; Kang, S.-O.; Park, S.; Park, H.S. Adsorption isotherms and kinetics of cationic and anionic dyes on three-dimensional reduced graphene oxide macrostructure. *J. Ind. Eng. Chem.* **2015**, *21*, 1191–1196. [CrossRef]
74. Zhan, W.; Gao, L.; Fu, X.; Siyal, S.H.; Sui, G.; Yang, X. Green synthesis of amino-functionalized carbon nanotube-graphene hybrid aerogels for high performance heavy metal ions removal. *Appl. Surf. Sci.* **2019**, *467*, 1122–1133. [CrossRef]

75. Mariana, M.; HPS, A.K.; Yahya, E.B.; Olaiya, N.; Alfatah, T.; Suriani, A.; Mohamed, A. Recent trends and future prospects of nanostructured aerogels in water treatment applications. *J. Water Process Eng.* **2022**, *45*, 102481. [CrossRef]
76. Wang, L.; Zhu, D.; Duan, L.; Chen, W. Adsorption of single-ringed N-and S-heterocyclic aromatics on carbon nanotubes. *Carbon* **2010**, *48*, 3906–3915. [CrossRef]
77. Yang, S.; Shen, C.; Chen, L.; Wang, C.; Rana, M.; Lv, P. Vapor–liquid Deposition Strategy to Prepare Superhydrophobic and Superoleophilic Graphene Aerogel for Oil–Water Separation. *ACS Appl. Nano Mater.* **2018**, *1*, 531–540. [CrossRef]
78. Gusain, R.; Kumar, N.; Ray, S.S. Recent advances in carbon nanomaterial-based adsorbents for water purification. *Coord. Chem. Rev.* **2020**, *405*, 213111. [CrossRef]
79. Yang, Q.; Lu, R.; Ren, S.; Chen, C.; Chen, Z.; Yang, X. Three dimensional reduced graphene oxide/ZIF-67 aerogel: Effective removal cationic and anionic dyes from water. *Chem. Eng. J.* **2018**, *348*, 202–211. [CrossRef]
80. Ye, Y.; Yin, D.; Wang, B.; Zhang, Q. Synthesis of three-dimensional Fe_3O_4/graphene aerogels for the removal of arsenic ions from water. *J. Nanomater.* **2015**, *2015*, 864864. [CrossRef]
81. Wu, S.; Zhang, K.; Wang, X.; Jia, Y.; Sun, B.; Luo, T.; Meng, F.; Jin, Z.; Lin, D.; Shen, W. Enhanced adsorption of cadmium ions by 3D sulfonated reduced graphene oxide. *Chem. Eng. J.* **2015**, *262*, 1292–1302. [CrossRef]
82. Kabiri, S.; Tran, D.N.; Azari, S.; Losic, D. Graphene-diatom silica aerogels for efficient removal of mercury ions from water. *ACS Appl. Mater. Interfaces* **2015**, *7*, 11815–11823. [CrossRef] [PubMed]
83. Liu, J.; Ge, X.; Ye, X.; Wang, G.; Zhang, H.; Zhou, H.; Zhang, Y.; Zhao, H. 3D graphene/δ-MnO_2 aerogels for highly efficient and reversible removal of heavy metal ions. *J. Mater. Chem. A* **2016**, *4*, 1970–1979. [CrossRef]
84. Platero, E.; Fernandez, M.E.; Bonelli, P.R.; Cukierman, A.L. Graphene oxide/alginate beads as adsorbents: Influence of the load and the drying method on their physicochemical-mechanical properties and adsorptive performance. *J. Colloid Interface Sci.* **2017**, *491*, 1–12. [CrossRef] [PubMed]
85. Yang, X.; Li, Y.; Du, Q.; Sun, J.; Chen, L.; Hu, S.; Wang, Z.; Xia, Y.; Xia, L. Highly effective removal of basic fuchsin from aqueous solutions by anionic polyacrylamide/graphene oxide aerogels. *J. Colloid Interface Sci.* **2015**, *453*, 107–114. [CrossRef]
86. Wu, Z.; Huang, W.; Shan, X.; Li, Z. Preparation of a porous graphene oxide/alkali lignin aerogel composite and its adsorption properties for methylene blue. *Int. J. Biol. Macromol.* **2020**, *143*, 325–333. [CrossRef] [PubMed]
87. Xu, J.; Du, P.; Bi, W.; Yao, G.; Li, S.; Liu, H. Graphene oxide aerogels co-functionalized with polydopamine and polyethylenimine for the adsorption of anionic dyes and organic solvents. *Chem. Eng. Res. Des.* **2020**, *154*, 192–202. [CrossRef]
88. Zhang, Y.; Li, K.; Liao, J. Facile synthesis of reduced-graphene-oxide/rare-earth-metal-oxide aerogels as a highly efficient adsorbent for Rhodamine-B. *Appl. Surf. Sci.* **2020**, *504*, 144377. [CrossRef]
89. Yao, Q.; Fan, B.; Xiong, Y.; Jin, C.; Sun, Q.; Sheng, C. 3D assembly based on 2D structure of cellulose nanofibril/graphene oxide hybrid aerogel for adsorptive removal of antibiotics in water. *Sci. Rep.* **2017**, *7*, 45914. [CrossRef] [PubMed]
90. Sun, X.F.; Guo, B.B.; He, L.; Xia, P.F.; Wang, S.G. Electrically accelerated removal of organic pollutants by a three-dimensional graphene aerogel. *AIChE J.* **2016**, *62*, 2154–2162. [CrossRef]
91. Chen, Z.; Jin, L.; Hao, W.; Ren, W.; Cheng, H.-M. Synthesis and applications of three-dimensional graphene network structures. *Mater. Today* **2019**, *5*, 100027. [CrossRef]
92. Wei, N.; Wang, Q.; Ma, Y.; Ruan, L.; Zeng, W.; Liang, D.; Xu, C.; Huang, L.; Zhao, J. Superelastic active graphene aerogels dried in natural environment for sensitive supercapacitor-type stress sensor. *Electrochim. Acta* **2018**, *283*, 1390–1400. [CrossRef]
93. Wang, M.; Duan, X.; Xu, Y.; Duan, X. Functional three-dimensional graphene/polymer composites. *ACS Nano* **2016**, *10*, 7231–7247. [CrossRef] [PubMed]
94. Zhao, J.; Pan, R.; Sun, R.; Wen, C.; Zhang, S.-L.; Wu, B.; Nyholm, L.; Zhang, Z.-B. High-conductivity reduced-graphene-oxide/copper aerogel for energy storage. *Nano Energy* **2019**, *60*, 760–767. [CrossRef]
95. Wu, Z.S.; Winter, A.; Chen, L.; Sun, Y.; Turchanin, A.; Feng, X.; Müllen, K. Three-dimensional nitrogen and boron co-doped graphene for high-performance all-solid-state supercapacitors. *Adv. Mater.* **2012**, *24*, 5130–5135. [CrossRef] [PubMed]
96. Wang, H.; Li, R.; Li, M.; Li, Z. Flower-like Fe_2O_3@ multiple graphene aerogel for high-performance supercapacitors. *J. Alloys Compd.* **2018**, *742*, 759–768. [CrossRef]
97. Xu, L.; Wang, S.; Zhang, X.; He, T.; Lu, F.; Li, H.; Ye, J. A facile method of preparing $LiMnPO_4$/reduced graphene oxide aerogel as cathodic material for aqueous lithium-ion hybrid supercapacitors. *Appl. Surf. Sci.* **2018**, *428*, 977–985. [CrossRef]
98. Sun, R.; Chen, H.; Li, Q.; Song, Q.; Zhang, X. Spontaneous assembly of strong and conductive graphene/polypyrrole hybrid aerogels for energy storage. *Nanoscale* **2014**, *6*, 12912–12920. [CrossRef]
99. Zhou, Y.; Hu, X.; Guo, S.; Yu, C.; Zhong, S.; Liu, X. Multi-functional graphene/carbon nanotube aerogels for its applications in supercapacitor and direct methanol fuel cell. *Electrochim. Acta* **2018**, *264*, 12–19. [CrossRef]
100. Wei, X.; Wan, S.; Gao, S. Self-assembly-template engineering nitrogen-doped carbon aerogels for high-rate supercapacitors. *Nano Energy* **2016**, *28*, 206–215. [CrossRef]
101. Tingting, Y.; Ruiyi, L.; Xiaohuan, L.; Zaijun, L.; Zhiguo, G.; Guangli, W.; Junkang, L. Nitrogen and sulphur-functionalized multiple graphene aerogel for supercapacitors with excellent electrochemical performance. *Electrochim. Acta* **2016**, *187*, 143–152. [CrossRef]
102. Kim, D.W.; Jung, S.M.; Jung, H.Y. Long term thermostable supercapacitor using in-situ SnO_2 doped porous graphene aerogel. *J. Power Sources* **2020**, *448*, 227422. [CrossRef]
103. Tian, W.; Cheng, D.; Wang, S.; Xiong, C.; Yang, Q. Phytic acid modified manganese dioxide/graphene composite aerogel as high-performance electrode materials for supercapacitors. *Appl. Surf. Sci.* **2019**, *495*, 143589. [CrossRef]

104. Liu, Z.; Lyu, J.; Fang, D.; Zhang, X. Nanofibrous Kevlar aerogel threads for thermal insulation in harsh environments. *ACS Nano* **2019**, *13*, 5703–5711. [CrossRef]
105. Wang, Y.-T.; Liao, S.-F.; Shang, K.; Chen, M.-J.; Huang, J.-Q.; Wang, Y.-Z.; Schiraldi, D.A. Efficient approach to improving the flame retardancy of poly (vinyl alcohol)/clay aerogels: Incorporating piperazine-modified ammonium polyphosphate. *ACS Appl. Mater. Interfaces* **2015**, *7*, 1780–1786. [CrossRef] [PubMed]
106. Cai, H.; Sharma, S.; Liu, W.; Mu, W.; Liu, W.; Zhang, X.; Deng, Y. Aerogel microspheres from natural cellulose nanofibrils and their application as cell culture scaffold. *Biomacromolecules* **2014**, *15*, 2540–2547. [CrossRef]
107. Chen, H.-B.; Chiou, B.-S.; Wang, Y.-Z.; Schiraldi, D.A. Biodegradable pectin/clay aerogels. *ACS Appl. Mater. Interfaces* **2013**, *5*, 1715–1721. [CrossRef] [PubMed]
108. Yue, C.; Feng, J.; Feng, J.; Jiang, Y. Ultralow-density and high-strength graphene aerogels composites for thermal insulation. *Mater. Lett.* **2017**, *188*, 169–171. [CrossRef]
109. Wang, Z.; Wei, R.; Gu, J.; Liu, H.; Liu, C.; Luo, C.; Kong, J.; Shao, Q.; Wang, N.; Guo, Z. Ultralight, highly compressible and fire-retardant graphene aerogel with self-adjustable electromagnetic wave absorption. *Carbon* **2018**, *139*, 1126–1135. [CrossRef]
110. Zhang, Q.; Lin, D.; Deng, B.; Xu, X.; Nian, Q.; Jin, S.; Leedy, K.D.; Li, H.; Cheng, G.J. Flyweight, superelastic, electrically conductive, and flame-retardant 3D multi-nanolayer graphene/ceramic metamaterial. *Adv. Mater.* **2017**, *29*, 1605506. [CrossRef] [PubMed]
111. Maleki, H.; Durães, L.; García-González, C.A.; del Gaudio, P.; Portugal, A.; Mahmoudi, M. Synthesis and biomedical applications of aerogels: Possibilities and challenges. *Adv. Colloid Interface Sci.* **2016**, *236*, 1–27. [CrossRef] [PubMed]
112. Zhang, B.; Wang, Y.; Zhai, G. Biomedical applications of the graphene-based materials. *Mater. Sci. Eng. C* **2016**, *61*, 953–964. [CrossRef]
113. Liu, S.; Zhou, C.; Mou, S.; Li, J.; Zhou, M.; Zeng, Y.; Luo, C.; Sun, J.; Wang, Z.; Xu, W. Biocompatible graphene oxide–collagen composite aerogel for enhanced stiffness and in situ bone regeneration. *Mater. Sci. Eng. C* **2019**, *105*, 110137. [CrossRef] [PubMed]
114. Ayazi, H.; Akhavan, O.; Raoufi, M.; Varshochian, R.; Motlagh, N.S.H.; Atyabi, F. Graphene aerogel nanoparticles for in-situ loading/pH sensitive releasing anticancer drugs. *Colloids Surf. B* **2020**, *186*, 110712. [CrossRef]
115. Luo, J.; Lai, J.; Zhang, N.; Liu, Y.; Liu, R.; Liu, X. Tannic acid induced self-assembly of three-dimensional graphene with good adsorption and antibacterial properties. *ACS Sustain. Chem. Eng.* **2016**, *4*, 1404–1413. [CrossRef]
116. McCallion, C.; Burthem, J.; Rees-Unwin, K.; Golovanov, A.; Pluen, A. Graphene in therapeutics delivery: Problems, solutions and future opportunities. *Eur. J. Pharm. Biopharm.* **2016**, *104*, 235–250. [CrossRef]
117. López-Dolado, E.; González-Mayorga, A.; Portolés, M.T.; Feito, M.J.; Ferrer, M.L.; Del Monte, F.; Gutiérrez, M.C.; Serrano, M.C. Subacute tissue response to 3D graphene oxide scaffolds implanted in the injured rat spinal cord. *Adv. Healthc. Mater.* **2015**, *4*, 1861–1868. [CrossRef]
118. Zheng, Q.; Lee, J.-h.; Shen, X.; Chen, X.; Kim, J.-K. Graphene-based wearable piezoresistive physical sensors. *Mater. Today* **2020**, *36*, 158–179. [CrossRef]
119. Ma, Y.; Zhao, M.; Cai, B.; Wang, W.; Ye, Z.; Huang, J. 3D graphene foams decorated by CuO nanoflowers for ultrasensitive ascorbic acid detection. *Biosen. Bioelectron.* **2014**, *59*, 384–388. [CrossRef]
120. Dong, X.; Wang, X.; Wang, L.; Song, H.; Zhang, H.; Huang, W.; Chen, P. 3D graphene foam as a monolithic and macroporous carbon electrode for electrochemical sensing. *ACS Appl. Mater. Interfaces* **2012**, *4*, 3129–3133. [CrossRef]
121. Sun, Y.; Lin, Y.; Sun, W.; Han, R.; Luo, C.; Wang, X.; Wei, Q. A highly selective and sensitive detection of insulin with chemiluminescence biosensor based on aptamer and oligonucleotide-AuNPs functionalized nanosilica@ graphene oxide aerogel. *Anal. Chim. Acta* **2019**, *1089*, 152–164. [CrossRef] [PubMed]

materials

MDPI

Review

The Synergistic Properties and Gas Sensing Performance of Functionalized Graphene-Based Sensors

Zandile Dennis Leve , Emmanuel Iheanyichukwu Iwuoha and Natasha Ross *

SensorLab, Chemistry Department, University of the Western Cape, Cape Town 7535, South Africa; 3135207@myuwc.ac.za (Z.D.L.); eiwuoha@uwc.ac.za (E.I.I.)
* Correspondence: nross@uwc.ac.za; Tel.: +27-842549220

Abstract: The detection of toxic gases has long been a priority in industrial manufacturing, environmental monitoring, medical diagnosis, and national defense. The importance of gas sensing is not only of high benefit to such industries but also to the daily lives of people. Graphene-based gas sensors have elicited a lot of interest recently, due to the excellent physical properties of graphene and its derivatives, such as graphene oxide (GO) and reduced graphene oxide (rGO). Graphene oxide and rGO have been shown to offer large surface areas that extend their active sites for adsorbing gas molecules, thereby improving the sensitivity of the sensor. There are several literature reports on the promising functionalization of GO and rGO surfaces with metal oxide, for enhanced performance with regard to selectivity and sensitivity in gas sensing. These synthetic and functionalization methods provide the ideal combination/s required for enhanced gas sensors. In this review, the functionalization of graphene, synthesis of heterostructured nanohybrids, and the assessment of their collaborative performance towards gas-sensing applications are discussed.

Keywords: graphene; graphene oxide; reduced graphene oxide; surface functionalization; gas sensor; metal oxide nanocomposites; gas sensing mechanism

1. Introduction

The globe has been faced with a burden of diseases linked to air pollution exposure which has had a massive toll on human health. The effects caused by exposure to air pollution have been estimated to cause millions of deaths and yearly losses of a healthy lifestyle. This burden has been reported to be on a par with other major global health risks, namely, unhealthy diets and tobacco smoking. Air pollutants are attributable as the single main environmental threat to the human health [1]. Air pollutants may be either released into the atmosphere, which may then be referred to as primary air pollutants, or formed within the atmosphere as secondary air pollutants. Primary air pollutants are composed of sulfur dioxide (SO_2), oxides of nitrogen, carbon monoxide (CO), volatile organic compounds (VOCs), and carbonaceous and non-carbonaceous primary particles. Secondary air pollutants are formed from chemical reactions of primary air pollutants, which may often involve natural environmental components such as oxygen and water. These include ozone (O_3), oxides of nitrogen, and secondary particulate matter (PM) [2].

Air pollution exposure is said to be largely determined by the concentration of air pollutants disposed in the environment to which people are exposed and the amount of time spent in that environment [2]. The World Health Organization (WHO) has, since 1987, periodically issued air quality guidelines (AQGs) based on health to better assist governments and civil society to reduce exposure to air pollution and its adverse effects. In 2005, WHO published AQGs for PM, O_3, NO_2, and SO_2 [1]. Carbon monoxide was assessed in 2000 and later, in 2010, as an indoor pollutant [3]. Table 1 shows the WHO AQGs set for health protection based on air pollutant concentrations and average times for short-term and long-term exposures. These were later updated, and the latest data showed that PM_{10} had a 15 μg/m^3 annual mean and a 45 μg/m^3 24-h mean; $PM_{2.5}$, a 5 μg/m^3 annual mean and 15 μg/m^3 24-h mean; O_3, a 100 μg/m^3 8-h daily maximum and 60 μg/m^3 8-h mean on a six-month basis; NO_2, a 10 μg/m^3 annual mean and 25 μg/m^3 24-h mean; SO_2, a 40 μg/m^3 24-h mean [3].

Table 1. Summary of short-term and long-term exposures to air pollutants shown by their mean concentrations and average exposures times following standards reported by WHO AQGs. Adapted from ref. [4].

Air Pollutant	Short-Term Exposure		Long-Term Exposure	
	Mean Concentration	Average Time	Mean Concentration	Average Time
O_3	100 μg/m^3	8 h	-	-
NO_2	200 μg/m^3	1 h	40 μg/m^3	1 year
CO	100 mg/m^3	15 min	60 mg/m^3 30 mg/m^3 10 mg/m^3	30 min 1 h 8 h
SO_2	500 μg/m^3	10 min	20 μg/m^3	24 h
PM_{10}	50 μg/m^3	24 h	20 μg/m^3	1 year
$PM_{2.5}$	25 μg/m^3	24 h	10 μg/m^3	1 year

The generation of toxic gases including nitrogen oxides (NOx), sulfur oxides (SOx), ammonia (NH_3), and CO has been stated as a major hazard to environmental security and individual health protection [5]. The detection of NO_2 has generated substantial attention as it is not only harmful for the respiratory system but also causes acid rain formation [6]. Hydrogen sulfide (H_2S), which is also a toxic gas produced from the process of oil and natural gas production, is highly dangerous for the human body with reported health effects following exposure including death and respiratory, ocular, neurological, cardiovascular, metabolic, and reproductive effects [7]. Ammonia (NH_3) is an irritant and corrosive gas, to the extent such that its low concentrations in air or liquid can lead to severe irritations and coughing in the case of skin or eye contact [6]. Carbon monoxide (CO) is also highly toxic to humans, amongst various gases, as it is an odorless, colorless, and tasteless gas which appears to be slightly denser than air, therefore making it difficult to recognize in a normal way [8]. Therefore, the detection of toxic gases and harmful chemical vapors within a limited time is of the utmost importance.

The importance of gas sensors has long been apparent in different aspects of certain fields since the first invention by Davy in 1815 [9,10] and a commercial catalytic combustion gas detector by Johnson in 1926 [11], for which first substantial studies began in the early 1970s and later rapidly expanded since 2002. Hence, simple, and accurate detection of toxic gas has become vital in our everyday lives, not only to industries but to all people [7]. Techniques such as optical, acoustic and gas chromatography, chemiluminescence ion chromatography, and spectrophotometry have been utilized for the detection toxic gases [12,13]. However, the methods mentioned are not cost-effective, complex, and are not suitable for implementation for widespread, continuous monitoring in ambient conditions. As a result, electrochemical sensing is the most widely used method for detecting dangerous gases. Electrochemical detection has several advantages over other approaches, including strong selectivity and repeatability, ppm level detection with high precision, low energy linear output with high resolution, and lower cost. In recent years, electrochemical sensors made of diverse functional materials have been the focus of harmful gas detection research [7].

Principles of Graphene-Based Gas Sensors

Graphene is described as a flat one-atom thick monolayer of sp^2-hybridized carbon atoms that are tightly stacked into a two-dimensional honeycomb lattice [14]. Its semimetal nature allows charge carriers to behave like Dirac fermions which results in extraordinary effects such as improved intrinsic mobility of up to ~200,000 $cm^2 \cdot V^{-1} \cdot s^{-1}$, with unique properties such as a higher thermal conductivity of ~5000 $W \cdot m^{-1} \cdot k^{-1}$, high mechanical stiffness of ~1060 GPa, an excellent optical transmittance of ~97.7%, and a large specific surface area of 2630 $m^2 \cdot g^{-1}$ [15]. Graphene is a basic building block for graphitic materials of all other dimensionalities [16].

Graphene oxide (GO) is an oxide form of graphene that is covered by a high density of oxygen functional groups such as hydroxyl, epoxy, and carboxyl on its basal plane and carboxyl on its edges, making it easy to suspend in water and other polar media [15]. Its carbon atoms are partially sp^3-hybridized and they can move above or below the graphene plane [17]. The ability of GO conduction depends on the degree of oxidation in the compound and the synthetic route proposed. Graphene-like sheets are produced by reduction of GO in which the oxygen functional groups are removed whilst recovering the π-conjugated network, which is the most fascinating property of GO [18].

A material related to GO is reduced graphene oxide (rGO), which possesses sheets that are regarded as chemically derived graphene [18]. Measurements of elemental analysis (atomic C/O ratio, ~10) for rGO performed by combustion reveal that a significant amount of oxygen exists in the structure, which indicates that rGO is not the same as pristine graphene (Park and Ruoff, 2009). Additionally, its conductance decreases by a magnitude of three orders when cooled to lower temperatures, which makes it exhibit non-metallic behavior whilst it is nearly metallic [19].

Graphene has emerged as a possible contender for sensing applications, among other things. Experimental and theoretical research has reported on the demonstrated monolayer graphene as a promising candidate to detect a variety of molecules, including gases, due to its appealing advantages [20]. Graphene oxide is a popular precursor of graphene because of its high water solubility, ease of functionalization, and simple processing [21]. In an as-oxidized state, GO has poor conductivity [22] as it is rendered too electrically insulating as a conductance-based sensor due to the disruption of the π-conjugated system by oxygen moieties [20,23]. Chemical reduction partially restores the conductivity by the removal of oxygen which then recovers the aromatic carbon double bonds. Yet, this still does not repair to pure graphene, as some oxygen groups remain in the network [24]. The rGO has intermediate conductivity and defect sites which make it attractive for sensor application [24].

Functional materials have been reported to be used in chemiresistive gas sensors. Volanti et al. [25] report on the development of CuO-based nanostructured chemiresistive gas sensors with different morphologies, which were exposed to oxidizing and reducing gases in the same test chamber over a range of temperatures and gas concentrations that were measured simultaneously [25]. However, chemiresistive gas sensors have been reported to have drawbacks such as a lack of selectivity, flexibility, high power consumption, safety risks, and a high operating temperature [26]. Nanostructured materials, such as conductive polymers (CPs), have been studied extensively around the world because they have unique and intriguing properties such as ease of synthesis, structural diversity, environmental stability, low cost, flexibility, and a sensitive response to chemical molecules at room temperature [27]. Conducting polymers possess a strong potential for producing enhanced sensor performance in comparison to their bulky opposite [28]. However, they lack stability at the nanoscale attributed to the nature of covalent bonds, also resulting in an unstable nanostructure [26]. Due to this factor, progress in the synthesis of CPs has been reported to be relatively slow with limited research when compared to inorganic nanomaterials [29].

Metal oxides have been used as a sensing material in low-cost sensors. Due to their good sensing capability, fast response, and recovery, gas sensing devices based on metal oxide sensors have been thoroughly investigated. These sensors do, however, have an operational limitation due to their failure to work at temperatures much higher than room temperature. Complex circuitry and high power consumption are required for operation or optimal response at higher temperatures [6]. There are some methods that have been successfully employed to enhance the selectivity of metal oxide sensors such as the optimization of temperature, bulk/surface doping, and the use of molecular filters [30]. The exceptional properties of graphene, GO, and rGO pose them as highly useful materials for various applications by the functionalization/doping of surfaces in a variety of ways and hence are widely investigated by researchers [18,22]. In this work, the synthesis and fabrication of GO/rGO/nanoscale metal oxide nanocomposites and assessing their performance for gas/vapor-sensing applications are reviewed.

2. Synthesis of Graphene-Based Inorganic Nanostructured Composites

Graphene has been synthesized using various methods as shown in Figure 1. These include mechanical exfoliation [31], chemical exfoliation [32], epitaxial growth on silicon carbide (SiC) [33,34], and chemical vapor deposition (CVD) [35]. Mechanical exfoliation is described as a simple peeling process involving a commercially available highly oriented pyrolytic graphite (HOPG) sheet that is dry etched in oxygen. It may produce graphene with exceptional properties, but it is limited by its low production which may not be sufficient for specific applications [31]. Chemical oxidation of graphite and its subsequent exfoliation lead to a greater amount of graphite monolayer, and the chemical treatment inevitably results in structural defects. The graphene can be used in industrial applications such as paint additives or composites [32,36].

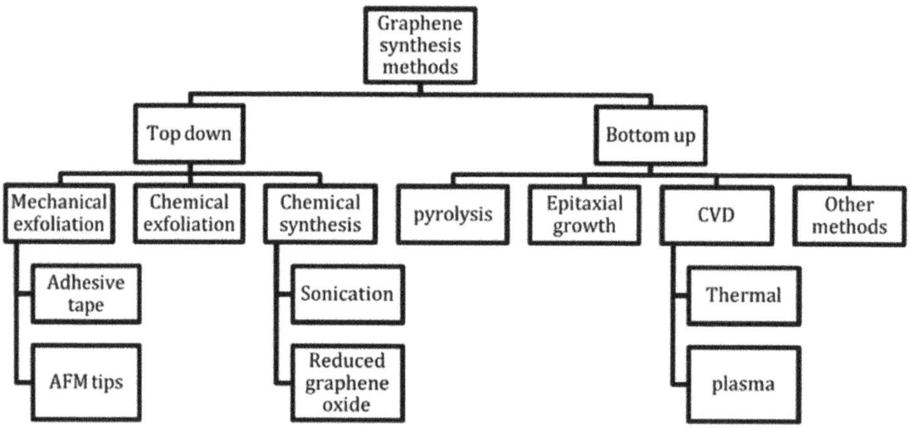

Figure 1. Summary of the graphene synthesis methods [14].

In thermal decomposition of SiC, the compound is heated under an ultrahigh vacuum (UHV) and the Si atoms sublimate from the substrate. The formed few-layer graphene (FLG) typically needs a short period of time to anneal at a temperature of 1200 °C. The graphene layers can be grown directly on a semiconducting substrate, but production is still not producible [37]. In CVD, gas species are placed in a reactor and passed through the hot zone in which hydrocarbon precursors are decomposed into carbon radicals at the surface of a metal substrate, and, thereafter, a single-layer and few-layer graphene are formed [35]. The method produces a large area and high-quality graphene, and it is also inexpensive [38]. Chemically derived graphene is achieved by synthesis of GO and its subsequent reduction into rGO [16]. In a typical procedure, graphite undergoes chemical oxidation into GO using a modified Hummers' method. The abundance of functional groups in GO results in hydrophilic behavior which is strongly dependent on the level of oxidation [39]. Reduction of GO follows thereafter to form rGO via several methods such as thermal or chemical reduction and electrochemical reduction which produce mass production of rGO [18]. Summary of the synthesis of graphene with different routes is shown in Figure 2.

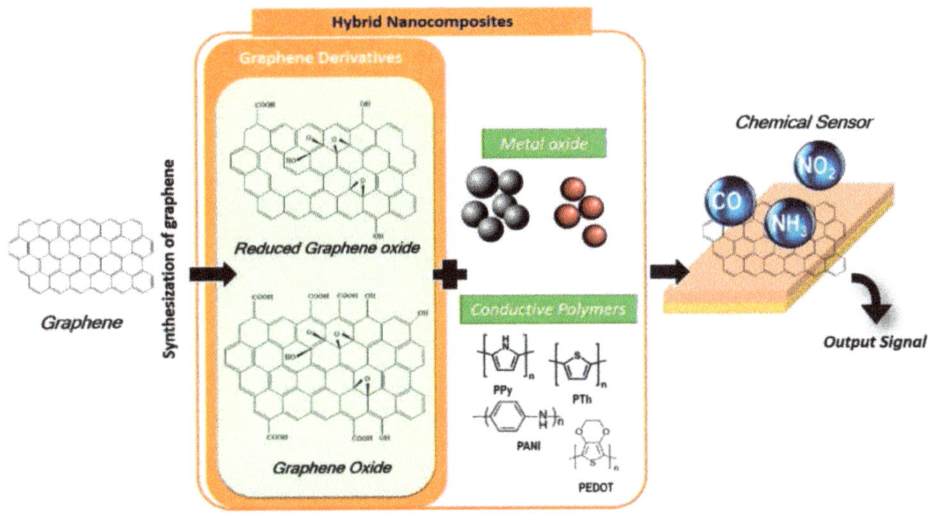

Figure 2. Schematic of graphene hybrid nanocomposites' fabrication into chemical sensors [39].

Graphene-semiconductor nanocomposites could open new possibilities for graphene-based catalytic and photocatalytic reactions [40]. To improve the properties of graphene-based composites, metal nanoparticles, metal oxides, and other inorganic compounds have been made into a structure with graphene [41]. Metals such as Au [41], Ag [42], Ni [43], Cu [44], Ru [41], Rh [45], and Co [46,47], along with metal oxides such as TiO_2 [48], ZnO [49], MnO_2 [50], Co_3O_4 [51], NiO [52], and Fe_3O_4 [53], and metal organic framework ZIF-8 [54] were among the materials used. There have also been graphene/MoS_2 and graphene/silica nanocomposites created both of which have fascinating electrical characteristics [55]. Coulombic charge transfer between noble metal NPs and graphene has been demonstrated during their interaction [56,57]. Through excited-state electron transport, graphene oxide interacts with NPs of semiconducting oxides [58,59]. Figure 2 shows a graphical representation of the creation of chemical sensors using graphene derivatives and metal NPs/metal oxides.

Inorganic nanostructures have been used to make graphene-based composites [42,49,51,60], and graphene has been employed as a novel and promising platform for the synthesis of graphene-based noble metal nanostructures [60,61]. Huge efforts have been undertaken in recent decades to create inorganic nanostructures with regulated shapes, size, crystallinity, and functionality [62,63]. Belts, tubes, rods, wires, particles, and polyhedrons are examples of inorganic components with a variety of morphologies [64]. By combining graphene and its derivatives with various types of functional materials in composites, it is possible to harness their desirable qualities [20,65–69]. The production of unique noble metal nanocrystals and their prospective applications in varied domains such as catalysis, electronics, sensing, and medicine have made significant development [70–75]. A novel class of material system for constructing unconventional inorganic electrical and optoelectronic devices is the hybrid heterostructure, which is made up of inorganic nanostructures grown directly atop graphene layers [15,76,77].

The high carrier mobility, radiative recombination rate, and long-term stability of inorganic semiconductor nanostructures enable the creation of high-performance optoelectronic and electrical devices [78–80]. Graphene–inorganic hybrid materials have been created in the last few years by inserting inorganic nanostructures between graphene sheets using the driving force of chemisorption interaction [81,82]. The methods for the synthesis of graphene-based inorganic hybrid materials can be divided into two categories: (i) graphene (oxides) assembly with generated inorganic nanostructures and (ii) graphene and inorganic nanostructure synthesis and assembly in one pot. The first method involves preparing inorganic nanostructures before mixing them with graphene or GO dispersion. The second technique, on the other hand, involves obtaining graphene and inorganic species in situ and then assembling them in a single-pot synthesis [83].

Alfano et al. [84] developed a sensitive material in which graphene was created by exfoliating graphite in its liquid phase, and then followed by microwave functionalization with ZnO. In comparison to equivalent devices made of bare graphene, chemiresistor devices made of hybrid materials showed that adding ZnO NPs to the graphene matrix can increase the sensing platform by increasing the sensing response and enhancing the selectivity. Wu et al. [85] created an inkjet-printed graphene-MOx-based sensor system that can be integrated onto tiny CMOS-compatible platforms to measure NH_3 selectively and accurately. Inkjet printing was employed to deposit ZnO-graphene functional inks directly onto the interdigitated Au electrodes (IDEs) on the Si_3N_4 membrane substrate (5 m finger width and gap; 250 m diameter) of the CMOS HP during the sensor fabrication process. The process described allowed for the automated manufacture of many devices at the same time. Kim et al. [86] developed a method for producing graphene/SnO_2 nanocomposites via explosive microwave synthesis for use in gas sensors. The fabrication technique was able to achieve rapid and large-scale production thanks to the fast surface chemical reactions enabled by microwave-generated plasmas, which could lead to the commercialization of semiconductor-type gas sensors by lowering production costs and improving sensing capabilities.

In comparison with graphene, GO presents advantages such as a low production cost, large-scale production, and easy processing [17,87]. A variety of materials have been created on GO or rGO nanosheets including inorganic nanostructures, organic crystals, polymers, metal organic frameworks (MOFs), biomaterials, and carbon nanotubes (CNTs) [60,61,88–95]. Kavinkumar and Manivannan [96] used chemical reduction with vitamin C in GO suspension to make AgNPs–GO composites with various $AgNO_3$ concentrations. Following that, an AgNPs–GO-coated fiber optic gas sensor for NH_3 detection was created. The composites were proven to have superior sensitivity and sensing performances to rGO. Jiang et al. [97] conducted a study in which nanoparticles and GO of SnO_2/NiO were produced. The GO/SnO_2/NiO materials were created using a hydrothermal growth composition in a neutral environment. A gas sensitivity test system was used to determine the composite's sensitivity, ideal working temperature, and selectivity.

Graphene oxide is typically used as a precursor in the preparation of rGO (Yu et al., 2020; Dreyer et al., 2010). Li et al. [98] used a one-pot solution approach at room temperature to successfully manufacture CuO/rGO nanohybrids, in which the reduction of GO and the synthesis of CuO occurred simultaneously. The sheet-like CuO produced was found to be consistently mixed with rGO nanosheets [98]. Gu et al. [99] stated that they have effectively manufactured n-type In_2O_3–rGO nanocomposites using a simple hydrothermal technique, with great selectivity, high response, and a quick response/recovery time. The experimental results showed that the In_2O_3–rGO nanocomposite-based gas sensor had a good chance of being a good candidate for NO_2 monitoring in the environment. Karthik et al. [100] used a spray pyrolysis technique combined with an annealing process to make rGO/TiO_2 thin films. The entire sensing nature of the rGO/TiO_2 sensor was allegedly owing to the design component of rGO, which reduced TiO_2 nanoparticle accumulation and advanced porosity and conductivity.

Zhang et al. [101] developed an rGO/SnO_2/Au sensor in which rGO/SnO_2 nanocomposites were decorated with Au nanoparticles at high concentrations of a GO precursor, which was obtained by adding $HAuCl_4$ and $SnCl_2$ to the reaction system. Following that, hybrid nanomaterials were used as gas sensors, and they performed well. Wang et al. [102] successfully manufactured a gas sensor based on AgNPs–SnO_2–rGO hybrids synthesized using the hydrothermal synthesis approach. When compared to SnO_2–rGO hybrids, the gas-sensing results showed that adding AgNPs to the SnO_2–rGO hybrids considerably improved the gas-sensing capability at room temperature. Ifterkhar Uddin et al. [103] used a simple chemical approach to construct a gas sensor based on a Ag-loaded ZnO–rGO hybrid to improve gas-sensing performance at low working temperatures. The as-synthesized sensing material was characterized and found to have a homogeneously dispersed and tightly adhered Ag–ZnO mixer on the surface of the reduced graphene oxide. The gas sensor performed best at 150 °C, with three wt% Ag-loaded ZnO–Gr exhibiting improved sensing properties as compared to individual equivalents.

3. Surface Functionalization of Graphene/GO/rGO with Metal Oxide Nanocomposites towards Gas Sensing

In general, different gases possess molecules with electron-withdrawing or -donating abilities that can adsorb onto the surface of graphene and thereby alter its conductivity. The sensing platform of this nature has intrinsically high sensitivity. This can be attributed to the conical band structure of graphene that ensures significant conductivity changes. Nonetheless, selectivity is an issue for a sensitive chemiresistor where many gases can result in large sensing responses. Therefore, functionalization of graphene surfaces has been proposed as a solution to this issue [104]. Chemical functionalization is a powerful tool for modifying structure and specific characteristics of graphene. This can be done via non-covalent as well as covalent routes according to the operation between the ligands and the sp^2 carbonaceous lattice [105]. In the former, the extended π-electron delocalization of the graphene sheet remains intact, whereas the latter takes place via the formation of covalent bonds between the graphene and different organic (inorganic) functional groups [105]. Combining graphene with newly added groups in the form of covalent bonds to improve and increase its performance is known as covalent bond functionalization [17,106–108]. By interacting of hydrogen bonds and the electrostatic forces between graphene and functional molecules, non-covalent bond functionalization of graphene or graphene oxide results in the formation of a composite material with a specific function, the greatest advantage of which is maintaining the bulk structure and excellent properties of graphene or graphene oxide, as well as improving the dispersibility and stability of graphene or graphene oxide [109,110].

Graphene on its own does not exhibit good sensing properties; however, derivatives have shown exceptional sensing ability due to better optical, mechanical, electrical, and chemical properties [99,111,112]. To combine different elements into graphene, element doping modification typically uses annealing heat treatment, ion bombardment, arc discharge, and other methods, resulting in the substitution of defects and vacancy defects in graphene while maintaining the intrinsic two-dimensional structure of graphene [113–115]. Generally, when the electrical conductivity and large surface of graphene are required, non-covalent modification methods are typically preferred. Likewise, when the stability and the strong mechanical properties of modified graphene are expected, covalent methods are ideal. Graphene sheets can be uniformly disseminated in aqueous and/or organic (inorganic) fluids by selectively adding functions to their surfaces [108,110,113,116,117]. Graphene oxide lacks chemical reactivity, which can be attributed to its homogeneity and highly delocalized electronic structure. In typical occurrence, chemical reactions are traced at locations that exhibit weak or labile bonds, highly localized orbitals, dangling bonds, or localized charges [117]. All these cannot be found in graphene, whilst in its honeycomb lattice structure, each sp^2 atom of carbon is characterized by a 3-fold symmetric electronic hybridization where the p-orbitals extend out of the atomic plane [118]. In this manner, a self-passivating and highly delocalized network is formed [113]. Disruption of this chemical structure is not only thermodynamically unfavorable, it also requires the formation of high-energy radicals on adjacent carbons which are difficult to support [117].

The oxides which are functionalized with various oxygen groups produce GO and rGO, which provide more adsorption sites for gases and so improve the sensitivity of the film. The presence of oxygen groups in GO renders it an insulating material [119] and since it is difficult to control the content of these groups during oxidation, GO is not an appropriate gas-sensing material. Hence, GO is reduced into rGO [120], and the generation of some oxygen functional groups that remain following reduction, coupled with some defects and vacancies, prove beneficial for gas adsorption [121]. These lead to electron transfer from rGO to the oxygen functional groups located at the surface of rGO. The holes are the main charge carriers and thus make rGO act as a p-type semiconductor [122]. Graphene or its derivatives provides faster carrier transport through barrier by opening of the sizable energy gap due to quantum confinement [123]. In this way, they can be exploited as catalytic active centers for covalent/non-covalent modification design, depending on the needs of particular application domains [17,32]. Furthermore, the presence of oxygen-containing groups broadens the graphene oxide interlayer gap. Small molecules or polymer intercalations can be used to functionalize it [124]. As a result, increasing the applications of graphene and graphene oxide requires functional modification [17]. Kumar and Kaur [125] described how thermal annealing reduced the electrical gas sensing of graphene oxide. The number of graphitic domains, as well as the specific surface area and pore volume, were found to increase the sample's gas-sensing ability when exposed to SO_2.

Chemical functionalization of graphene using synthetic chemistry methods allows for the creation of p- and n-doped graphene by selecting electron-donating or electron-withdrawing complexes that are covalently bound to the graphene carbon network. The doping concentration could often control the electrical characteristics [126]. The bandgap would open at the Fermi level of graphene as a result of successful doping, and graphene's 'metallic' character would be transformed to a 'semiconductor' one [106,127–129]. For n-type (p-type) doping, electrons must be released into (or extracted from) the graphene layer, which is commonly accomplished by adsorbing atoms and/or molecules on its surface, i.e., surface transfer doping [126,128,130–132]. P-type doping for graphene is a lot more difficult [133]. For strong dimer bonds, many elements with a high electronegativity, such as nitrogen, oxygen, or fluorine, are used. On the graphene surface, they are unlikely to form a stable layer.

To generate p-type doping in graphene, several chemicals such as NO_2, H_2O, NH_3, or charge transfer complexes have been utilized. However, NO_2, H_2O, and NH_3 are highly reactive compounds that should not be used in electronic materials [128,133–136]. The heavier elements, which are less reactive than oxygen or fluorine, offer feasible options. Bismuth and antimony, while having a lower electron affinity than atomic carbon, are able to pull electrons from the graphene sheet, which is not immediately apparent. The functionalization of graphene and graphene oxide is achieved by altering their intrinsic structure further [137]. The correct functionalization of graphene and graphene oxide prevents agglomeration and protects their inherent properties during the reduction process. The functional modification of graphene and graphene oxide preserves their remarkable properties while also introducing additional functional groups to offer them new properties [128].

Hybridization with metal oxide nanostructures improves graphene-based sensors even more [123] by offering a higher surface-to-volume ratio with good adsorption of gas molecules on the sensor surface at numerous active sites [6,138]. This causes variations in the carrier concentration of graphene-based metal oxide composite film and, thus, the resistance of the film. The change in resistivity allows sensors to identify the target gas as an electron donor or acceptor [99,103,111,112]. Furthermore, graphene's properties prevent metal oxide agglomeration, whilst in turn the metal oxides prevent graphene fossilization [6]. There have been several reports on the functionalization of graphene for gas-sensing applications throughout literature [139]. Graphene-based gas sensors can overcome the limitations on which traditional sensors fall short, such as sensitivity and selectivity coupled with power consumption, temperature-dependency that is significantly large, and safety issues [140]. Wicaksono et al. [141] exhibited the gas-sensing characteristics of graphene–TiO_2/TiSiO-coated fabrics towards various gases. Martinez-Orozco et al. [142] presented a study on the preparation of a hydrogen-gas sensor, synthesized by the microwave method, based on Pd nanoparticles which were dispersed and anchored on graphene layers. The methodology utilized allows for the synthesis of functional Pd–graphene nanostructures.

A wet chemical route through Hummers' method for the synthesis of functionalized rGO was achieved by Panda et al. [8], for which the sensing properties (~71% sensitivity against 30 ppm CO) were used to detect low-concentration CO at room temperature and atmospheric pressure under ambient humidity. Muda et al. [143] also reported on the fabrication of a gas sensor based on a vacuum method to achieve a homogenous and uniform thin film of multi-layer rGO on a plastic substrate. The fabricated sensor was used to detect NO_2, and a sensitivity of ~25% over 50 ppm gas concentration was reported. However, the recovery was slow as it took time to fully recover to its baseline resistance before the exposure timescales were tested.

In regard to the response time, Jia and Wang [5] reported on a novel NO_2 gas sensor where rGO adsorbs NO_2 gas molecules as well as transports electrical signals, and the AgNPs act as catalysts to enhance the sensing response. Kang et al. [144] designed an rGO gas sensor functionalized with SnO_2 nanoclusters in order to improve the recovery performance. The sensor was used to measure NO_2 under UV illumination, and it was discovered that the rGO device that was functional near the percolation threshold had the best recovery and then the best sensitivity in subsequent cycles. The rGO–metal-oxide semiconductor nanocomposites, however, were reported to be not suitable to detect other gases, such as H_2, CO, and C_2H_5OH, from a power consumption point of view and may not be favorable due to their operation at high temperatures [141].

4. Morphological Influence of Graphene-Based Metal Oxide Nanocomposite in a Gas Sensing Mechanism

Karthik et al. [100] studied the fabrication of rGO nanosheets functionalized with titanium dioxide (TiO_2) towards CO_2 gas. The gas-sensing mechanism involved CO reacting when it came into contact with ionized oxygen. As reaction products, CO_2 and surplus electrons were to be emitted and the extra electrons contributed to the material's increased conductivity. When compared to the air atmosphere, the nanocomposite material showed a decrease in resistance, which led to an increase in conductivity owing to the synergistic effect of GO and TiO_2 nanoparticles. This was attributed to the higher charge-carrier density at the nanocomposite material surfaces due to adsorbed CO. When these two come together, the n-n intersection is detected, and the charge carriers are transferred from TiO_2 to GO. Due to the tall depletion layer of TiO_2, handle electrons in GO were rapidly incremented at the interface. The chemisorbed interaction between oxygen atoms from GO and gas atoms was mostly responsible for the change in resistance. The sensing nature of the rGO/TiO_2 sensors was attributed to the design component of rGO, which reduces TiO_2 nanoparticle accumulation and therefore advances porosity and conductivity.

A technique used by Tadeusz Pisarkiewicz and co-workers [145] demonstrates Fe_2O_3 as an n-type semiconductor, but in the rGO/Fe_2O_3 hybrid structure it behaves similarly to p-type rGO. Both chemisorbed O_2 and NO_2 act as electron traps, decreasing the concentration of electrons, with a decrease of resistance (hole density increases). Pure Fe_2O_3 is nearly insensitive to NO_2, but, within an rGO/Fe_2O_3 composite, NO_2 reacts with O^{2-} adsorbed on the Fe_2O_3 surface, resulting in the formation of an intermediate NO_3 complex. The unbalanced charges on the Fe_2O_3 surface are compensated by the transfer of additional electrons from rGO to Fe_2O_3, which results in additional holes in the rGO and increased conductivity, as shown in Figure 3. At the interface between rGO flakes and Fe_2O_3 grains, the p-n heterojunctions can be formed. The concentration of holes in the accumulation layer increases after interaction with NO_2, leading to the increased conductivity of GO flakes in the presence of NO_2 gas.

A tin oxide (SnO_2)/rGO/polyaniline (PANI) sensor constructed by Zhang et al. [146] displayed superior H_2S-gas-detection capabilities. The SnO_2/rGO/PANI nanocomposites, with PANI, rGO, and SnO_2, were tightly wrapped together to form a porous nanostructure. In the heterostructure, two different types of depletion layers were seen during the gas sensing process. The adsorption of oxygen (O) species at the surface of SnO_2 was attributed to the first depletion layer, whereas the other depletion layer was linked to the SnO_2 and

PANI heterojunction. The performance of chemiresistive gas sensors has been shown to be influenced by the sensing properties of metal-oxide-based surface reactions between chemically adsorbed oxygen species [147]. The electron depletion layer was formed on the SnO_2 surface by the chemisorption of oxygen species [146,148]. Adsorbed oxygen species (O^{2}-(ads)) adhered to the SnO_2 surface of the SnO_2/rGO/PANI film [149]. As a result, a thicker electron depletion layer formed on the sensing film's surface, causing the sensor to have a high resistance state in the air. A substantial number of electrons were liberated into the conduction band of metal oxide when the adsorbed oxygen species reacted with H_2S. The sensor resistance was reduced when the thickness of the electron depletion layer was reduced [150]. The porous nanostructure of the SnO_2/rGO/PANI heterojunction was shown to contribute significantly towards enhancing the H_2S sensing properties [146].

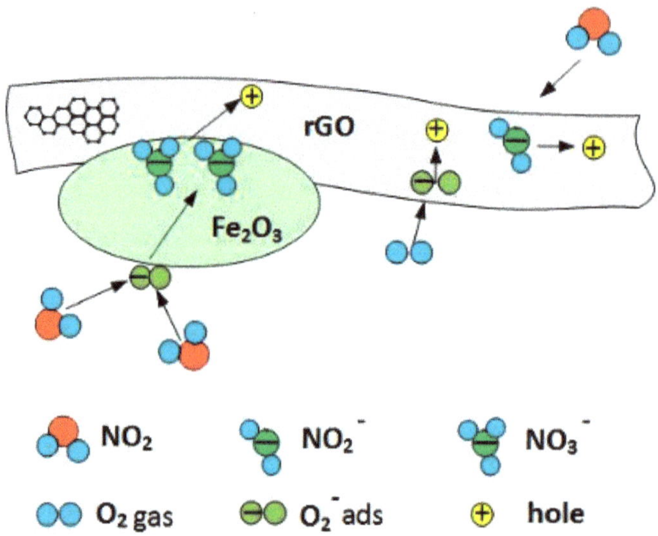

Figure 3. Interaction of NO_2 gas with oxygen adsorbed on Fe_2O_3 surface can effectively increase the concentration of holes in rGO [145].

To detect SO_2 gas, Zhang et al. [151] used TiO_2/rGO nanocomposite metal organic frameworks (MOFs) as sensor platforms. The sensing process involved SO_2 molecules dissociating and adsorbing on the accessible apertures. O_2 around the sensing material was surely adsorbed on the surface of the non-stoichiometric TiO_2 and transformed to an O species by releasing a hole due to the gap of the oxygen atom [152]. When a significant number of holes on the p-type TiO_2 surface were facilitated as the majority of carriers, the resistance of the MOFs TiO_2 decreased. When the sensor was exposed to air, a portion of the oxygen molecules decomposed into O ions, releasing holes (h^+) inside the Debye length. Upon exposure of the sensor in a SO_2 atmosphere, the amount of O ions on the surface of the MOFs TiO_2 and its positive-charge hole (h^+) were reduced compared to when the sensor was in a dry gas, which resulted in a decrease in the majority carrier concentration, and thus the resistance of the TiO_2 film climbed. When the sensor was exposed to SO_2, O ions adsorbed on the surface of the MOFs TiO_2 interacted with SO_2 (reducing gas), resulting in the production of unstable SO_3 via trapping holes in the surface of the MOFs TiO_2. The number of positively charged carriers began to reduce, resulting in a decrease in the majority carrier concentration, causing the resistance of the MOFs TiO_2/rGO film sensor to rapidly increase [153]. In the MOFs TiO_2/rGO nanocomposite, p-type semiconductor activity was observed. The resistance of p-type nanomaterials was affected by the thickness of the hole accumulation layer (HAL), which resulted in poor resistance [151].

Revolved around the discussed works, graphene-based resistive gas sensors possess the advantages of rapid responsivity, outstanding sensitivity, excellent repeatability, and stability. Zhang et al. [146] compared the H_2S sensing properties of SnO_2, SnO_2/PANI, SnO_2/rGO, and in situ polymerized SnO_2/rGO/PANI sensors at 25 °C, observing that the responses of the in situ polymerized SnO_2/rGO/PANI sensor were about 3.18%, 8.34%, 24.07%, 32.16%, 44.91%, 60.11%, 76.25%, and 91.11% toward the corresponding H_2S concentration of 50 ppb, 100 ppb, 200 ppb, 500 ppb, 1 ppm, 2 ppm, 5 ppm, and 10 ppm, indicating that the sensor can achieve a sub-ppm-level detection of 50 ppb H_2S gas. The sensor also showed long-term stability where it was measured every day over a period of a month; the little change in response confirmed its good long-term stability. The rGO/TiO_2 sensor outperformed the bare rGO and TiO_2 sensors in terms of sensitivity, with the rGO/TiO_2 sensor having a maximum sensitivity of 77% compared to 38% for the bare rGO and TiO_2 sensors with over 1500 ppm H_2S, respectively. Similar results were achieved for CO_2 gas sensitivities, which were 42% and 92% for the bare TiO_2 and rGO/TiO_2 sensors, respectively, and these were also measured with a gas concentration of against 1500 ppm [100]. The MOF-derived rGO/TiO_2 sensor demonstrated good reproducibility for SO_2 sensing at 1, 3, and 5 ppm over a time span of 0 to 1600 s [151]. Hence, graphene-based gas sensors are considered to be among the most ideal for toxic gas detection. There are however still the concerns of poor selectivity and high operating temperatures. To reduce the working temperature of the resistive gas sensor, it is necessary to innovatively develop high-performance, low-temperature gas-sensing materials and further clarify their working mechanisms. Concomitantly, the causes of selective behavior are not completely understood to date [6]. At present, to advance gas-sensing performances, the best option is to optimize the gas-sensitive materials via doping, heterostructures, and composites. These methods can adjust the grain size, porosity, and specific surface area of the material, improve the electron transport characteristics, and increase the surface adsorption at the active sites, thereby improving the sensitivity and selectivity of the gas sensors.

5. Conclusions

A major cause of the rising environmental hazards is toxins in the atmosphere. The scientific community continually examines new sensing materials for environmental gases at a laboratory level (concentrations at ppm and ppb scales; absorption of target gas molecules at low and high ranges) which shows good performances. In this review, we focused on the design and optimization strategies of graphene surfaces, in particular the synthesis of graphene composites and the assessment of their performance potential towards their use in resistive gas sensors. Functionalization aspects are discussed, highlighting the properties of the graphene surface interaction with the target toxic gases, whilst also observing the shortcomings regarding response time, full recovery to baseline of the gas sensor, power consumption, and elevated temperatures. Modified graphene surfaces and their derivatives achieved by doping/functionalization have been displayed to overcome these issues by adopting metal oxide and/or heterostructured nanohybrids in effective synthetic routes towards sensitive and selective gas sensing. The synergistic effect of the graphene–metal oxide combination and proposed mechanisms have indicated that there are still more investigations to be done towards developing the next generation of graphene-based gas sensors. Therefore, it is pivotal for scientists to develop novel innovative materials which are reported at the lab level to compete with existing commercial technology in terms of good stability and the ability to operate for long times without any need for re-calibration.

Author Contributions: N.R. and Z.D.L. conceived the presented idea and drafted the original version of manuscript. E.I.I. assisted with developing the theory and interpretation of data. Z.D.L. verified the functionalization methods of previously published research and revised it critically for important intellectual content. All authors have read and agreed to the published version of the manuscript.

Funding: This research received no external funding.

Data Availability Statement: Not applicable.

Acknowledgments: We would like to thank the SensorLab of the Chemistry Department at the University of the Western Cape for providing resources and expertise in support of this output.

Conflicts of Interest: The authors declare no conflict of interest.

References

1. Nielsen, P. WHO global air quality guidelines. In *Coastal And Estuarine Processes*; World Scientific Publishing Company: Singapore, 2009; pp. 1–360.
2. WHO. Particulate matter, ozone, nitrogen dioxide and sulfur dioxide. In *Air Quality Guidelines: Global Update 2005*; WHO: Geneva, Switzerland, 2006. [CrossRef]
3. WHO. Ambient Air Pollution. In *Pediatric Clinical Practice Guidelines & Policies*; WHO: Geneva, Switzerland, 2019; pp. 1001–1002. [CrossRef]
4. WHO. World Health Organization: European Environment and Health Process. In *WHO Expert Consultation: Available Evidence for the Future Update of the WHO Global Air Quality Guidelines (AQGs)*; WHO: Geneva, Switzerland, 2015; p. 50.
5. Jia, X.; Wang, X. Mosaic-like Micropatterned Monolayer RGO/AgNPs Film Gas Sensor with Enhanced Room-Temperature NO_2 Response/Recovery Properties. *J. Microelectromech. Syst.* **2019**, *28*, 833–840. [CrossRef]
6. Wang, C.; Wang, Y.; Yang, Z.; Hu, N. Review of recent progress on graphene-based composite gas sensors. *Ceram. Int.* **2021**, *47*, 16367–16384. [CrossRef]
7. Khan, M.A.H.; Rao, M.V.; Li, Q. Recent advances in electrochemical sensors for detecting toxic gases: NO_2, SO_2 and H_2S. *Sensors* **2019**, *19*, 905. [CrossRef] [PubMed]
8. Panda, D.; Nandi, A.; Datta, S.K.; Saha, H.; Majumdar, S. Selective detection of carbon monoxide (CO) gas by reduced graphene oxide (rGO) at room temperature. *RSC Adv.* **2016**, *6*, 47337–47348. [CrossRef]
9. Fioravanti, A.; Carotta, M.C. Year 2020: A snapshot of the last progress in flexible printed gas sensors. *Appl. Sci.* **2020**, *10*, 1741. [CrossRef]
10. Thomas, J.M. Sir Humphry Davy and the coal miners of the world: A commentary on Davy (1816) "An account of an invention for giving light in explosive mixtures of fire-damp in coal mines". *Philos. Trans. R. Soc. A Math. Phys. Eng. Sci.* **2015**, *373*, 2039. [CrossRef]
11. Sarf, F. Metal Oxide Gas Sensors by Nanostructures. *Gas Sens.* **2020**, 1–17. Available online: https://www.intechopen.com/chapters/68941 (accessed on 10 January 2022).
12. Jin, Y.; Huang, S.; Zhang, M.; Jia, M.; Hu, D. A green and efficient method to produce graphene for electrochemical capacitors from graphene oxide using sodium carbonate as a reducing agent. *Appl. Surf. Sci.* **2013**, *268*, 541–546. [CrossRef]
13. Zaaba, N.I.; Foo, K.L.; Hashim, U.; Tan, S.J.; Liu, W.W.; Voon, C.H. Synthesis of Graphene Oxide using Modified Hummers Method: Solvent Influence. *Procedia Eng.* **2017**, *184*, 469–477. [CrossRef]
14. Bastani, S.; Darani, M.K. Carbon Nanotube-Based UV-Curable Nanocomposite Coatings. *Carbon Nanotub. Curr. Prog. Polym. Compos.* **2016**, 275–296. [CrossRef]
15. Geim, A.K.; Novoselov, K.S. The rise of graphene. *Nanosci. Technol. A Collect. Rev. Nat. J.* **2007**, *6*, 11–19. [CrossRef] [PubMed]
16. Singh, V.; Joung, D.; Zhai, L.; Das, S.; Khondaker, S.I.; Seal, S. Graphene based materials: Past, present and future. *Prog. Mater. Sci.* **2011**, *56*, 1178–1271. [CrossRef]
17. Yu, W.; Sisi, L.; Haiyan, Y.; Jie, L. Progress in the functional modification of graphene/graphene oxide: A review. *RSC Adv.* **2020**, *10*, 15328–15345. [CrossRef]
18. Sharma, N.; Sharma, V.; Jain, Y.; Kumari, M.; Gupta, R.; Sharma, S.K.; Sachdev, K. Synthesis and Characterization of Graphene Oxide (GO) and Reduced Graphene Oxide (rGO) for Gas Sensing Application. *Macromol. Symp.* **2017**, *376*, 1700006. [CrossRef]
19. Park, S.; Ruoff, R.S. Chemical methods for the production of graphenes. *Nat. Nanotechnol.* **2009**, *4*, 217–224. [CrossRef]
20. Li, X.; Zhang, G.; Bai, X.; Sun, X.; Wang, X.; Wang, E.; Dai, H. Highly conducting graphene sheets and Langmuir-Blodgett films. *Nat. Nanotechnol.* **2008**, *3*, 538–542. [CrossRef]
21. Yavari, F.; Koratkar, N. Graphene-based chemical sensors. *J. Phys. Chem. Lett.* **2012**, *3*, 1746–1753. [CrossRef]
22. Georgakilas, V.; Otyepka, M.; Bourlinos, A.B.; Chandra, V.; Kim, N.; Kemp, K.C.; Hobza, P.; Zboril, R.; Kim, K.S. Functionalization of graphene: Covalent and non-covalent approaches, derivatives and applications. *Chem. Rev.* **2012**, *112*, 6156–6214. [CrossRef]
23. Lu, C.; Huang, P.J.J.; Liu, B.; Ying, Y.; Liu, J. Comparison of Graphene Oxide and Reduced Graphene Oxide for DNA Adsorption and Sensing. *Langmuir* **2016**, *32*, 10776–10783. [CrossRef]
24. Robinson, J.T.; Perkins, F.K.; Snow, E.S.; Wei, Z.; Sheehan, P.E. Reduced graphene oxide molecular sensors. *Nano Lett.* **2008**, *8*, 3137–3140. [CrossRef]
25. Volanti, D.P.; Felix, A.A.; Orlandi, M.O.; Whitfield, G.; Yang, D.J.; Longo, E.; Tuller, H.L.; Varela, J.A. The role of hierarchical morphologies in the superior gas sensing performance of CuO-based chemiresistors. *Adv. Funct. Mater.* **2013**, *23*, 1759–1766. [CrossRef]
26. Yoon, H. Current trends in sensors based on conducting polymer nanomaterials. *Nanomaterials* **2013**, *3*, 524–549. [CrossRef] [PubMed]
27. Kailasa, S.; Sai Bhargava Reddy, M.; Geeta Rani, B.; Maseed, H.; Venkateswara Rao, K. Twisted Polyaniline Nanobelts @ rGO for Room Temperature NO_2 Sensing. *Mater. Lett.* **2019**, *257*, 126687. [CrossRef]

28. Yoon, H.; Jang, J. Conducting-polymer nanomaterials for high-performance sensor applications: Issues and challenges. *Adv. Funct. Mater.* **2009**, *19*, 1567–1576. [CrossRef]
29. Yoon, H.; Choi, M.; Lee, K.J.; Jang, J. Versatile strategies for fabricating polymer nanomaterials with controlled size and morphology. *Macromol. Res.* **2008**, *16*, 85–102. [CrossRef]
30. Nemade, K.R. Gas sensors based on inorganic materials: An overview. *Sens. Transducers* **2011**, *132*, 1–13.
31. Novoselov, K.S.; Geim, A.K.; Morozov, S.V.; Jiang, D.; Zhang, Y.; Dubonos, S.V.; Grigorieva, I.V.; Firsov, A.A. Electric Field Effect in Atomically Thin Carbon Films. *Science* **2004**, *146*, 666–669. [CrossRef]
32. Stankovich, S.; Dikin, D.A.; Piner, R.D.; Kohlhaas, K.A.; Kleinhammes, A.; Jia, Y.; Wu, Y.; Nguyen, S.B.T.; Ruoff, R.S. Synthesis of graphene-based nanosheets via chemical reduction of exfoliated graphite oxide. *Carbon N. Y.* **2007**, *45*, 1558–1565. [CrossRef]
33. Berger, C.; Song, Z.; Li, T.; Li, X.; Ogbazghi, A.Y.; Feng, R.; Dai, Z.; Marchenkov, A.N.; Conrad, E.H.; First, P.N.; et al. Ultrathin epitaxial graphite: 2D electron gas properties and a route toward graphene-based nanoelectronics. *J. Phys. Chem. B* **2004**, *108*, 19912–19916. [CrossRef]
34. Berger, C.; Song, Z.; Li, X.; Wu, X.; Brown, N.; Naud, C.; Mayou, D.; Li, T.; Hass, J.; Marchenkov, A.N. Electronic confinement and coherence in patterned epitaxial graphene. *Science* **2016**, *312*, 1191–1196. [CrossRef]
35. Zhang, Y.; Zhang, L.; Zhou, C. Review of chemical vapor deposition of graphene and related applications. *Acc. Chem. Res.* **2013**, *46*, 2329–2339. [CrossRef]
36. Eda, G.; Fanchini, G.; Chhowalla, M. Large-area ultrathin films of reduced graphene oxide as a transparent and flexible electronic material. *Nat. Nanotechnol.* **2008**, *3*, 270–274. [CrossRef] [PubMed]
37. Penuelas, J.; Ouerghi, A.; Lucot, D.; David, C.; Gierak, J.; Estrade-Szwarckopf, H.; Andreazza-Vignolle, C. Surface morphology and characterization of thin graphene films on SiC vicinal substrate. *Phys. Rev. B Condens. Matter Mater. Phys.* **2009**, *79*, 1–4. [CrossRef]
38. Saeed, M.; Alshammari, Y.; Majeed, S.A.; Al-Nasrallah, E. Chemical Vapour Deposition of Graphene Synthesis, Characterisation, and Application: A Review. *Molecules* **2020**, *25*, 3856. [CrossRef] [PubMed]
39. Nurazzi, N.M.; Abdullah, N.; Demon, S.Z.N.; Halim, N.A.; Azmi, A.F.M.; Knight, V.F.; Mohamad, I.S. The frontiers of functionalized graphene-based nanocomposites as chemical sensors. *Nanotechnol. Rev.* **2021**, *10*, 330–369. [CrossRef]
40. Das, B.; Choudhury, B.; Gomathi, A.; Manna, A.K.; Pati, S.K.; Rao, C.N.R. Interaction of inorganic nanoparticles with graphene. *ChemPhysChem* **2011**, *12*, 937–943. [CrossRef]
41. Mitra, S.; Banerjee, S.; Datta, A.; Chakravorty, D. A brief review on graphene/inorganic nanostructure composites: Materials for the future. *Indian J. Phys.* **2016**, *90*, 1019–1032. [CrossRef]
42. Zhou, X.; Huang, X.; Qi, X.; Wu, S.; Xue, C.; Boey, F.Y.; Yan, Q.; Chen, P.; Zhang, H. In situ synthesis of metal nanoparticles on single-layer graphene oxide and reduced graphene oxide surfaces. *J. Phys. Chem. C* **2009**, *113*, 10842–10846. [CrossRef]
43. Wang, H.; Robinson, J.T.; Diankov, G.; Dai, H. Supporting Information Nanocrystal Growth on Graphene with Various Degrees of Oxidation. *J. Am. Chem. Soc.* **2010**, *132*, 3270–3271. [CrossRef]
44. Hassan, H.M.A.; Abdelsayed, V.; Khder, A.E.R.S.; Abouzeid, K.M.; Terner, J.; El-Shall, M.S.; Al-Resayes, S.I.; El-Azhary, A.A. Microwave synthesis of graphene sheets supporting metal nanocrystals in aqueous and organic media. *J. Mater. Chem.* **2009**, *19*, 3832–3837. [CrossRef]
45. Marquardt, D.; Vollmer, C.; Thomann, R.; Steurer, P.; Mülhaupt, R.; Redel, E.; Janiak, C. The use of microwave irradiation for the easy synthesis of graphene-supported transition metal nanoparticles in ionic liquids. *Carbon N. Y.* **2011**, *49*, 1326–1332. [CrossRef]
46. Akhtar, A.J.; Gupta, A.; Kumar Shaw, B.; Saha, S.K. Unusual dielectric response in cobalt doped reduced graphene oxide. *Appl. Phys. Lett.* **2013**, *103*, 242902. [CrossRef]
47. Mandal, S.; Saha, S.K. Anomalous magnetic behavior at the graphene/Co interface. *Appl. Phys. Lett.* **2014**, *105*, 022402. [CrossRef]
48. Liu, J.; Bai, H.; Wang, Y.; Liu, Z.; Zhang, X.; Sun, D.D. Self-assembling TiO_2 nanorods on large graphene oxide sheets at a two-phase interface and their anti-recombination in photocatalytic applications. *Adv. Funct. Mater.* **2010**, *20*, 4175–4181. [CrossRef]
49. Yin, Z.; Wu, S.; Zhou, X.; Huang, X.; Zhang, Q.; Boey, F.; Zhang, H. Electrochemical deposition of ZnO nanorods on transparent reduced graphene oxide electrodes for hybrid solar cells. *Small* **2010**, *6*, 307–312. [CrossRef]
50. Chen, S.; Zhu, J.; Wu, X.; Han, Q.; Wang, X. Graphene Oxide MnO_2. *ACS Nano* **2010**, *4*, 2822–2830. [CrossRef]
51. Wu, Q.; Xu, Y.; Yao, Z.; Liu, A.; Shi, G. Supercapacitors based on flexible graphene/polyaniline nanofiber composite films. *ACS Nano* **2010**, *4*, 1963–1970. [CrossRef]
52. Son, J.Y.; Shin, Y.H.; Kim, H.; Jang, H.M. NiO resistive random access memory nanocapacitor array on graphene. *ACS Nano* **2010**, *4*, 2655–2658. [CrossRef]
53. Shen, J.; Hu, Y.; Shi, M.; Li, N.; Ma, H.; Ye, M. One step synthesis of graphene oxide-magnetic nanoparticle composite. *J. Phys. Chem. C* **2010**, *114*, 1498–1503. [CrossRef]
54. Kumar, R.; Jayaramulu, K.; Maji, T.K.; Rao, C.N.R. Hybrid nanocomposites of ZIF-8 with graphene oxide exhibiting tunable morphology, significant CO_2 uptake and other novel properties. *Chem. Commun.* **2013**, *49*, 4947–4949. [CrossRef]
55. Mitra, S.; Singha, A.; Chakravorty, D. Non-linear temperature variation of resistivity in graphene/silicate glass nanocomposite. *J. Phys. D Appl. Phys.* **2013**, *46*, 375306. [CrossRef]
56. Subrahmanyam, K.S.; Manna, A.K.; Pati, S.K.; Rao, C.N.R. A study of graphene decorated with metal nanoparticles. *Chem. Phys. Lett.* **2010**, *497*, 70–75. [CrossRef]

57. Kamat, P.V. Graphene-based nanoarchitectures. Anchoring semiconductor and metal nanoparticles on a two-dimensional carbon support. *J. Phys. Chem. Lett.* **2010**, *1*, 520–527. [CrossRef]
58. Williams, G.; Kamat, P.V. Graphene-semiconductor nanocomposites: Excited-state interactions between ZnO nanoparticles and graphene oxide. *Langmuir* **2009**, *25*, 13869–13873. [CrossRef]
59. Williams, G.; Seger, B.; Kamat, P. V UV-Assisted Photocatalytic Reduction of Graphene Oxide. *ACS Nano* **2008**, *2*, 1487–1491. [CrossRef]
60. Huang, X.; Qi, X.; Boey, F.; Zhang, H. Graphene-based composites. *Chem. Soc. Rev.* **2012**, *41*, 666–686. [CrossRef]
61. Huang, X.; Yin, Z.; Wu, S.; Qi, X.; He, Q.; Zhang, Q.; Yan, Q.; Boey, F.; Zhang, H. Graphene-based materials: Synthesis, characterization, properties, and applications. *Small* **2011**, *7*, 1876–1902. [CrossRef]
62. Daniel, M.C.; Astruc, D. Gold Nanoparticles: Assembly, Supramolecular Chemistry, Quantum-Size-Related Properties, and Applications Toward Biology, Catalysis, and Nanotechnology. *Chem. Rev.* **2004**, *104*, 293–346. [CrossRef]
63. Xia, B.Y.; Yang, P.; Sun, Y.; Wu, Y.; Mayers, B.; Gates, B.; Yin, Y.; Kim, F.; Yan, H. One-Dimensional Nanostructures: Synthesis, Characterization, and Applications. *Adv. Mater.* **2003**, *15*, 353–389. [CrossRef]
64. Li, N.; Cao, M.; Hu, C. Review on the latest design of graphene-based inorganic materials. *Nanoscale* **2012**, *4*, 6205–6218. [CrossRef]
65. Lee, S.; Fan, C.; Wu, T.; Anderson, S.L. CO Oxidation on Aun/TiO_2 Catalysts Produced by Size-Selected Cluster Deposition. *J. Am. Chem. Soc.* **2004**, *126*, 5682–5683. [CrossRef] [PubMed]
66. Yu, K.M.K.; Yeung, C.M.Y.; Shik, C.T. Carbon dioxide fixation into chemicals (methyl formate) at high yields by surface coupling over a Pd/Cu/ZnO nanocatalyst. *J. Am. Chem. Soc.* **2007**, *129*, 6360–6361. [CrossRef] [PubMed]
67. Li, B.; Lu, G.; Zhou, X.; Cao, X.; Boey, F.; Zhang, H. Controlled assembly of gold nanoparticles and graphene oxide sheets on dip pen nanolithography-generated templates. *Langmuir* **2009**, *25*, 10455–10458. [CrossRef] [PubMed]
68. Zhou, W.; Du, G.; Hu, P.; Yin, Y.; Li, J.; Yu, J.; Wang, G.; Wang, J.; Liu, H.; Wang, J.; et al. Nanopaper based on Ag/TiO_2 nanobelts heterostructure for continuous-flow photocatalytic treatment of liquid and gas phase pollutants. *J. Hazard. Mater.* **2011**, *197*, 19–25. [CrossRef]
69. Zeng, Z.; Zhou, X.; Huang, X.; Wang, Z.; Yang, Y.; Zhang, Q.; Boey, F.; Zhang, H. Electrochemical deposition of Pt nanoparticles on carbon nanotube patterns for glucose detection. *Analyst* **2010**, *135*, 1726–1730. [CrossRef]
70. Huang, X.; Qi, X.; Huang, Y.; Li, S.; Xue, C.; Gan, C.L.; Boey, F. Photochemically Controlled Synthesis of Nanorods, Platelet-like Au. *ACS Nano* **2010**, *4*, 6196–6202. [CrossRef]
71. Tao, A.R.; Huang, J.; Yang, P. Langmuir—Blodgettry of Nanocrystals and Nanowires. *Langmuir* **2008**, *41*, 1662–1673. [CrossRef]
72. Murray, R.W. Nanoelectrochemistry: Metal Nanoparticles, Nanoelectrodes, and Nanopores. *Chem. Rev.* **2008**, *108*, 2688–2720. [CrossRef]
73. Guo, S.; Wang, E. Noble metal nanomaterials: Controllable synthesis and application in fuel cells and analytical sensors. *Nano Today* **2011**, *6*, 240–264. [CrossRef]
74. Jones, M.R.; Osberg, K.D.; MacFarlane, R.J.; Langille, M.R.; Mirkin, C.A. Templated techniques for the synthesis and assembly of plasmonic nanostructures. *Chem. Rev.* **2011**, *111*, 3736–3827. [CrossRef]
75. Cobley, C.M.; Chen, J.; Chul Cho, E.; Wang, L.V.; Xia, Y. Gold nanostructures: A class of multifunctional materials for biomedical applications. *Chem. Soc. Rev.* **2011**, *40*, 44–56. [CrossRef] [PubMed]
76. Bonaccorso, F.; Sun, Z.; Hasan, T.; Ferrari, A.C. Graphene photonics and optoelectronics. *Nat. Photonics* **2010**, *4*, 611–622. [CrossRef]
77. Zhu, Y.; Murali, S.; Cai, W.; Li, X.; Suk, J.W.; Potts, J.R.; Ruoff, R.S. Graphene and graphene oxide: Synthesis, properties, and applications. *Adv. Mater.* **2010**, *22*, 3906–3924. [CrossRef] [PubMed]
78. Yi, G.C.; Wang, C.; Park, W. Il ZnO nanorods: Synthesis, characterization and applications. *Semicond. Sci. Technol.* **2005**, *20*, S22. [CrossRef]
79. Thelander, C.; Agarwal, P.; Brongersma, S.; Eymery, J.; Feiner, L.F.; Forchel, A.; Scheffler, M.; Riess, W.; Ohlsson, B.J.; Samuelson, L. Nanowire-based one-dimensional electronics. *Mater. Today* **2006**, *9*, 28–35. [CrossRef]
80. Yan, R.; Gargas, D.; Yang, P. Nanowire photonics. *Nat. Photonics* **2009**, *3*, 569–576. [CrossRef]
81. Compton, O.C.; Nguyen, S.T. Graphene oxide, highly reduced graphene oxide, and graphene: Versatile building blocks for carbon-based materials. *Small* **2010**, *6*, 711–723. [CrossRef]
82. Sun, Y.; Wu, Q.; Shi, G. Graphene based new energy materials. *Energy Environ. Sci.* **2011**, *4*, 1113–1132. [CrossRef]
83. Zhou, D.; Cui, Y.; Han, B.H. Graphene-based hybrid materials and their applications in energy storage and conversion. *Chin. Sci. Bull.* **2012**, *57*, 2983–2994. [CrossRef]
84. Alfano, B.; Miglietta, M.L.; Polichetti, T.; Massera, E.; Bruno, A.; Di Francia, G.; Veneri, P.D. Improvement of NO_2 Detection: Graphene Decorated with ZnO Nanoparticles. *IEEE Sens. J.* **2019**, *19*, 8751–8757. [CrossRef]
85. Wu, T.C.; De Luca, A.; Zhong, Q.; Zhu, X.; Ogbeide, O.; Um, D.S.; Hu, G.; Albrow-Owen, T.; Udrea, F.; Hasan, T. Inkjet-printed CMOS-integrated graphene–metal oxide sensors for breath analysis. *NPJ 2D Mater. Appl.* **2019**, *3*, 42. [CrossRef]
86. Kim, H.W.; Na, H.G.; Kwon, Y.J.; Kang, S.Y.; Choi, M.S.; Bang, J.H.; Wu, P.; Kim, S.S. Microwave-Assisted Synthesis of Graphene-SnO_2 Nanocomposites and Their Applications in Gas Sensors. *ACS Appl. Mater. Interfaces* **2017**, *9*, 31667–31682. [CrossRef] [PubMed]
87. Dreyer, D.R.; Park, S.; Bielawski, C.W.; Ruoff, R.S. The chemistry of graphene oxide. *Chem. Soc. Rev.* **2010**, *39*, 228–240. [CrossRef] [PubMed]

88. Tan, C.; Huang, X.; Zhang, H. Synthesis and applications of graphene-based noble metal nanostructures. *Mater. Today* **2013**, *16*, 29–36. [CrossRef]
89. He, Q.; Wu, S.; Yin, Z.; Zhang, H. Graphene-based electronic sensors. *Chem. Sci.* **2012**, *3*, 1764–1772. [CrossRef]
90. Huang, X.; Zeng, Z.; Fan, Z.; Liu, J.; Zhang, H. Graphene-based electrodes. *Adv. Mater.* **2012**, *24*, 5979–6004. [CrossRef]
91. Qi, X.; Li, H.; Lam, J.W.Y.; Yuan, X.; Wei, J.; Tang, B.Z.; Zhang, H. Graphene oxide as a novel nanoplatform for enhancement of aggregation-induced emission of silole fluorophores. *Adv. Mater.* **2012**, *24*, 4191–4195. [CrossRef]
92. Cao, X.; Shi, Y.; Shi, W.; Lu, G.; Huang, X.; Yan, Q.; Zhang, Q.; Zhang, H. Preparation of novel 3D graphene networks for supercapacitor applications. *Small* **2011**, *7*, 3163–3168. [CrossRef]
93. Wu, S.; Yin, Z.; He, Q.; Lu, G.; Yan, Q.; Zhang, H. Nucleation mechanism of electrochemical deposition of Cu on reduced graphene oxide electrodes. *J. Phys. Chem. C* **2011**, *115*, 15973–15979. [CrossRef]
94. He, Q.; Wu, S.; Gao, S.; Cao, X.; Yin, Z.; Li, H.; Chen, P.; Zhang, H. Transparent, flexible, all-reduced graphene oxide thin film transistors. *ACS Nano* **2011**, *5*, 5038–5044. [CrossRef]
95. Liu, Y.; Dong, X.; Chen, P. Biological and chemical sensors based on graphene materials. *Chem. Soc. Rev.* **2012**, *41*, 2283–2307. [CrossRef] [PubMed]
96. Kavinkumar, T.; Manivannan, S. Uniform decoration of silver nanoparticle on exfoliated graphene oxide sheets and its ammonia gas detection. *Ceram. Int.* **2016**, *42*, 1769–1776. [CrossRef]
97. Jiang, L.; Tu, S.; Xue, K.; Yu, H.; Hou, X. Preparation and gas-sensing performance of GO/SnO_2/NiO gas-sensitive composite materials. *Ceram. Int.* **2021**, *47*, 7528–7538. [CrossRef]
98. Li, Z.; Liu, Y.; Guo, D.; Guo, J.; Su, Y. Room-temperature synthesis of CuO/reduced graphene oxide nanohybrids for high-performance NO_2 gas sensor. *Sens. Actuators B Chem.* **2018**, *271*, 306–310. [CrossRef]
99. Gu, F.; Nie, R.; Han, D.; Wang, Z. In_2O_3-graphene nanocomposite based gas sensor for selective detection of NO_2 at room temperature. *Sens. Actuators B Chem.* **2015**, *219*, 94–99. [CrossRef]
100. Karthik, P.; Gowthaman, P.; Venkatachalam, M.; Rajamanickam, A.T. Propose of high performance resistive type H_2S and CO_2 gas sensing response of reduced graphene oxide/titanium oxide (rGO/TiO_2) hybrid sensors. *J. Mater. Sci. Mater. Electron.* **2020**, *31*, 3695–3705. [CrossRef]
101. Zhang, H.; Wang, L.; Zhang, T. Reduced graphite oxide/SnO_2/Au hybrid nanomaterials for NO_2 sensing performance at relatively low operating temperature. *RSC Adv.* **2014**, *4*, 57436–57441. [CrossRef]
102. Wang, Z.; Zhang, Y.; Liu, S.; Zhang, T. Preparation of Ag nanoparticles-SnO_2 nanoparticles-reduced graphene oxide hybrids and their application for detection of NO_2 at room temperature. *Sens. Actuators B Chem.* **2016**, *222*, 893–903. [CrossRef]
103. Iftekhar Uddin, A.S.M.; Phan, D.T.; Chung, G.S. Low temperature acetylene gas sensor based on Ag nanoparticles-loaded ZnO-reduced graphene oxide hybrid. *Sens. Actuators B Chem.* **2015**, *207*, 362–369. [CrossRef]
104. Alzate-Carvajal, N.; Luican-Mayer, A. Functionalized Graphene Surfaces for Selective Gas Sensing. *ACS Omega* **2020**, *5*, 21320–21329. [CrossRef]
105. Saha, B.; Bhattacharyya, P.K. Adsorption of amino acids on boron and/or nitrogen doped functionalized graphene: A Density Functional Study. *Comput. Theor. Chem.* **2016**, *1086*, 45–51. [CrossRef]
106. Kiang Chua, C.; Pumera, M. Covalent chemistry on graphene. *Chem. Soc. Rev.* **2013**, *42*, 3222–3233. [CrossRef] [PubMed]
107. Yan, L.; Zheng, Y.B.; Zhao, F.; Li, S.; Gao, X.; Xu, B.; Weiss, P.S.; Zhao, Y. Chemistry and physics of a single atomic layer: Strategies and challenges for functionalization of graphene and graphene-based materials. *Chem. Soc. Rev.* **2012**, *41*, 97–114. [CrossRef] [PubMed]
108. Gao, X.; Wei, Z.; Meunier, V.; Sun, Y.; Zhang, S.B. Opening a large band gap for graphene by covalent addition. *Chem. Phys. Lett.* **2013**, *555*, 1–6. [CrossRef]
109. Bai, H.; Xu, Y.; Zhao, L.; Li, C.; Shi, G. Non-covalent functionalization of graphene sheets by sulfonated polyaniline. *Chem. Commun.* **2009**, 1667–1669. [CrossRef]
110. Xu, Y.; Bai, H.; Lu, G.; Li, C.; Shi, G. Flexible Graphene Films via the Filtration of Water-Soluble. *JACS* **2008**, *130*, 5856–5857. [CrossRef]
111. Song, N.; Fan, H.; Tian, H. PVP assisted in situ synthesis of functionalized graphene/ZnO (FGZnO) nanohybrids with enhanced gas-sensing property. *J. Mater. Sci.* **2015**, *50*, 2229–2238. [CrossRef]
112. Zhang, Z.; Zou, X.; Xu, L.; Liao, L.; Liu, W.; Ho, J.; Xiao, X.; Jiang, C.; Li, J. Hydrogen gas sensor based on metal oxide nanopaerticles decorated graphene transistor. *R. Soc. Chem.* **2015**, *3*, 10715–10722. [CrossRef]
113. Wang, X.; Shi, G. An introduction to the chemistry of graphene. *Phys. Chem. Chem. Phys.* **2015**, *17*, 28484–28504. [CrossRef]
114. Yao, B.; Li, C.; Ma, J.; Shi, G. Porphyrin-based graphene oxide frameworks with ultra-large d-spacings for the electrocatalyzation of oxygen reduction reaction. *Phys. Chem. Chem. Phys.* **2015**, *17*, 19538–19545. [CrossRef]
115. Duan, X.; Indrawirawan, S.; Sun, H.; Wang, S. Effects of nitrogen-, boron-, and phosphorus-doping or codoping on metal-free graphene catalysis. *Catal. Today* **2015**, *249*, 184–191. [CrossRef]
116. Quintana, M.; Vazquez, E.; Prato, M. Organic functionalization of graphene in dispersions. *Acc. Chem. Res.* **2013**, *46*, 138–148. [CrossRef] [PubMed]
117. Johns, J.E.; Hersam, M.C. Atomic covalent functionalization of graphene. *Acc. Chem. Res.* **2013**, *46*, 77–86. [CrossRef] [PubMed]

118. Bekyarova, E.B.; Niyogi, S.; Sarkar, S.; Tian, X.; Chen, M.; Moser, M.L.; Ayub, K.; Mitchell, R.H.; Haddon, R.C. Stereochemical effect of covalent chemistry on the electronic structure and properties of the carbon allotropes and graphene surfaces. *Synth. Met.* **2015**, *210*, 80–84. [CrossRef]
119. Gómez-Navarro, C.; Weitz, R.T.; Bittner, A.M.; Scolari, M.; Mews, A.; Burghard, M.; Kern, K. Electronic transport properties of individual chemically reduced graphene oxide sheets. *Nano Lett.* **2007**, *7*, 3499–3503, Erratum in *Nano Lett.* **2009**, *9*, 2206. [CrossRef]
120. Sun, D.; Luo, Y.; Debliquy, M.; Zhang, C. Graphene-enhanced metal oxide gas sensors at room temperature: A review. *Beilstein J. Nanotechnol.* **2018**, *9*, 2832–2844. [CrossRef]
121. Ko, G.; Jung, Y.; Lee, K.Y.; Lee, K.; Kim, J. Improved sorption characteristics of NH_3 molecules on the solution-processed graphene sheets. *J. Cryst. Growth* **2011**, *326*, 208–211. [CrossRef]
122. Gilje, S.; Han, S.; Wang, M.; Wang, K.L.; Kaner, R.B. A chemical route to graphene for device applications. *Nano Lett.* **2007**, *7*, 3394–3398. [CrossRef]
123. Chatterjee, S.G.; Chatterjee, S.; Ray, A.K.; Chakraborty, A.K. Graphene-metal oxide nanohybrids for toxic gas sensor: A review. *Sens. Actuators B Chem.* **2015**, *221*, 1170–1181. [CrossRef]
124. Joshi, N.; Hayasaka, T.; Liu, Y.; Liu, H.; Oliveira, O.N.; Lin, L. A review on chemiresistive room temperature gas sensors based on metal oxide nanostructures, graphene and 2D transition metal dichalcogenides. *Microchim. Acta* **2018**, *185*, 213. [CrossRef]
125. Kumar, R.; Kaur, A. Chemiresistive gas sensors based on thermally reduced graphene oxide for sensing sulphur dioxide at room temperature. *Diam. Relat. Mater.* **2020**, *109*, 108039. [CrossRef]
126. Ristein, J. Surface Transfer Doping of Semiconductors. *Science* **2006**, *313*, 1057–1058. [CrossRef] [PubMed]
127. Bangert, U.; Zan, R. Electronic functionalisation of graphene via external doping and dosing. *Int. Mater. Rev.* **2015**, *60*, 133–149. [CrossRef]
128. Gierz, I.; Riedl, C.; Starke, U.; Ast, C.R.; Kern, K. Atomic Hole Doping of Graphene. *Nano Lett.* **2008**, *8*, 4603–4607. [CrossRef]
129. Ohta, T.; Bostwick, A.; Seyller, T.; Horn, K.; Rotenberg, E. Controlling the electronic structure of bilayer graphene. *Science* **2006**, *313*, 951–954. [CrossRef]
130. Maier, F.; Riedel, M.; Mantel, B.; Ristein, J.; Ley, L. Origin of surface conductivity in diamond. *Phys. Rev. Lett.* **2000**, *85*, 3472–3475. [CrossRef]
131. Chakrapani, V.; Angus, J.C.; Anderson, A.B.; Wolter, S.D.; Stoner, B.R.; Sumanasekera, G.U. Charge transfer equilibria between diamond and an aqueous oxygen electrochemical redox couple. *Science* **2007**, *318*, 1424–1430. [CrossRef]
132. Sque, S.J.; Jones, R.; Briddon, P.R. The transfer doping of graphite and graphene. *Phys. Status Solidi Appl. Mater. Sci.* **2007**, *204*, 3078–3084. [CrossRef]
133. Chen, W.; Chen, S.; Dong, C.Q.; Xing, Y.G.; Wee, A.T.S. Surface transfer p-type doping of epitaxial graphene. *J. Am. Chem. Soc.* **2007**, *129*, 10418–10422. [CrossRef]
134. Hwang, E.H.; Adam, S.; Das Sarma, S. Transport in chemically doped graphene in the presence of adsorbed molecules. *Phys. Rev. B Condens. Matter Mater. Phys.* **2007**, *76*, 195421. [CrossRef]
135. Wehling, T.O.; Novoselov, K.S.; Morozov, S.V.; Vdovin, E.E.; Katsnelson, M.I.; Geim, A.K.; Lichtenstein, A.I. Molecular doping of graphene. *Nano Lett.* **2008**, *8*, 173–177. [CrossRef] [PubMed]
136. Zhou, S.Y.; Siegel, D.A.; Fedorov, A.V.; Lanzara, A. Metal to insulator transition in epitaxial graphene induced by molecular doping. *Phys. Rev. Lett.* **2008**, *101*, 086402. [CrossRef] [PubMed]
137. Zhang, W.; Wu, L.; Li, Z.; Liu, Y. Doped graphene: Synthesis, properties and bioanalysis. *RSC Adv.* **2015**, *5*, 49521–49533. [CrossRef]
138. Chowdhury, N.K.; Bhowmik, B. Role of graphene-metal oxide composite for performance improvement of chemical sensor: Study for various analytes. In *AIP Conference Proceedings*; AIP Publishing LLC.: Melville, NY, USA, 2021; Volume 2341. [CrossRef]
139. Paulchamy, B.; Arthi, G.; Lignesh, B.D. A Simple Approach to Stepwise Synthesis of Graphene Oxide Nanomaterial. *J. Nanomed. Nanotechnol.* **2015**, *6*, 1–4. [CrossRef]
140. Korotcenkov, G. Current trends in nanomaterials for metal oxide-based conductometric gas sensors: Advantages and limitations. part 1: 1D and 2D nanostructures. *Nanomaterials* **2020**, *10*, 1392. [CrossRef]
141. Wicaksono, D.H.B.; Utari, L.; Wulan, N.; Engel, D.J.; Widjaja, S.T.; Jovinka, X.; Genilar, L.A.; Setiawan, S.A.; Yuliarto, B.; Dipojono, H.K.; et al. Preliminary study on graphene/metal oxide nanoparticles-coated cotton fabrics for flexible gas sensor. In *AIP Conference Proceedings*; AIP Publishing LLC.: Melville, NY, USA, 2018; Volume 2024. [CrossRef]
142. Martínez-Orozco, R.D.; Antaño-López, R.; Rodríguez-González, V. Hydrogen-gas sensors based on graphene functionalized palladium nanoparticles: Impedance response as a valuable sensor. *New J. Chem.* **2015**, *39*, 8044–8054. [CrossRef]
143. Muda, M.R.; Ramli, M.M.; Isa, S.S.M.; Jamlos, M.F.; Murad, S.A.Z.; Norhanisah, Z.; Isa, M.M.; Kasjoo, S.R.; Ahmad, N.; Nor, N.I.M.; et al. Fundamental study of reduction graphene oxide by sodium borohydride for gas sensor application. In *AIP Conference Proceedings*; AIP Publishing LLC.: Melville, NY, USA, 2017; Volume 1808. [CrossRef]
144. Kang, I.S.; So, H.M.; Bang, G.S.; Kwak, J.H.; Lee, J.O.; Won Ahn, C. Recovery improvement of graphene-based gas sensors functionalized with nanoscale heterojunctions. *Appl. Phys. Lett.* **2012**, *101*, 123504. [CrossRef]
145. Pisarkiewicz, T.; Maziarz, W.; Małolepszy, A.; Stobiński, L.; Michoń, D.A.; Szkudlarek, A.; Pisarek, M.; Kanak, J.; Rydosz, A. Nitrogen dioxide sensing using multilayer structure of reduced graphene oxide and α-Fe_2O_3. *Sensors* **2021**, *21*, 1011. [CrossRef]

146. Zhang, D.; Wu, Z.; Zong, X. Flexible and highly sensitive H_2S gas sensor based on in-situ polymerized SnO_2/rGO/PANI ternary nanocomposite with application in halitosis diagnosis. *Sens. Actuators B Chem.* **2019**, *289*, 32–41. [CrossRef]
147. Song, X.Z.; Qiao, L.; Sun, K.M.; Tan, Z.; Ma, W.; Kang, X.L.; Sun, F.F.; Huang, T.; Wang, X.F. Triple-shelled ZnO/$ZnFe_2O_4$ heterojunctional hollow microspheres derived from Prussian Blue analogue as high-performance acetone sensors. *Sens. Actuators B Chem.* **2018**, *256*, 374–382. [CrossRef]
148. Zhang, D.; Liu, A.; Chang, H.; Xia, B. Room-temperature high-performance acetone gas sensor based on hydrothermal synthesized SnO_2-reduced graphene oxide hybrid composite. *RSC Adv.* **2015**, *5*, 3016–3022. [CrossRef]
149. Koo, W.T.; Yu, S.; Choi, S.J.; Jang, J.S.; Cheong, J.Y. and Kim I.D. Nanoscale PdO Catalyst Functionalized Co_3O_4 Hollow Nanocages Using MOF Templates for Selective Detection of Acetone Molecules in Exhaled Breath. *ACS Appl. Mater. Interfaces* **2017**, *9*, 8201–8210. [CrossRef] [PubMed]
150. Zhang, D.; Jiang, C.; Li, P.; Sun, Y. Layer-by-Layer Self-assembly of Co_3O_4 Nanorod-Decorated MoS_2 Nanosheet-Based Nanocomposite toward High-Performance Ammonia Detection. *ACS Appl. Mater. Interfaces* **2017**, *9*, 6462–6471. [CrossRef] [PubMed]
151. Zhang, D.; Wu, D.; Zong, X.; Yang, Z. Enhanced SO_2 gas sensing properties of metal organic frameworks-derived titanium dioxide/reduced graphene oxide nanostructure. *J. Mater. Sci. Mater. Electron.* **2019**, *30*, 11070–11078. [CrossRef]
152. Bhowmik, B.; Dutta, K.; Hazra, A.; Bhattacharyya, P. Low temperature acetone detection by p-type nano-titania thin film: Equivalent circuit model and sensing mechanism. *Solid State Electron.* **2014**, *99*, 84–92. [CrossRef]
153. Bai, S.; Du, L.; Sun, J.; Luo, R.; Li, D.; Chen, A.; Liu, C.C. Preparation of reduced graphene oxide/Co_3O_4 composites and sensing performance to toluene at low temperature. *RSC Adv.* **2016**, *6*, 60109–60119. [CrossRef]

materials

MDPI

Article

Magnetite Nanoparticles In-Situ Grown and Clustered on Reduced Graphene Oxide for Supercapacitor Electrodes

Yue Jiang [1], Jinxun Han [2], Xiaoqin Wei [2], Hanzhuo Zhang [2,*], Zhihui Zhang [1,*] and Luquan Ren [1]

1 Key Laboratory of Bionic Engineering of Ministry of Education, College of Biological and Agricultural Engineering, Jilin University, Changchun 130022, China
2 School of Materials and Physics, China University of Mining & Technology, Xuzhou 221116, China
* Correspondence: zhz@cumt.edu.cn (H.Z.); zhzh@jlu.edu.cn (Z.Z.)

Abstract: Fe_3O_4 nanoparticles with average sizes of 3–8 nm were in-situ grown and self-assembled as homogeneous clusters on reduced graphene oxide (RGO) via coprecipitation with some additives, where RGO sheets were expanded from restacking and an increased surface area was obtained. The crystallization, purity and growth evolution of as-prepared Fe_3O_4/RGO nanocomposites were examined and discussed. Supercapacitor performance was investigated in a series of electrochemical tests and compared with pure Fe_3O_4. In 1 M KOH electrolyte, a high specific capacitance of 317.4 F g^{-1} at current density of 0.5 A g^{-1} was achieved, with the cycling stability remaining at 86.9% after 5500 cycles. The improved electrochemical properties of Fe_3O_4/RGO nanocomposites can be attributed to high electron transport, increased interfaces and positive synergistic effects between Fe_3O_4 and RGO.

Keywords: magnetite; supercapacitors; graphene; coprecipitation; electrochemistry

1. Introduction

Since the last two decades, various energy devices, such as batteries, fuel cells and supercapacitors, have been developed in order to alleviate the energy crisis and environmental problems [1–3]. Among them, pseudocapacitors have intrigued much attention due to their exceptional advantages, including higher power density, high rates of charge-discharge, cost reduction and relative safety [4]. Transition metal oxides are considered as ideal electrode materials for pseudocapacitors because of their high theoretical specific capacitance and high electrochemical activity. RuO_2 has been revealed as the first pseudocapacitive electrode material with an excellent pseudocapacitance of about 720 F g^{-1} [5]. However, its prohibitive cost and toxic nature has motivated the search for economical and environment-friendly alternatives with equivalent performance. Other transition metal oxides, such as Co_3O_4, NiO, MnO_2 and Fe_3O_4, have been studied extensively as substitutes for Ru-based oxide electrodes [6–9]. Among them, Fe_3O_4 has been identified as the promising material, considering its high theoretical specific capacitance (2299 F g^{-1}), low cost and large potential window. In an early study, Fe_3O_4 hollow spheres possessed a capacity of 294 F g^{-1} and 90.8% capacity retention after 500 cycles [9]. However, research on Fe_3O_4 as electrodes have shown relatively low capacitances below 400 F g^{-1} [9–11]. The reason might be related to its limited electrical conductivity that hinders ion diffusions. So far, exploring novel Fe_3O_4-based materials for supercapacitors application is still challenging.

Nanostructured Fe_3O_4 composited with carbonaceous materials has been reported as an effective strategy that may offer higher energy density with larger specific capacitance. For instance, Fe_3O_4 composited with carbon nanotubes, exhibited improved cyclic stability and energy density in contrast to those of pure Fe_3O_4 [12]. The specific capacity of 275.9 F g^{-1} was acquired at 0.5 A g^{-1} in Na_2SO_4 aqueous solution for carbon-coated Fe_3O_4 composites [13]. Recently, reduced graphene oxide (RGO) has attracted much attention on account of its tremendous electronic conductivity and high surface area. Considerable efforts have been made to design and synthesize Fe_3O_4/RGO composites for supercapacitor applications. Kumar et al. prepared 3D network of Fe_3O_4/RGO composites by one-pot microwave approach and acquired specific capacity of 297.0 F g^{-1} at 4.4 A g^{-1} [14]. Another piece of research on Fe_3O_4 nanorods decorated on RGO surfaces in two-step procedures achieved 95% specific capacity retention after 2000 cycles [15]. The improved electrochemical performance with the presence of RGO is found to be prevalent and strongly dependent on the synergistic effects between Fe_3O_4 and RGO [15]. Therefore, it is highly desirable to design Fe_3O_4/RGO composites with combined interfaces for specific capacitance enhancement.

In this paper, Fe_3O_4/RGO nanocomposites with unique microstructures were prepared by a facile coprecipitation process. By introducing certain additives, Fe_3O_4 nanoparticles could be anchored on RGO surfaces and formed as nanoclusters without aggregation. The interlayer spacing between RGO sheets was expanded by these Fe_3O_4 nanoclusters, which contributed to ion transport in the electrolyte. In a series of electrochemical tests, the as-prepared Fe_3O_4/RGO nanocomposites exhibited high specific capacitance, good rate capability and long cyclic stability.

2. Materials and Methods

2.1. Preparation of Fe_3O_4/RGO Nanocomposites

In a typical procedure, Fe_3O_4/RGO nanocomposites were synthesized as follows: First, 0.20 g GO was dispersed into 200 mL distilled water and ultrasonicated for 5 h to form a homogeneous suspension. Then, it was put into a five-neck flask with a thermostatic water bath around 80 °C. Second, 5.56 g $FeSO_4 \cdot 7H_2O$ was dissolved and added into the flask with constant stirring while Ar was forced into the mixed solution to obtain an oxygen-free environment. Subsequently, 14.47 g $NH_4Fe(SO_4)_2 \cdot 12H_2O$ and 20 mL polyethylene glycol (PEG-400) were dissolved, respectively, and added into the flask successively. Third, 4.80 g NaOH dissolved in 100 mL distilled water was dropwised into the flask for 1 h, while the resultant solution was maintained for another 1 h to complete the coprecipitation. Corresponding reactions can be described as:

$$Fe^{2+} + 2NaOH = Fe(OH)_2\downarrow + 2Na^+ \tag{1}$$

$$Fe^{3+} + 3NaOH = Fe(OH)_3\downarrow + 3Na^+ \tag{2}$$

$$Fe(OH)_2 + 2Fe(OH)_3 = Fe_3O_4 + 4H_2O \tag{3}$$

Finally, 4 mL $N_2H_4 \cdot H_2O$ (85%) was dropwised into the flask for 1 h. Then, the reaction system was kept for 3 h to accomplish the reduction from GO to RGO. After the reaction was completed, the black products were washed several times, then dried at 50 °C for 9 h. According to the above synthesis procedures, the molar ratio of Fe_3O_4 to RGO in the as-prepared composites was estimated to be 2:1. The schematic for synthesizing Fe_3O_4/RGO nanocomposites was illustrated in Figure 1. For comparison, Fe_3O_4 nanoparticles were prepared under the similar experimental condition where GO was removed from the flask.

2.2. Material Characterization

The as-prepared GO, Fe_3O_4 nanoparticles, and Fe_3O_4/RGO nanocomposites were examined by X-ray diffraction (XRD, Rigaku D/max 2500PC, Tokyo, Japan) and Fourier transform infrared (FT-IR, Bruker Tensor 27, Billerica, MA, USA) spectroscopy, respectively. Microstructure and morphology analysis were conducted on transmission electron microscope and high-resolution transmission electron microscopy (TEM & HRTEM, FEI TecnaiG^{2}F20, Lincoln, NE, USA). To quantify the chemical compositions, X-ray photoelectron spectroscopy (XPS, Thermo Fisher ESCALAB 250Xi, Waltham, MA, USA) measurements were performed by using 300 W Al Kα radiations as the X-ray source for excitation.

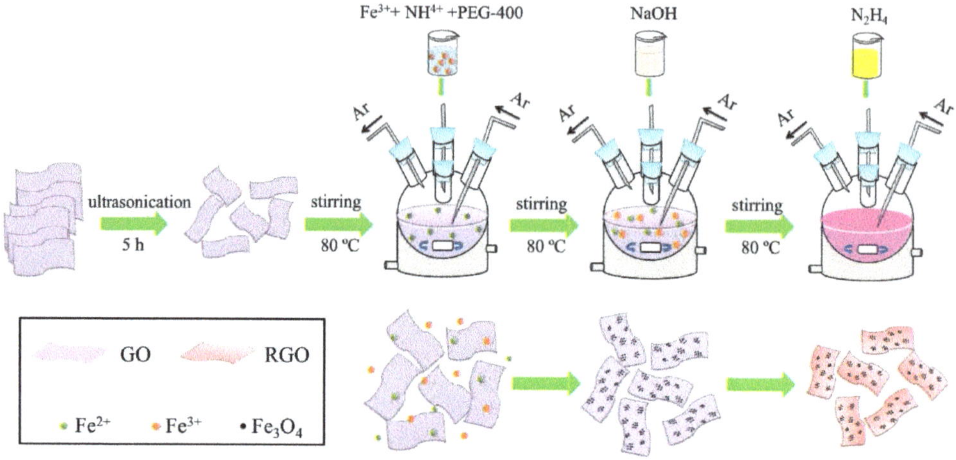

Figure 1. Schematic illustration of the synthesis process and growth evolution of Fe_3O_4/RGO nanocomposites.

2.3. Electrochemical Measurements

The working electrodes were prepared by mixing 80 wt.% active material, 10 wt.% acetylene black and 10 wt.% polyvinylidene fluoride dissolved to form a slurry and coated on nickel foam (10 mm × 15 mm). The electrochemical investigations were employed in 1 M KOH electrolyte. Platinum and Hg/HgO were used as counter and reference electrodes, respectively. Electrochemical performances were detected by Chi660e electrochemical workstation.

3. Results and Discussion

XRD results of as-prepared GO, Fe3O4 nanoparticles and Fe3O4/RGO nanocomposites are shown in Figure 2a. The as-prepared GO presents a strong diffraction peak around $2\theta = 10.5°$, corresponding to its characteristic (001) plane. The layer-to-layer distance of GO is calculated to be 0.83 nm by Scherrer equation, which is much larger than that of graphite. The introduction of the oxygen-containing functional groups between GO sheets should be responsible for it [14]. For Fe3O4 nanoparticles, their sharp and strong diffraction peaks are indexed to spinel phase (JCPDS: No.65-3107), indicating a preferable crystallization and high purity. The diffraction peak positions of Fe3O4/RGO nanocomposites are well-matched with Fe3O4 nanoparticle, where the decreased peak intensities and relatively broadened peaks suggest their reduced crystallinity [16]. FT-IR analysis results of the as-prepared samples were collected by attenuated total reflection method and preprocessed by OMNIC software (version number 9.2). As is shown in Figure 2b, distinct peaks around 3430 cm^{-1} are prevalent in all spectra, which is related to the stretching vibrations of O-H bands. The other three characteristic peaks in the spectrum of GO should be ascribed to C=C (1643 cm^{-1}), CO-H (1388 cm^{-1}) and C-O (1082 cm^{-1}) stretching vibrations [17]. In the spectra of Fe3O4 nanoparticles and Fe3O4/RGO nanocomposites, stretching vibrations of Fe-O bands around 580 cm^{-1} are conspicuous [15]. For Fe3O4/RGO nanocomposites, another peak around 1646 cm^{-1} is accurate and confirmed the existence of RGO. Compared with GO, the spectrum of Fe3O4/RGO has no infrared peaks around 1082 cm^{-1}, implying that GO has been completely reduced into RGO after hydrothermal processes.

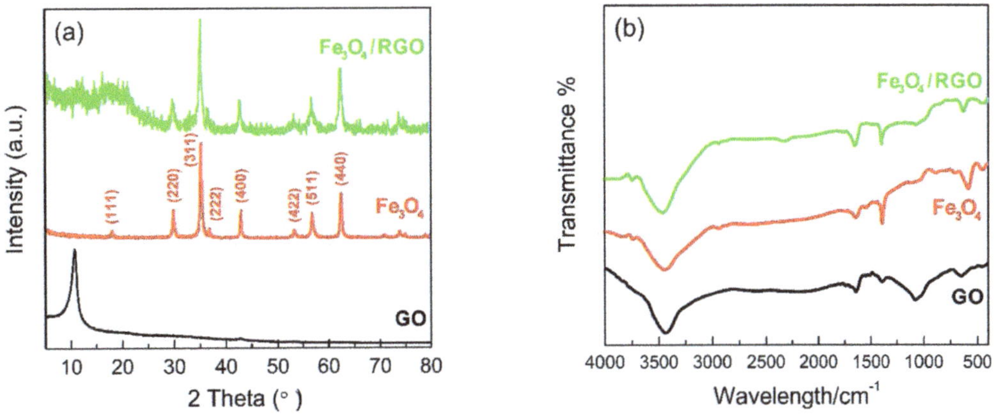

Figure 2. (**a**) XRD patterns and (**b**) FT-IR spectrum of as-prepared GO, Fe_3O_4 nanoparticles and Fe_3O_4/RGO nanocomposites.

Chemical compositions and oxidation states of as-prepared GO and Fe_3O_4/RGO nanocomposites were determined by XPS analysis. C 1s spectrum of GO in Figure 3a can be deconvoluted into three different peaks. The peaks centered at 284.7 eV, 286.7 eV and 288.7 eV should be indexed to C–C/C=C, C–O and C=O groups, respectively [18]. The survey spectrum of Fe_3O_4/RGO nanocomposites in Figure 3b confirms the existence of Fe, C, and O elements, while no other elemental signals can be detected. Compared with Figure 3a, the peak of sp^2 hybridized C–C /C=C bonds is increased, while the peaks of both C–O and C=O bonds are significantly decreased in Figure 3c, implying the adequate reduction from GO to RGO. For Fe 2p spectrum in Figure 3d, two major peaks at 711.5 eV and 724.7 eV were identified as Fe $2p_{3/2}$ and Fe $2p_{1/2}$, respectively, demonstrating Fe^{2+} and Fe^{3+} ions in Fe_3O_4 [16].

Figure 4 displays the TEM images of Fe_3O_4 nanoparticles prepared without the addition of GO. All the particles are approximately spherical with uniform sizes around 10 nm while agglomeration among these nanoparticles is inevitable. Similar experimental results have been reported in previous studies [18]. The TEM & HRTEM images of Fe_3O_4/RGO nanocomposites are given in Figure 5. From relatively low magnification, it can be seen that Fe_3O_4 nanoclusters with irregular ellipsoid shapes are dispersed uniformly and wrapped by RGO sheets, where the sizes of nanoclusters are in range of 45–110 nm. Accordingly, the interlayer spacing between RGO sheets is expanded by these magnetite nanoclusters, which avoids RGO from restacking. Figure 5c,d indicate that each Fe_3O_4 nanocluster is composed of dozens of tiny Fe_3O_4 nanoparticles without agglomeration. These self-assembled nanoparticles are anchored on RGO sheets with sizes around 3–8 nm, and some foldings can be observed at the edges of RGO sheets, which increase the surface area. Figure 5e illustrates the distinct interfaces between Fe_3O_4 nanoparticles and RGO sheets. HRTEM lattice fringes with d-spacing distances of 0.25 nm and 0.29 nm were determined from the top and bottom right corners of Figure 5e, corresponding to (311) and (220) planes of Fe_3O_4, respectively. Some RGO sheets are overlapped and corrugated due to the embedded and stabilized Fe_3O_4 nanoclusters on RGO support. Such unique microstructures could hinder the agglomeration of Fe_3O_4 nanoparticles and facilitate ion transport inside RGO sheets. Therefore, improved electrochemical performance of Fe_3O_4/RGO nanocomposites is anticipated in the following studies.

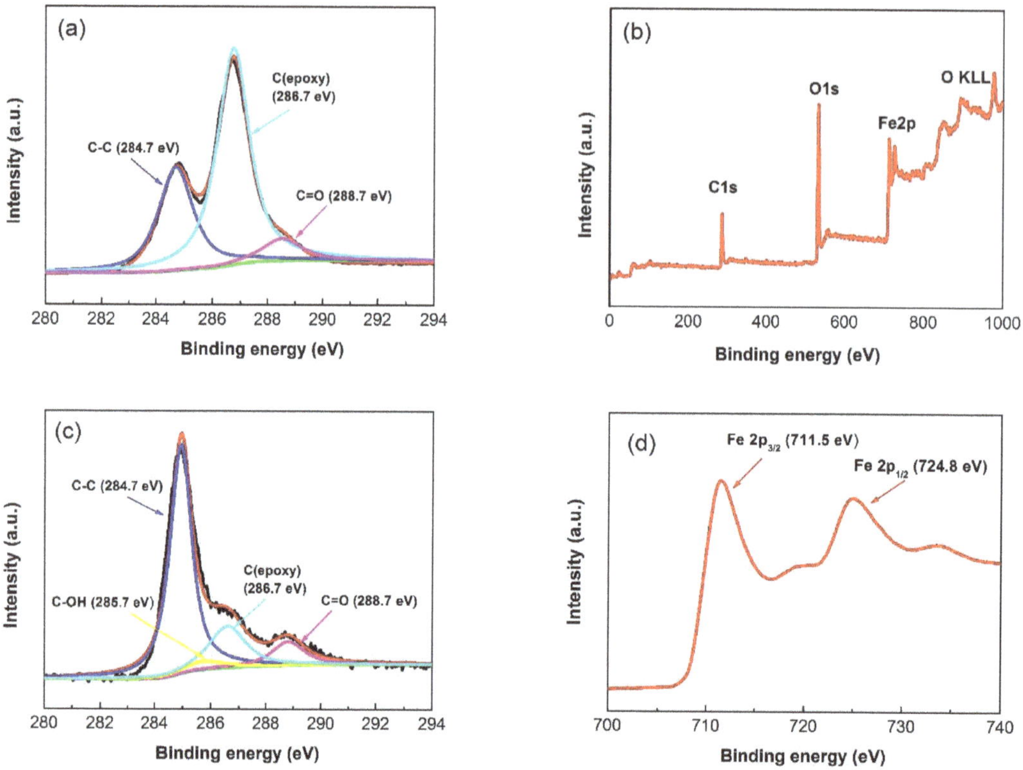

Figure 3. XPS spectrum of (**a**) C 1s for as-prepared GO, (**b**) wide scanning for Fe_3O_4/RGO nanocomposites with (**c**) C 1s and (**d**) Fe 2p spectral profiles.

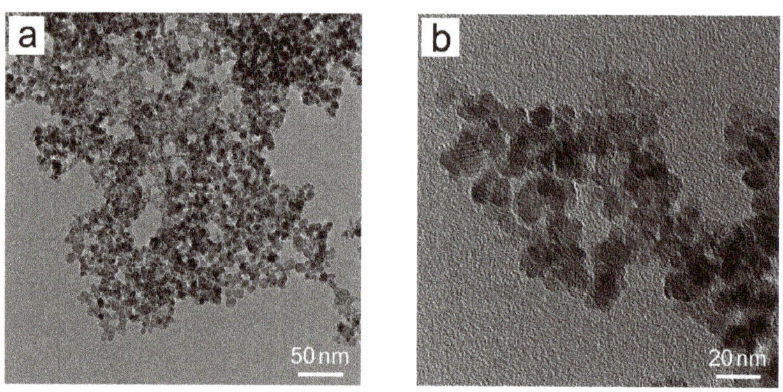

Figure 4. TEM images in (**a**) low and (**b**) high magnifications of Fe_3O_4 nanoparticles.

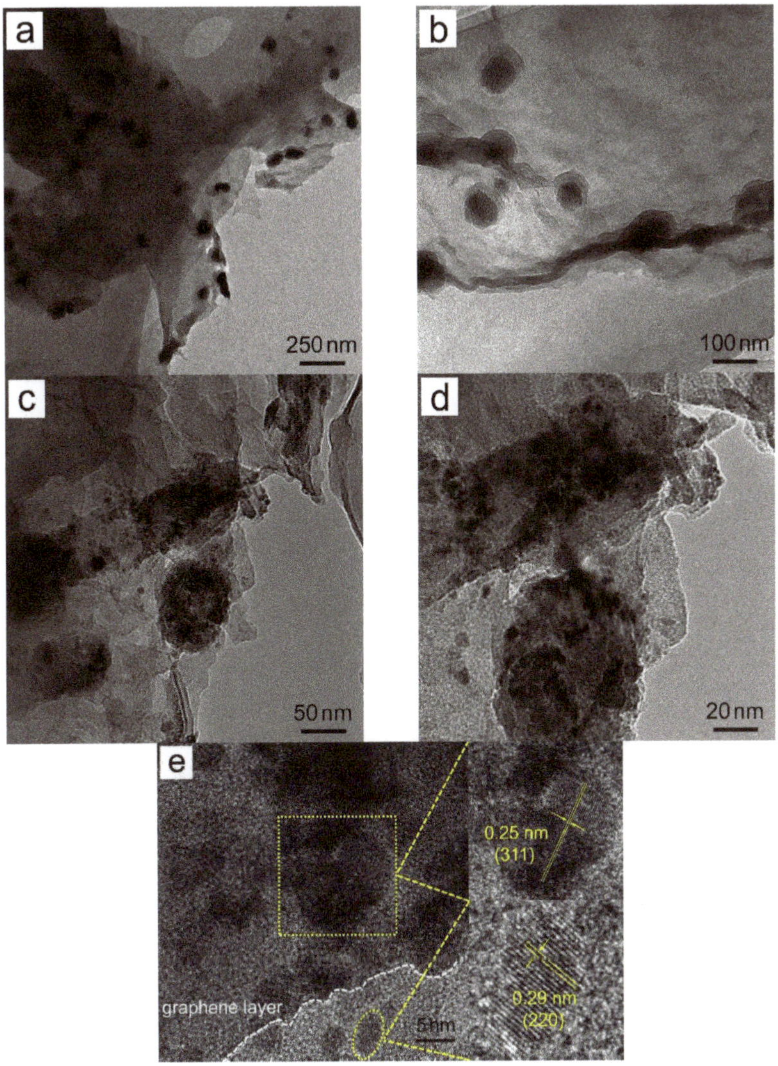

Figure 5. (**a–d**) TEM and (**e**) HRTEM images of Fe_3O_4/RGO nanocomposites where the selected areas in (**e**) are enlarged and shown in the top and bottom right corners, respectively.

The growth evolution of Fe_3O_4/RGO nanocomposites is based on the synthesis processes with the effects of additives. GO sheets achieved from the expanded graphite generally contain abundant hydroxyl, epoxy, carboxyl and carbonyl functional groups [18]. Fe^{3+} and Fe^{2+} ions added into GO suspension tend to be attached by these functional groups owing to electrostatic attraction, which might be served as nucleation centers in coprecipitation reactions. The introduction of NH^{4+} and PEG-400 makes Fe_3O_4 nucleus grow selectively and inhibits agglomeration [19,20]. After the reduction from GO to RGO by hydrazine hydrate, Fe_3O_4/RGO nanocomposites with unique microstructures could be obtained. Compared with hydrothermal [21] and sol–gel routes [22], coprecipitation processes in this study have several advantages, including high yield, simple equipment, relative low temperature and efficiency.

Figure 6a illustrates the cyclic voltammetry (CV) curves of Fe_3O_4/RGO nanocomposites at various scan speeds. A lack of symmetry in the pairs of cathodic and anodic peaks is evident, which demonstrates pseudocapacitance. These redox peaks are assigned to reversible faradic reactions between Fe^{2+} and Fe^{3+} [23]. By increasing scan rates, distance between oxidation peak and reduction peak increases, which is due to the resistance of electrode. Additionally, the well-preserved shapes of CV curves at various scan rates imply the efficient electrochemical transportation and reversibility [24]. For comparison, the CV curves of Fe_3O_4 nanoparticles (Figure 6b) indicate much lower specific capacitance than that of the Fe_3O_4/RGO nanocomposites. The specific capacitance (C_s) of Fe_3O_4 nanoparticles and Fe_3O_4/RGO nanocomposites can be calculated by the following equation:

$$C_s = \frac{1}{sm(V_a - V_b)} \int_{V_b}^{V_a} idV \tag{4}$$

where s is scan rate, m is mass of electrode materials, i is respond current density, V_a and V_b are the integration limits of the voltammetric curves, respectively. At the scanning speed of 5 mV s^{-1}, the calculated specific capacitance of Fe_3O_4/RGO nanocomposites and Fe_3O_4 nanoparticles are 572.8 F g^{-1} and 418.3 F g^{-1}, respectively. The higher specific capacitance of Fe_3O_4/RGO nanocomposites is attributed to the existence of RGO providing high electrical conductivity and inhibiting the aggregation of Fe_3O_4 nanoparticles [25].

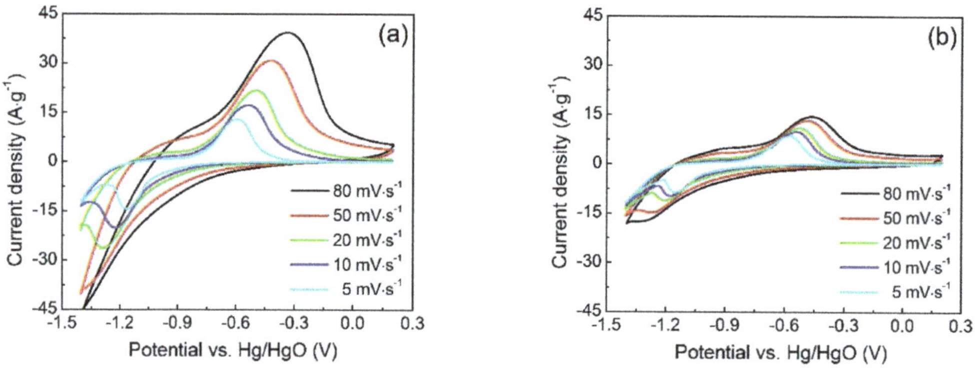

Figure 6. CV curves of (**a**) Fe_3O_4/RGO nanocomposites and (**b**) Fe_3O_4 nanoparticles at various scan rates in 1 M KOH solution.

Typical galvanostatic charge–discharge (GCD) curves of Fe_3O_4/RGO nanocomposites and Fe_3O_4 nanoparticles are shown in Figure 7a,b, respectively. The charge and discharge curves of both electrode materials are approximately symmetrical in shape with a slight curvature, indicating a pseudocapacitor characteristic [26]. In contrast to Fe_3O_4 nanoparticles, Fe_3O_4/RGO nanocomposites have much longer discharge times at the same current density, which illustrates improved capacitive performance. The specific capacity can also be calculated by following equation: $C'_s = I\Delta t/m\Delta v$, where I, Δt and Δv referred to discharge current, discharge time and voltage window, respectively. C'_s value of Fe_3O_4/RGO nanocomposites is calculated to be 317.4 F g^{-1} at scan rate of 0.5 A g^{-1}, which is almost twice that of Fe_3O_4 nanoparticles (169.1 F g^{-1}). The specific capacitances of Fe_3O_4/RGO nanocomposites are 264.5, 223.7 and 198.1 F g^{-1}, respectively, at current densities of 1.0, 2.0 and 5.0 A g^{-1}. Even with a 10-fold increase in current density, 62.4% of the initial specific capacitance can still be maintained, showing the good rate performance of Fe_3O_4/RGO nanocomposites.

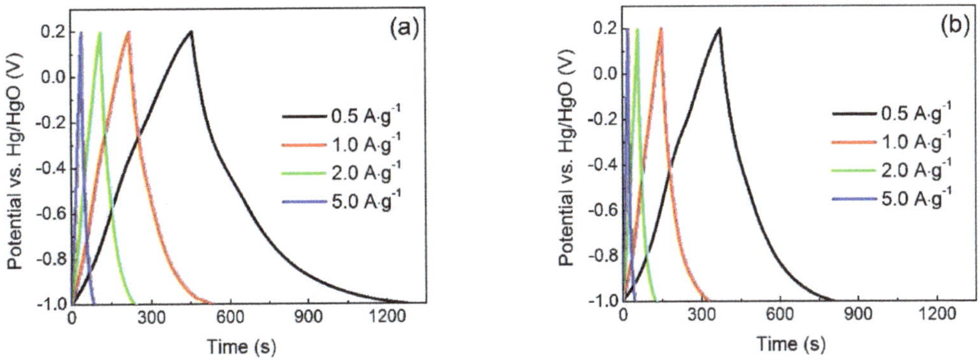

Figure 7. GCD curves of (**a**) Fe_3O_4/RGO nanocomposites and (**b**) Fe_3O_4 nanoparticles at various discharge current densities in 1 M KOH solution.

The charge transport and ionic diffusion characteristics of Fe_3O_4/RGO nanocomposites and Fe_3O_4 nanoparticles were determined by electrochemical impedance spectroscopy (EIS) measurements. As shown in Figure 8a, Nyquist plots of both electrode materials include semicircle parts at higher frequencies and straight-lined parts in lower frequencies. Generally, the semicircle parts represent charge transfer resistance (R_{ct}) while the straight-lined parts attribute to Warburg impedance [27]. Corresponding circuit diagrams were matched with the help of Zview software and given in the inset of Figure 8a. Fe_3O_4/RGO nanocomposites exhibit a smaller semicircle in higher frequency ranges due to the existence of RGO [28]. Accordingly, R_{ct} values for Fe_3O_4/RGO nanocomposites is around 1.45 Ω, much lower than that of Fe_3O_4 particles (2.57 Ω), demonstrating their preferable electron transference. The straight line at lower frequencies close to 90° indicates ideal capacitive behavior. The oblique line of the Fe_3O_4/RGO nanocomposites is more vertical than Fe_3O_4 particles, revealing more efficient access of electrolyte ions. The intrinsic resistance (R_s) of Fe_3O_4/RGO nanocomposites is about 1.18 Ω, lower than that of the Fe_3O_4 particles (1.41 Ω), which confirms the improved ion transport [28]. The cycling stability of both Fe_3O_4/GO nanocomposites and Fe_3O_4 nanoparticles were examined at the current density of 5.0 A g^{-1} for 5500 cycles, and the results were presented in Figure 8b. It can be figured out that the capacitance of both materials drops rapidly within 1000 cycles, then for Fe_3O_4 nanoparticles it maintains about 69.2% of their initial capacitance after 5500 cycles, while the capacitance retention of Fe_3O_4/RGO nanocomposite is as high as 86.9%.

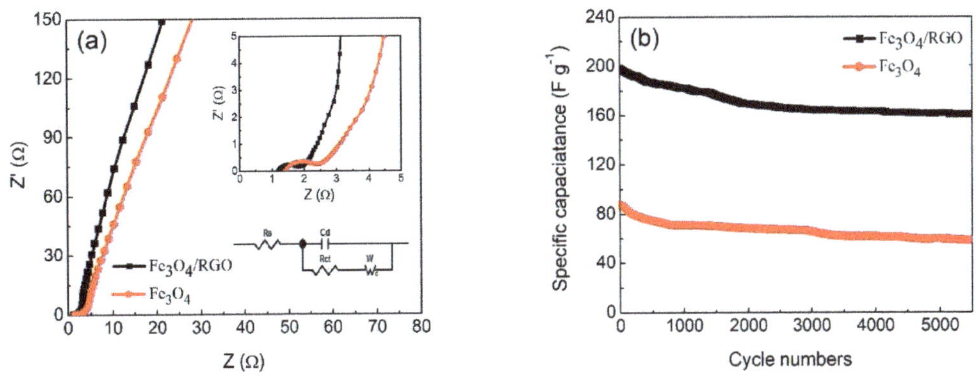

Figure 8. (**a**) Nyquist impedance plots of Fe_3O_4 nanoparticles and Fe_3O_4/RGO nanocomposites with insets of high-frequency regions and matched circuit diagram, (**b**) Cyclic performance of Fe_3O_4 nanoparticles and Fe_3O_4/RGO nanocomposites in 1 M KOH solution at a current density of 5.0 A g^{-1}.

Table 1 lists the electrochemical performance of Fe_3O_4 and related carbonaceous composites as supercapacitor electrodes in the present study and literature survey. It can be seen that pure Fe_3O_4 [9–11] has limited specific capacitance and cycle stability than its carbonaceous composites, owing to its limited electrical conductivity and agglomeration characteristics. Moreover, Fe_3O_4 composited with RGO [14,15,21,26] generally exhibits better performance, since RGO possesses a larger specific surface area than other carbonaceous forms [12,13]. Considering the complicated procedures to synthesize samples in the literature, the as-prepared Fe_3O_4/RGO nanocomposites are easy to obtain with comparable capacitance and stability.

Table 1. Comparisons on electrochemical performance of Fe_3O_4 and related carbonaceous composites as supercapacitor electrodes in the present study and literature survey.

Materials	Morphology	Specific Capacitance (F g^{-1})	Electrolyte	Stability	Ref.
Fe_3O_4	Hollow microspheres	294 (0.5 A g^{-1})	8 M KOH	90.8% after 500 cycles	[9]
Fe_3O_4	Microflowers	183 (1.0 A g^{-1})	0.5 M Na_2SO_3	65.0% after 5000 cycles	[10]
Fe_3O_4	Nanoparticles	383.2 (0.5 A g^{-1})	1 M Na_2SO_3	83.6% after 2000 cycles	[11]
Fe_3O_4	Nanoparticles	169.1 (0.5 A g^{-1})	1 M KOH	69.2% after 5500 cycles	This work
Fe_3O_4/C	Nanoparticles/Nanotubes	187.1 (1.0 A g^{-1})	1 M Na_2SO_3	80.2% after 1000 cycles	[12]
Fe_3O_4/C	Nanorods/Nanoparticles	275.9 (0.5 A g^{-1})	1 M Na_2SO_3	81.2% after 500 cycles	[13]
Fe_3O_4/RGO	Nanosheets/Nanosheets	297.0 (4.4 A g^{-1})	2 M KOH	91.4% after 9600 cycles	[14]
Fe_3O_4/RGO	Nanorods/Nanosheets	315 (5.0 A g^{-1})	1 M KOH	95.0% after 2000 cycles	[15]
Fe_3O_4/RGO	Nanodiscs/Nanosheets	1149 (1.5 A g^{-1})	6 M KOH	97.5% after 10,000 cycles	[21]
Fe_3O_4/RGO	Nanoflowers/Nanosheets	454.3 (1.0 A g^{-1})	2 M KOH	94.0% after 10,000 cycles	[26]
Fe_3O_4/RGO	Nanoparticles/Nanosheets	317.4 (0.5 A g^{-1})	1 M KOH	86.9% after 5500 cycles	This work

Based on the above discussions, the improved electrochemical properties of as-prepared Fe_3O_4/RGO nanocomposites can be ascribed to the following aspects. Firstly, the existence of RGO not only provides plenty of electron transport channels for supercapacitor electrode but also partially accommodates the volume change during cycling. Secondly, the dispersion and intercalation of Fe_3O_4 nanoclusters separate RGO sheets and inhibit their restacking, leading to an increased surface area and ensuring the utilization of magnetite. Finally, the unique microstructures of Fe_3O_4 can shorten the diffusion path of electrolyte ions and contribute the synergistic effects to RGO. Therefore, the as-prepared Fe_3O_4/RGO nanocomposites are considered as prospective candidates for future electrode materials in energy storage.

4. Conclusions

In this paper, Fe_3O_4/RGO nanocomposites were prepared by coprecipitation in an oxygen-free environment with some additives. Fe_3O_4 nanoparticles were dispersed homogeneously with average sizes of 3–8 nm, which were self-assembled as irregular oval-shaped nanoclusters and wrapped by RGO sheets. XRD and XPS analysis confirmed their preferable crystallization and high purity. Compared with pure Fe_3O_4, Fe_3O_4/RGO nanocomposites deliver improved specific capacitance of 572.8 F g^{-1} at the scan rate of 5 mV s^{-1} and better cycling stability remaining 86.9% after 5500 cycles. The excellent electrochemical properties were ascribed to high electron transport, increased interfaces and positive synergistic effects between Fe_3O_4 and RGO, which enlightened an effective strategy in producing supercapacitor electrodes for energy storage devices.

Author Contributions: Conceptualization, H.Z. and Y.J.; methodology, J.H. and X.W.; formal analysis, J.H. and X.W.; writing—original draft preparation, J.H. and Y.J.; writing—review and editing, Y.J., H.Z., and Z.Z.; validation, X.W. and L.R.; supervision, H.Z., Z.Z. and L.R.; funding acquisition, Y.J., H.Z., Z.Z. and L.R. All authors have read and agreed to the published version of the manuscript.

Funding: This research was funded by the National Natural Science Foundation of China (Grant Nos. 52105303 and 52025053), China Postdoctoral Science Foundation Funded Project (2021M691215), Foundation for Innovative Research Groups of the National Natural Science Foundation of China (No. 52021003), and Fundamental Research Funds for the Central Universities of China (No. 2019ZDPY22).

Institutional Review Board Statement: Not applicable.

Informed Consent Statement: Not applicable.

Data Availability Statement: The data presented in this study are available from the corresponding authors upon reasonable request.

Conflicts of Interest: The authors declare no conflict of interest.

References

1. Dong, C.; Zhou, H.; Jin, B.; Gao, W.; Lang, X.; Li, J.; Jiang, Q. Enabling high-performance room-temperature sodium/sulfur batteries with few-layer 2H-MoSe 2 embellished nitrogen-doped hollow carbon spheres as polysulfide barriers. *J. Mater. Chem. A* **2021**, *9*, 3451–3463. [CrossRef]
2. Zhu, X.; Sun, M.; Zhao, R.; Li, Y.; Zhang, B.; Zhang, Y.; Lang, X.; Zhu, Y.; Jiang, Q. 3D hierarchical self-supported NiO/Co_3O_4@C/CoS_2 nanocomposites as electrode materials for high-performance supercapacitors. *Nanoscale Adv.* **2020**, *2*, 2785–2791. [CrossRef]
3. Zhou, S.; Yang, Y.; Chen, H.; Ling, Y. In situ exsolved Co–Fe nanoparticles on the Ruddlesden-Popper-type symmetric electrodes for intermediate temperature solid oxide fuel cells. *Ceram. Int.* **2020**, *46*, 18331–18338. [CrossRef]
4. Lokhande, P.E.; Chavan, U.S.; Pandey, A. Materials and Fabrication Methods for Electrochemical Supercapacitors: Overview. *Electrochem. Energy Rev.* **2019**, *3*, 155–186. [CrossRef]
5. Hu, C.-C.; Chang, K.-H.; Lin, A.M.-C.; Wu, Y.-T. Design and Tailoring of the Nanotubular Arrayed Architecture of Hydrous RuO_2 for Next Generation Supercapacitors. *Nano Lett.* **2006**, *6*, 2690–2695. [CrossRef] [PubMed]
6. Jang, G.-S.; Ameen, S.; Akhtar, M.S.; Shin, H.-S. Cobalt oxide nanocubes as electrode material for the performance evaluation of electrochemical supercapacitor. *Ceram. Int.* **2018**, *44*, 588–595. [CrossRef]
7. Shi, X.; Zhang, S.; Chen, X.; Tang, T.; Klingeler, R.; Mijowska, E. Ultrathin NiO confined within hollow carbon sphere for efficient electrochemical energy storage. *J. Alloys Compd.* **2019**, *797*, 702–709. [CrossRef]
8. Yin, B.; Zhang, S.; Jiang, H.; Qu, F.; Wu, X. Phase-controlled synthesis of polymorphic MnO_2 structures for electrochemical energy storage. *J. Mater. Chem. A* **2015**, *3*, 5722–5729. [CrossRef]
9. Zeng, X.; Yang, B.; Li, X.; Li, R.; Yu, R. Solvothermal synthesis of hollow Fe3O4 sub-micron spheres and their enhanced electrochemical properties for supercapacitors. *Mater. Des.* **2016**, *101*, 35–43. [CrossRef]
10. Manikandan, N.; Lakshmi, B.; Shivakumara, S. Preparation of self-assembled porous flower-like nanostructured magnetite (Fe_3O_4) electrode material for supercapacitor application. *J. Solid State Electrochem.* **2022**, *26*, 887–895. [CrossRef]
11. Ma, J.; Guo, X.; Yan, Y.; Xue, H.; Pang, H. FeOx-based materials for electrochemical energy storage. *Adv. Sci.* **2018**, *5*, 1700986. [CrossRef] [PubMed]
12. Park, S.K.; Sure, J.; Vishnu, D.S.M.; Jo, S.J.; Lee, W.C.; Ahmad, I.A.; Kim, H.K. Nano-Fe_3O_4/carbon nanotubes composites by one-pot microwave solvothermal method for supercapacitor applications. *Energies* **2021**, *14*, 2908. [CrossRef]
13. Liu, J.; Liu, S.; Zhuang, S.; Wang, X.; Tu, F. Synthesis of carbon-coated Fe_3O_4 nanorods as electrode material for supercapacitor. *Ionics* **2013**, *19*, 1255–1261. [CrossRef]
14. Kumar, R.; Singh, R.K.; Vaz, A.R.; Savu, R.; Moshkalev, S.A. Self-assembled and one-step synthesis of interconnected 3D network of Fe_3O_4/reduced graphene oxide nanosheets hybrid for high-performance supercapacitor electrode. *ACS Appl. Mater. Interfaces* **2017**, *9*, 8880–8890. [CrossRef]
15. Das, A.K.; Sahoo, S.; Arunachalam, P.; Zhang, S.; Shim, J.J. Facile synthesis of Fe_3O_4 nanorod decorated reduced graphene oxide (RGO) for supercapacitor application. *RSC Adv.* **2016**, *6*, 107057–107064. [CrossRef]
16. Satapathy, S.; Prabakaran, P.; Prasad, E. Augmenting Photoinduced Charge Transport in a Single-Component Gel System: Controlled In Situ Gel–Crystal Transformation at Room Temperature. *Chem. Eur. J.* **2018**, *24*, 6217–6230. [CrossRef]
17. Satapathy, S.; Prasad, E. Charge Transfer Modulated Self-Assembly in Poly (aryl ether) Dendron Derivatives with Improved Stability and Transport Characteristics. *ACS Appl. Mater. Interfaces* **2016**, *8*, 26176–26189. [CrossRef]
18. Zhang, H.; Zhao, J.; Ou, X. Facile synthesis of Fe_3O_4 nanowires at low temperature (80 C) without autoclaves and their electromagnetic performance. *Mater. Lett.* **2017**, *209*, 48–51. [CrossRef]
19. Shu, T.; Gao, H.; Li, Q.; Wei, F.; Ren, Y.; Sun, Z.; Qi, J.; Sui, Y. One-step phosphating synthesis of CoP nanosheet arrays combined with Ni_2P as a high-performance electrode for supercapacitors. *Nanoscale* **2020**, *12*, 20710–20718. [CrossRef]
20. Sheng, S.; Liu, W.; Zhu, K.; Cheng, K.; Ye, K.; Wang, G.; Cao, D.; Yan, J. Fe_3O_4 nanospheres in situ decorated graphene as high-performance anode for asymmetric supercapacitor with impressive energy density. *J. Colloid Interface Sci.* **2018**, *536*, 235–244. [CrossRef]
21. Khan, A.J.; Khan, A.; Javed, M.S.; Arshad, M.; Asim, S.; Khalid, M.; Siyal, S.H.; Hussain, S.; Hanif, M.; Liu, Z. Surface assembly of Fe_3O_4 nanodiscs embedded in reduced graphene oxide as a high-performance negative electrode for supercapacitors. *Ceram. Int.* **2020**, *46*, 19499–19505. [CrossRef]

22. Ma, J.; Shi, N.; Jia, J. Fe_3O_4 nanospheres decorated reduced graphene oxide as anode to promote extracellular electron transfer efficiency and power density in microbial fuel cells. *Electrochimica Acta* **2020**, *362*, 137126. [CrossRef]
23. Zhang, H.; Han, J.; Xu, J.; Ling, Y.; Ou, X. Self-assembled $NiCo_2O_4$ microspheres for hybrid supercapacitor applications. *J. Mater. Sci.* **2022**, *57*, 5566–5576. [CrossRef]
24. Zhao, X.; Jia, Y.; Liu, Z.-H. GO-graphene ink-derived hierarchical 3D-graphene architecture supported Fe_3O_4 nanodots as high-performance electrodes for lithium/sodium storage and supercapacitors. *J. Colloid Interface Sci.* **2019**, *536*, 463–473. [CrossRef] [PubMed]
25. Madhuvilakku, R.; Alagar, S.; Mariappan, R.; Piraman, S. Green one-pot synthesis of flowers-like Fe_3O_4/rGO hybrid nanocomposites for effective electrochemical detection of riboflavin and low-cost supercapacitor applications. *Sens. Actuators B Chem.* **2017**, *253*, 879–892. [CrossRef]
26. Rosaiah, P.; Zhu, J.; Hussain, O.M.; Qiu, Y. Facile and cost-effective synthesis of flower-like RGO/Fe_3O_4 nanocomposites with ultra-long cycling stability for supercapacitors. *Ionics* **2019**, *25*, 655–664. [CrossRef]
27. Qi, M.; Zhu, W.; Lu, Z.; Zhang, H.; Ling, Y.; Ou, X. Synthesis of nickel sulfide–graphene oxide composite microflower structures to enhance supercapacitor performance. *J. Mater. Sci. Mater. Electron.* **2020**, *31*, 12536–12545. [CrossRef]
28. Prabakaran, P.; Satapathy, S.; Prasad, E.; Sankararaman, S. Architecting pyrediyne nanowalls with improved inter-molecular interactions, electronic features and transport characteristics. *J. Mater. Chem. C* **2018**, *6*, 380–387. [CrossRef]

materials

Review

A Brief Review: The Use of L-Ascorbic Acid as a Green Reducing Agent of Graphene Oxide

Mariano Palomba, Gianfranco Carotenuto and Angela Longo *

Institute for Polymers, Composites, and Biomaterials, National Research Council, SS Napoli/Portici, Piazzale Enrico Fermi 1, 80055 Portici, Italy
* Correspondence: angela.longo@cnr.it; Tel.: +39-081-775-8824

Abstract: The reduced form of graphene oxide (r-GO) represents a versatile precursor to obtain graphene derivatives. Graphene oxide (GO) consists of a layered material based on a carbon skeleton functionalized by different oxygen-containing groups, while r-GO is obtained by the almost complete removal of these oxygen-containing functional groups. The r-GO has mechanical, electrical, and optical properties quite similar to graphene, thus, it proves to be a convenient 2D material useful for many technological applications. Nowadays, the most important aspects to consider in producing r-GO are: (i) the possibility of obtaining the highest reduction grade; (ii) the possibility of improving the dispersion stability of the resulting graphene using surfactants; (iii) the use of environmentally friendly and inexpensive reducing agents. Consequently, the availability of effective soft-chemistry approaches based on a green reducing agent for converting GO to r-GO are strongly needed. Among the green reductants, the most suitable is L-ascorbic acid (L-aa). Different studies have revealed that L-aa can achieve C/O ratio and conductivity values comparable to those obtained by hydrazine, a typical reducing agent. These aspects could promote an effective application strategy, and for this reason, this review summarizes and analyzes, in some detail, the up-to date literature on the reduction of GO by L-aa. The results are organized according to the two most important approaches, which are the reduction in liquid-phase, and the reduction in gel-phase. Reaction mechanisms and different experimental parameters affecting the processes were also compared.

Keywords: graphene oxide; reduced graphene oxide; L-ascorbic acid; liquid-phase reduction; gel-phase reduction

1. Introduction

Nowadays, the development of new advanced devices is strictly related to the availability of 2D materials, such as graphene-based materials, because of their unique mechanical, electrical, thermal, and optical properties [1–10]. Therefore, there is a strong need to develop new schemes for a massive synthesis of these materials, characterized by low-cost and sustainability [11]. Consequently, there is a growing interest in graphene oxide (GO) and its reduced form (r-GO), which is an inexpensive, versatile, printable, and biocompatible precursor of graphene-like materials [12]. GO consists of a layered material based on a carbon skeleton functionalized by different oxygen-containing groups (typically the C/O atomic ratio is less than 3) [13], having physical and chemical properties depending on different parameters, such as the type of oxygen-containing groups, the oxidation level, and the type of graphite used as precursor [14]. The most acceptable structural model proposed for GO is the Lerf–Klinowski model in which the basal planes of GO are decorated by hydroxyl and epoxide groups, whereas the edges are mainly occupied by carboxyl and carbonyl groups in a random manner [14–16]. The reduced graphene oxide (r-GO) obtained from the almost complete removal of the oxygen functional groups has mechanical, electrical, and optical properties quite similar to graphene, thus, resulting in a convenient 2D material useful for many technological applications [17–26].

According to the chemical structures of these layered materials, it is understood that interactions between GO or r-GO sheets and the environment (i.e., molecules, solvents, substrates, embedding polymers, etc.) are strictly related to the presence of oxygen-containing groups and their interaction (i.e., Van der Waals and electrostatic forces). Clearly, GO is predominantly hydrophilic while r-GO is hydrophobic, and the electrostatic interactions are stronger for GO, whereas the Van der Waals interactions play a major role in the case of r-GO because of the increase in unfunctionalized regions [27]. Moreover, the π–π stacking is another important interaction type among the sheets. In fact, each carbon atom of the basal plane bonds with three adjacent carbon atoms with sp^2 hybridized orbitals forming robust σ bonds, and the remaining electrons in the p atomic orbitals are delocalized all over the basal plane of the sheet forming a strong π bond, which makes possible the π–π stacking interaction [28].

Generally, the conversion from GO to r-GO requires some reducing agents [29–32], thermal treatment [33–36], laser-radiation [37–46], or bacterial methods [47–49]. Nowadays, different aspects are taken into consideration when selecting the best way to develop a massive and low-cost production of r-GO. Among the most important aspects, we can include the following: (i) the prospect of obtaining the highest reduction grade, (ii) the possibility of improving the dispersion stability of the resulting graphene using surfactants, (iii) the use of environmentally friendly and inexpensive reducing agents and solvents. So far, it has been proven that commonly used chemical reductants, such as hydrazine or hydrazine hydrate, are highly toxic and explosive, which can potentially induce environmental and safety risks [31–33,50]. Consequently, the availability of effective soft-chemistry approaches using a green reducing agent for converting GO to r-GO are strongly needed. Recently, green reductants, such as L-ascorbic acid (L-aa) [51–54], L-cysteine [55–57], glycine [58,59], green tea [60–62], various plant extracts [63–65], etc., have been studied as reductants for GO. Different studies have revealed that among all cited green chemical reductants, L-aa appears to be the outstanding candidate for achieving a C/O ratio and conductivity values that are comparable to those produced by hydrazine [54], having a mild activity and non-toxic properties. L-aa has proven to have a broad range of applications not only in bio-medical devices but also in electronic devices [51–54].

In summary, the advantages obtained by using L-aa as a reductant are the following:

- an environmentally friendly and non-toxic reductant;
- a highly efficient removal of the oxygen-groups;
- a low risk of introducing heteroatoms in the reaction products, because L-aa is composed only of carbon, oxygen, and hydrogen atoms.

In addition, preserving the environmentally friendly condition, the reduction process can be carried out in water, the most common and eco-friendly solvent [51]. However, this approach has the disadvantage of the irreversible formation of graphene agglomerates that are not useful for technological applications, and it is difficult to separate r-GO from the solvent and the by-products. Nevertheless, it is well known that a large amount of L-aa can stabilize the r-GO, thus, avoiding the addition of polymeric surface stabilizers and/or surfactants [66].

Recently, an innovative gel-phase technique has been developed to reduce a thin film of GO deposited on a substrate using the L-aa as reductant with the important advantages of avoiding the use of surfactants and being easily available for technological applications [67–70]. This approach overcomes the limitation of thermal reduction in the presence of a substrate. This limitation is represented by the operating temperature to be applied, which is dependent on the thermal stability of the substrate.

This review analyzes recent studies available in the literature concerning the reduction of GO by the use of an efficient and inexpensive green reductant (L-aa) under different experimental conditions. In addition, some possible mechanisms for the reduction of GO were also discussed. This review could be a useful reference for those scholars involved in this research area proposing its application as the most convenient new green method to reduce GO.

2. Reaction Conditions of Reduction of Graphene Oxide by L-aa

This section presents a concise description of the most important experimental results obtained for the reduction of GO by L-aa, their interpretation, as well as the possible filed application of the obtained material. To describe the evolution of this study clearly, the experimental results are presented chronologically and gathered according to the technique used for reduction.

2.1. Liquid-Phase Reduction

Zhang et al. (2009) first verified the possibility of developing the reduction of GO via L-aa [51]. The method developed was performed in an aqueous solution at room temperature under vigorous stirring. The study showed the monitoring of the reduction progress by optical absorption spectroscopy. The UV–vis spectrum of GO is typically characterized by the typical π–π^* transition peak at 233 nm from the C=C bond, and the n–π^* transition peak at around 300 nm from the C=O bond [34]. The red-shift of the π–π^* transition of GO originates related to the extension of its π-conjugated structure in the r-GO and the decreasing intensity of peaks centered at 300 nm, which is caused by the decrease in the C=O bond, allowing analysis of the reduction. The reduction that occurred was confirmed by FT-IR, Raman, and AFM results. The obtained r-GO, which showed a strongly restacked sheet arrangement with a wrinkled texture, provided a low specific surface area of 11.8 m^2/g and specific capacity of 128 F/g at a current density of 50 mA/g [51].

In the same period, Gao et al. (2010) presented a "green" reduction of GO using L-aa, analyzing the possible use of L-tryptophan as a stabilizer, to produce a stable dispersion of r-GO in an aqueous solution [66]. In brief, an aqueous solution of GO, L-aa, L-tryptophan, and NaOH was treated by ultrasonication at 80 °C for 24 h. After that, the mixture was cooled to room temperature, followed by another 1 h of ultrasonication. Thus, a large amount of stably dispersed aqueous r-GO was easily obtained. The experimental results showed the efficiency of L-tryptophan as a stabilizer to avoid the agglomeration and precipitation of the resulting r-GO sheets. L-tryptophan contains an electron-rich aromatic group that can function as an electron donor and be absorbed onto the r-GO sheet, based on the π–π interaction. In the meantime, the terminal carboxylic acid can supply enough negative charge, and the electrostatic repulsion can make the r-GO dispersions stable [66].

Successively, Fernández-Merino et al. (2010) [54] verified the use of L-aa to obtain stable suspensions of highly reduced r-GO in some common organic solvents, such as N,N-dimethylformamide (DMF) and N-methyl-2-pyrrolydone (NMP). In addition, the comparison of the deoxygenation efficiency of GO by different reductants (i.e., hydrazine, sodium borohydride, and pyrogallol) showed that only L-aa was found to yield highly reduced suspensions in a way that was comparable to those provided by hydrazine [54].

Sui et al. (2011) analyzed the effect of the amount of L-aa to reduced GO hydrogel and demonstrated that the mass ratio of L-aa to GO, temperature, and pH value of the reaction mixtures play a significant role in the formation of irreversible r-GO agglomerates in the form of hydrogels [71]. The hydrogels consist of a 3D cross-linked network of r-GO sheets self-assembling into a well-defined and interconnected 3D porous network through π–π interaction during gelation. Furthermore, this study demonstrated that r-GO in the form of a hydrogel with an excess of L-aa as a bioactive component can easily be used to release it in a diffusion-controlled manner. After the complete release of L-aa, the hydrogel exhibits excellent mechanical and electrical properties, and for this reason, it can be advantageously used in the fields of tissue engineering, drug delivery, soft machines, regenerative medicine, biosensors, etc. [71].

A complete study about the pH effect on the morphology of the as-prepared hydrogel r-GO was presented by Ha et al. (2019) [72]. This work shows that the reduction of GO in the liquid phase was completed in 1 h. The morphological characterization performed by scanning electron microscopy (SEM) demonstrated the formation of the 3D cross-linked spherical structure of r-GO by the hydrogel process with a diameter from 4 to 2 μm and with a specific surface area of 150 m^2/g, when the pH of the solution was 2. The experimental measurement showed that more spherical and compact structures were obtained at pH 10 with an improved specific surface area of 216 m^2/g.

The effect of pH on the degree of dispersion, packed with r-GO, can be explained by zeta potential analysis. The colloid of GO dispersion has a zeta potential −43 mV when the pH is 10 [73]. It is well kwon that Zeta potential values greater than −30 mV are generally considered to exhibit sufficient mutual repulsion to ensure the stability of dispersion [74]. When the droplets are generated from the GO colloid fabricated at pH 10, a more densely compacted structure of r-GO can be formed by the capillary-force-driven self-assembly of well dispersed GO sheets in the droplet during the solvent evaporation. Similar experimental results for the formation of the r-GO hydrogels were confirmed by Kondratowicz et al. (2017) [75]. This study explained that the by-products of reduction, such as dehydroascorbic acid and water molecules, may form additional hydrogen bonds with residual carboxylic groups on r-GO planes and contribute to the final structure of hydrogels [75].

Successively, the influence of pH and surfactants used for reduction by L-aa on the adsorption mechanism of organic contaminants, such as phenanthrene (a representative nonpolar, nonionic, and aromatic contaminant) and 1-naphthol (a representative polar, aromatic contaminant) was studied by Wang et al. in 2019 [76]. The study shows that the pH of the solution had a negligible effect on phenanthrene adsorption for both GO and r-GO, and it inhibited 1-naphthol adsorption at high pH because of the electronic repulsion. This was mainly attributed to the hydrophobic interaction, π–π interaction, and H-bonding between graphene sheets and organic contaminants. The same study shows that the use of the surfactants had different effects on the adsorption of polar and nonpolar aromatics onto graphene materials. For sodium dodecyl benzene sulfonate (SDBS), the exfoliation effect could enhance the adsorption affinity; thus, it could counteract the inhibition effect caused by competition. Cetyltrimethylammonium bromide (CTAB) could form hemimicelles on reduced graphene oxide, which may provide a favorable media for organic contaminant partitioning. In addition, this study shows that the r-GO could be regenerated and reused with high recyclability over five cycles. These findings could provide a promising material for wastewater treatment and the understanding of the fate and transport of organic contaminants in aquatic environments.

De Silva et al. (2018) [77] monitored the reduction of GO by L-aa in an aqueous solution in 10-min intervals up to 1 h in order to study how structural and morphological changes would take place. The reduced products obtained at different time periods were characterized in detail by UV–visible spectroscopy, X-ray diffraction (XRD), X-ray photoelectron spectroscopy (XPS), attenuated total reflectance Fourier transform infrared (ATR-FT-IR) spectroscopy, Raman spectroscopy, thermogravimetric analysis (TGA), atomic force microscopy (AFM), and scanning electron microscopy (SEM). The UV–visible spectra displayed a complete removal of the GO peak by 50 min, while other characterization techniques revealed the presence of residual oxygen functionalities. In particular, the XPS results showed that the decline in the oxygen atomic percentage was mainly due to the removal of the hydroxyl and epoxy groups located at the basal planes of the GO sheets and, to a small extent, due to edge carbonyl groups. AFM characterization indicated that at the intermediate stages of reduction, both GO and r-GO coexist in the material, as confirmed by XRD results.

The study of the impact of ultrasounds on the rate of GO reduction in L-aa aqueous solutions was carried out by Abulizi et al., 2014 [78]. They found that the r-GO formation under ultrasound treatment was accelerated in comparison with the conventional mechanical mixing treatment. To understand the effects of ultrasound, the authors compared the experimental results on the trend rates of r-GO formation, as a function of temperature, under ultrasound and mixing treatment, showing that this rate was increased by ultrasound treatment. The authors proposed that physical effects such as shear forces, microjets, and shock waves during acoustic cavitation enhanced the mass transfer and reaction of L-aa with GO to form r-GO, as well as the change in the surface morphology of GO. Furthermore, the rates of r-GO formation were suggested to be affected by local high temperatures of cavitation bubbles [78].

Similarly, the increase in kinetics and in the GO reduction degree under UV irradiation were analyzed by Go et al., 2018 [79]. They demonstrated that when the reactant solution was placed under UV irradiation (254 nm), L-aa improved its chemical activity caused by its UV-sensitive oxidation property [80]. In particular, the reduction was performed under various conditions, (i) without any reducing agent, (ii) using L-aa, (iii) using L-aa under UV irradiation (254 nm), and monitored by using UV–visible spectroscopy up to 24 h to explore the effect of UV irradiation on the reduction of GO by L-aa. The evolution of UV–Visible spectra for these different conditions, confirmed that the UV irradiation (254 nm) improves the activity of L-aa.

In fact, the UV-visible spectrum of GO showed the typical π–π^* transition peak at 233 nm from the C=C bond, and the n–π^* transition peak at around 300 nm from the C=O bond [33]. The red-shift of the GO π–π^* transition can be used as an indicator of its reduction [51]. Consequently, analyzing the trend of this shift for all three different reduction conditions, it is possible to deduce the advantageous effect of UV radiation on a reduction of GO [79,80].

To further understand the commonality of the procedures developed and presented in this review, Table 1 lists the main information about the parameters used in all reported studies. In particular, Table 1 lists the solvent used, the weight ratio between GO and L-aa, the pH value of solution, the presence of a stabilizer, the temperature and time of reaction, and the main experimental characterizations.

Table 1. Summary of the principal parameters used in the cited methods of the liquid-phase reduction of GO.

Solvent	GO ÷ L-aa	Base for pH	Stabilizer	Temperature and Time	Characterization	Reference
water	0.05 ÷ 50	—	—	23 °C for 48 h under stirring	UV–Visible, AFM, TEM, Raman, FT-IR	Zhang et al., 2009 [51]
water	0.2 > 10	pH 9–10 with NaOH	L-tryptophan	80 °C for 24 h under sonication for 0.5 h	AFM, XPS, Raman	Gao et al., 2010 [66]
water, NMP, DMF	—	in case of water pH 9–10 with 25% NH_4OH	—	95 °C for 15, 30, 180, 240 min	ATR-FTIR, XPS, TGA, UV–Vis	Fernández-Merino et al. [54]
water	From 1:832 To 8:1	pH 7–8 with several drops of 0.1 M HCl solution	—	Without stirring, for a few hours, from 25 °C to 80 °C	XPS, SEM	Sui et al. [71]
water	1.2	pH 2–10 was adjusted using $NH_3{\cdot}H_2O$		various reaction times of 5, 15, 30, and 60 min, T 95 °C	SEM, UV–Vis	Ha et al. [72]
water	—	—	SDBS, CTAB	stirred at 500 rpm for 72 h	XPS, FT-IR, Raman	Wang et al. [76]
water	—	pH to 10 with $NH_3{\cdot}H_2O$	—	Under stirring, 65 °C, 1 h	XRD, SEM, XPS	De Silva et al., 2018 [77]
water	—	pH to 11 with NaOH	—	High-power horn-type ultrasound sonicator	UV–Vis, FT-IR, SEM	Abulizi et al., 2014 [78]
water	—	pH to 10 with $NH_3{\cdot}H_2O$	—	UV light (254 nm) was irradiated at 2 mW/cm^2 power for 24 h	UV–Vis, FT-IR, Raman, XPS	Go et al., 2018 [79]

In many cases, the approach to reduce GO in an aqueous solution by L-aa was used to prepare functional and advantageous nanocomposite materials based on r-GO for many application fields. A few examples are shown below.

Ding et al. in 2015 [81] used L-aa to realize r-GO–TiO_2 composite films as a photoanode in DSSC. The experimental results demonstrated a 30% increase in conversion efficiency compared to that of the pure TiO_2 photoanode [81].

Dan et al. in 2018 [82] examined three types of r-GO/polyhydroxy butyrate (PHB) composites by employing three reducing agents: sodium borohydride, hydrazine, and L-ascorbic acid. The electrical properties of the r-GO/PHB composites achieved by L-aa were comparable to the best values known in the bio-composite field.

As far as we know, studies associated with the utilization of r-GO in osteogenesis indices by electrical stimulation have rarely been reported. Xiong et al. in 2017 [83] presented the study on the preparation of reduced graphene oxide/zinc silicate/calcium silicate (r-GO/ZS/CS) by L-aa with an optimal surface electroconductivity. The conductive biocomposite obtained was analyzed in vitro osteogenesis of mouse bone mesenchymal stem cells.

2.2. Gel-Phase Reduction

Recently, a gel-phase technique has achieved a growing interest. This technique is based on the reduction of a thin film of GO deposited on a substrate by the diffusion of L-aa molecules in it. The peculiarity of this approach is that GO is not dispersed in a solvent, but it is swollen by water, and it persists in the form of coating stacked to the substrate, and the chemical interaction with L-aa takes place by permeation of the reductant in these lamellar structures. This approach represents an important technological breakthrough because the development of many functional devices can be made directly by the reduction of a large area of GO coating deposited on selected substrates (i.e., polymers, glass, etc.) [52,68–70,84]. In addition, considering that this approach allows the GO reduction by L-aa at a low temperature, it is possible to overcome the temperature limitation due to the thermal stability of the substrate. This method preserves the properties of substrates, and it avoids the use of surfactants.

According to the literature, only a few methods have been developed for the reduction of GO thin films. Two approaches for reducing GO film deposited on a substrate were described: the first one is based on the dipping of supported GO in a reducing L-aa solution, and the second method requires the exposure of supported GO to vapors of a reducing L-aa solution. Both procedures, which could need a controlled temperature, are shown in Figure 1.

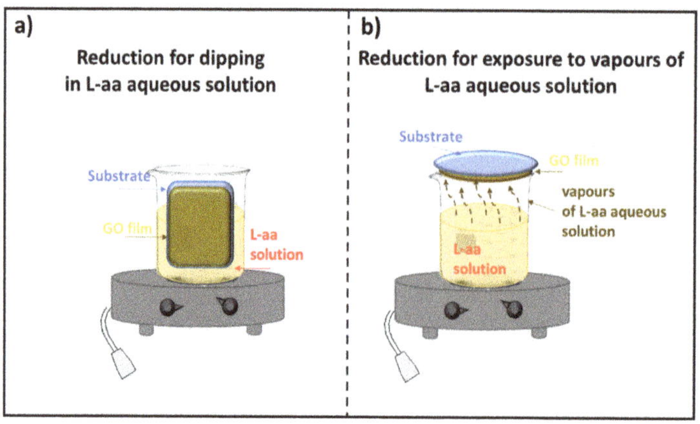

Figure 1. Scheme of the reduction of a GO film deposited on a substrate (**a**) by dipping; (**b**) by exposure to vapors of a L-aa solution.

Liu et al. in 2015 [52] first reported the use of L-ascorbic acid/water vapor as a reducing agent for GO films. In this paper, GO on cellulose was placed on the top of a glass bottle in a Teflon-lined autoclave containing different concentrations of an aqueous solution of L-aa. Finally, the autoclave was heated at 100 °C for 48 h. The same procedure was used to prepare r-GO–Ag composites that can be used for active substrate surface-enhanced Raman scattering and as antibacterial material.

In 2016, Li et al. [70] published one article concerning the preparation of porous r-GO membranes reducing GO on copper hydroxide nano-strand freestanding membranes by dipping in a 60 mL L-aa aqueous solution heated at 90 °C for 4 h. The results confirmed that a porous r-GO membrane, fabricated from a graphene oxide sheet via etching copper hydroxide nano-strands by L-aa reduction, provides an effective structural configuration for enhancing its gauge factor.

Tas et al. in 2019 [69] proposed the reduction of graphene oxide thin films deposited on glass by dipping in an opportune solution of the L-aa at low-temperature. To compare the effectiveness of the reduction process, hydrazine hydrate was also used as a chemical reducing agent following the same method. The results have shown that this reduction process, which does not contain heavy toxic chemicals and does not require nitrogen, argon, etc., is more successful.

Chen et al. in 2020 [84] reported on the preparation of cellulose/r-GO aerogels for the development of chemical vapor sensors. For the GO reduction, the cellulose/GO hydrogels were put in an aqueous solution of L-aa at 95 °C for 2 h. Sensors based on these aerogels exhibited fast response, good recovery, high sensitivity, and excellent reproducibility. The inexpensive, easy, green, and scalable preparation of this new type of vapor sensor could be expected to lead to new sensing and biomedical applications.

In the same year, Longo et al. [67,68] published a new method for a green gel-state chemical reduction of GO supported on cellulose substrates. The possibility of having an effective mass transport of the reductant inside the swollen GO deposit was ensured by spraying a reducing solution of L-aa on the GO film, allowing it to reflux for 48 h in a closed microenvironment at 50 °C. A scheme of the apparatus used for reduction is shown in Figure 2.

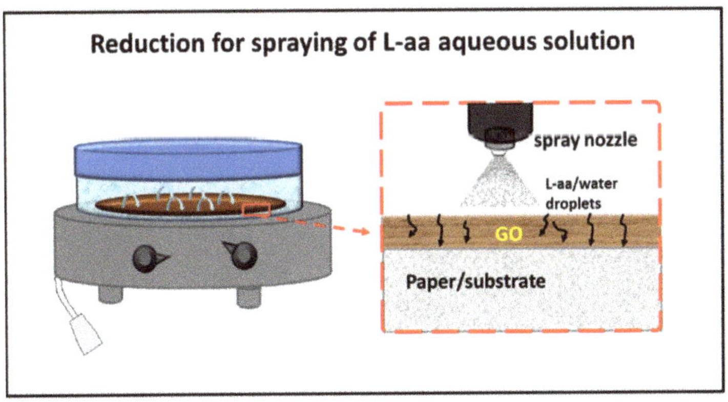

Figure 2. Scheme of the reduction of GO film deposited on paper/substrate by spraying of a L-aa aqueous solution [67].

The principal information on the parameters selected in the above-described gel-phase approach are summarized in Table 2. In particular, the method used to reduce the GO on the substrate, the concentration of the L-aa in water, the selected substrate, the temperature and time of reaction, and the characterization techniques are given.

Table 2. Summary of principal parameters used in the cited methods of the gel-phase reduction of GO.

Methods	Concentration of L-aa	Substrate	Temperature and Time	Characterizations	Reference
Exposure to vapors	Different concentrations	Cellulose	100 °C for 48 h	UV–Visible, FT-IR, Raman, AFM, XPS	Lui et al., 2015 [52]
Dipping	30 mg/mL of L-aa in water	Freestanding membrane	90 °C for 4 h	UV–Vis	Li et al., 2016 [70]
Dipping	Different concentrations	Glass	95 °C for 15 min	XRD, Raman, XPS	Tas et al. [69]
Dipping	30 mg/mL of L-aa in water	Paper	95 °C for 2 h	XRD, Raman, XPS	Chen et al. [84]
Spraying		Paper	50 °C for 48 h	XRD, FT-IR, SEM	Longo et al., 2020 [67] Palomba et al., 2021 [68]

The Reduction of GO Film Deposited on Paper/Substrate Spraying of L-aa Aqueous Solution

This gel-state reduction technique, based on spraying an L-aa aqueous solution, represents a convenient approach for a complete reduction of the GO layers supported on thermally unstable substrates [67,68]. The most important experimental results published in previous manuscripts [67,68] can be summarized as follows. In addition, in order to improve the previous results, a quantitative analysis of the degree of oxidation of GO before and after the reduction is presented.

According to the thermogravimetric investigation, the process temperature selected (i.e., 50 °C) is necessary to increase the mobility of the water and L-aa molecules in the GO inter-layers [68]. The SEM investigation confirmed a structural modification of the GO coating after the treatment, mainly consisting of an increase in the coating flatness. In addition, SEM confirmed a strong interfacial adhesion between the GO/r-GO coating and the fibrous substrate. This micro-structural characteristic, due to an excellent adhesion at the GO–paper interface, is relevant for achieving a highly flexible r-GO layer supported on paper, and it is vitally necessary for industrial exploitation [68].

The XRD results of r-GO/paper show the presence of the main peaks of the r-GO pattern combined with the XRD signals of residual GO. In particular, the results showed that the obtained r-GO coating is composed of platelets with an average thickness of ca. 27 nm and a width of ca. 40 nm, which are aligned parallel to the interfacial plane and show good graphitic quality [67]. These experimental results confirmed the obtained reduction of GO on paper only qualitatively.

To establish the degree of oxidation of GO before and after the reduction, the Fourier transform infra-red spectroscopy was recorded in attenuated total reflectance (ATR) mode in the 4000–700 cm^{-1} range by using a spectrophotometer (PerkinElmer Frontier NIR, Milan, Italy).

The ATR spectra (see Figure 3) showed that most of the peaks referring to oxygen-containing groups are present in the GO layer. In addition, the broad peak in the spectra in the region 3740–3100 $cm^{-1}$and the peak at $\approx$1640 cm^{-1} can be referring to the water molecules absorbed on the interlayers of GO. Table 3 details the main peaks measured by spectra.

To perform a quantitative analysis of the degree of oxidation of GO before and after the reduction by ATR spectra, the following procedure proposed by Guerrero-Contreras et al. [85,86] was used:

- A polynomial baseline was calculated and subtracted from the raw spectra.
- The resulting spectra were multiplied by -1 in order to have positive bands.
- The peak deconvolution was obtained by Gaussian fit to achieve the peak area.

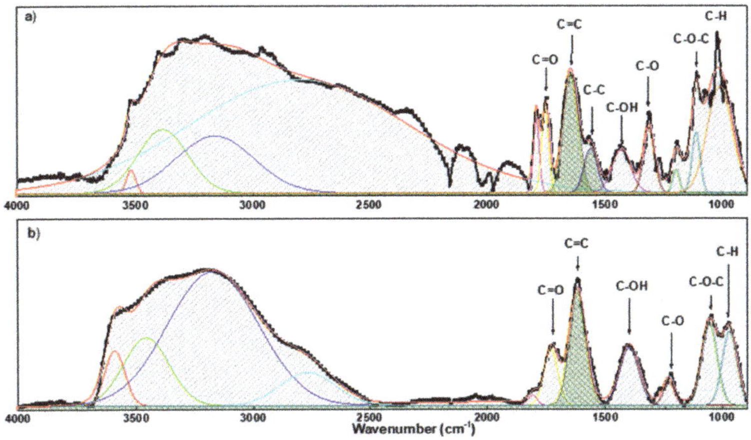

Figure 3. FT-IR spectrum of GO on paper/substrate (**a**) and r-GO on paper/substrate (**b**).

Table 3. ATR peaks assignments for GO and r-GO on paper.

Vibration Mode	GO Position	r-GO Position
O-H stretching, water molecules	3740–3100	3740-3100
C-H stretching	2800–2600	2766
C=O stretching carbonyl functional groups located on the edge of the sheets (yet COOH and C=O); C=C aromatic skeletal stretching	1753 1647	1726 1620
CH_2 deformation stretching	1433	1415
C-OH stretching bending mode of hydroxyl groups	1431	1404
C-OH stretching	1239	1227
C-O-C stretching epoxy groups	1150	1147
C-H bending aromatic skeletal	971	965

The comparison between the spectrum before and after reduction shows a general decrease in intensity of the peaks related to oxygen functional groups. At this point, the degree of oxidation of GO was evaluated by calculating the relative percentage of oxygen-containing functional groups (*RPox*) compared to the presence of all functional groups observed in the wavenumber range of 900−1850 cm^{-1} (for all peaks in Figure 3 [85,86]). *RPox* was calculated using the following formula:

$$RPox = \left[\frac{sum\ of\ area\ of\ all\ peaks - area\ od\ C = C\ peak}{sum\ of\ area\ of\ all\ peaks}\right] * 100 \quad (1)$$

The analysis reveals that the *RPox* decreases from 68% to 37% after the reduction.

3. Reduction of GO by L-aa

Nowadays, the mechanism of the chemical reaction between GO and L-aa is not completely understood. This section provides a detailed description of the mechanism given in the literature for the GO reduction by L-aa and describes the effect of UV light on this mechanism.

3.1. Possible Mechanism of Reduction

According to the literature [29,66], two different reactions are involved in the reduction of GO steps. The first reaction involves the reduction of GO vicinal-hydroxyls by L-a. The second one involves the reduction of epoxy groups by L-aa. Both reaction pathways are shown in Figure 4. In particular, the electron density withdrawing from the five-membered ring of L-aa makes the hydroxyls contained in this molecule much more acidic, consequently the L-aa can dissociate, providing two protons, which are transferred to GO, while nucleophilic species (i.e., the oxyanion of L-aa: $C_6H_7O_6^-$) are generated. Figure 4 shows that both reacting sites require the formation of a good leaving group, which is a hydroxyl group in the case of epoxies and a water molecule in the case of vicinal-hydroxyl groups. These types of leaving group are generated by protonation of the cited GO reactive groups (i.e., vicinal hydroxyls and epoxy groups). After this preliminary acid–base reaction, a SN2 reaction step follows. In this reaction, the nucleophilic agent attacks the Sp^2-carbon of the epoxy group or the α-Sp^3-carbon of the alcoholic groups, and hydroxyl or water results as a by-product, respectively. In the case of the reaction involving epoxy, a further condensation by the SN2 mechanism follows. Then, this intermediate undergoes a thermally induced red-ox reaction, which leads to the formation of: reduced graphene oxide (r-GO), and as by-products, dehydroascorbic acid and water molecules [29,66]. Some studies report that dehydroascorbic acid ($C_6H_6O_6$) can be further converted to guluronic and oxalic acids [59,86], and then CO_2, CO, and water are generated during GO reduction to r-GO.

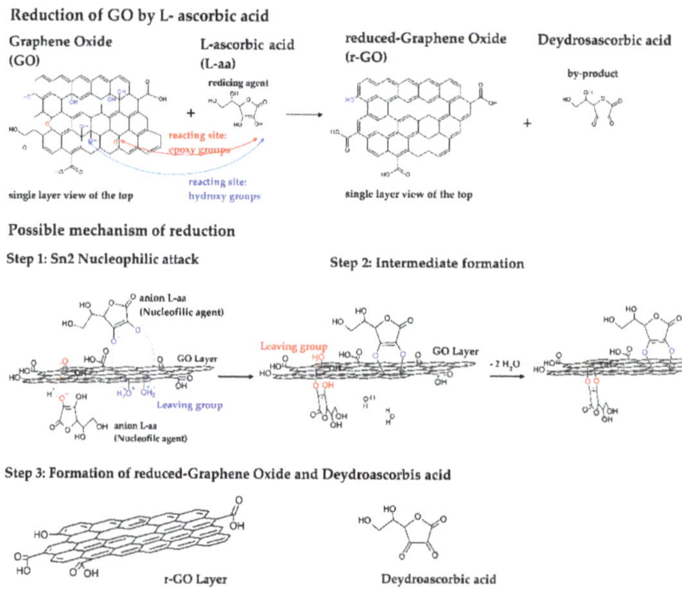

Figure 4. Schematic representation of the reaction pathway [67].

A similar chemical mechanism, proposed by Longo et al., 2020 [67], took place when a L-aa aqueous solution is sprayed on a GO thin film. The presence of a large amount of oxygenated functional groups on GO makes the material very hydrophilic and, thus, provides it with the capability to absorb water and swell [87,88]. In fact, the absorption of water molecules on the GO layer by physical interaction with the epoxide and/or hydroxyl groups is a well-known phenomenon (see Figure 5). This absorption gives the layered stacked GO a natural tendency to swell as a consequence of the enlarged inter-layer spacing. At room temperature, a slow diffusion rate of L-aa molecules, which always characterizes a gel-phase reduction, and consequently, a lower value of the reaction rate, is expected. To increase both the mobility of water and the corresponding mobility of L-aa molecules in the GO channels, the process temperature was increased to 50 °C.

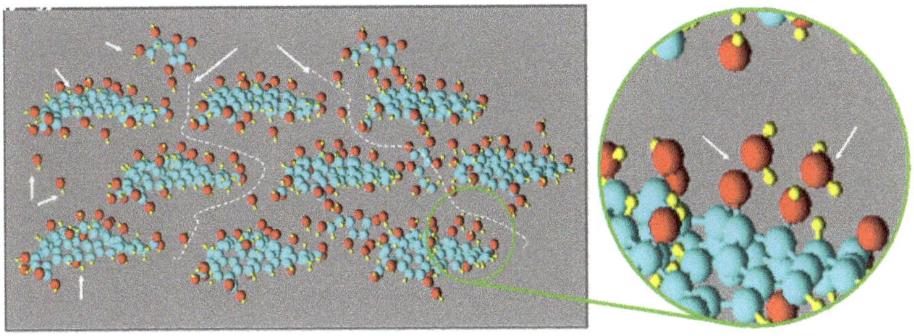

Figure 5. Representation of swollen GO and the diffusion pathway of a L-aa aqueous solution.

3.2. Possible Influence of UV-Irradiation

This eco-friendly approach can be improved by using L-aa as a photosensitive reducing agent. Go et al. demonstrated that when L-aa was excited by UV irradiation at 254 nm, it oxidized with simultaneous deprotonation and this proton-coupled electron transfer was capable of inducing a chemical reduction of GO [79,88]. Therefore, UV irradiation of L-aa is expected to accelerate the reduction of GO. This method can be advantageously used for developing eco-friendly and scalable processes [79,88].

4. Discussion, Conclusions, and Future Perspectives

The aim of this review is to highlight the potential for using L-aa as a green reducing agent to improve eco-friendly and large-scale production of r-GO. Liquid-phase and gel-phase reductions were briefly discussed here. As far as the first approach is concerned, experimental results have demonstrated the advantageous use of some factors in improving the reduction process shown in Figure 6. Higher temperature, the use of sonication, and exposure under UV radiation are factors that (separately or simultaneously) allow an increase in reduction. Furthermore, this review presents an analysis of the role of the pH of the reaction mixture on the formation of irreversible r-GO agglomerates and the use of surfactants to modify the adsorption properties of r-GO. Many studies have demonstrated that this approach can be advantageously used to obtained functional nanomaterials based on r-GO

More recently, there has been a growing interest in the potential for developing reduction processes in the gel-phase. This approach overcomes some limitations of the reduction of GO in the liquid-phase, such as the isolation of r-GO from the solution. In addition, this approach can be used to reduce a large area of the GO coating deposited on selected substrates (i.e., polymers, glasses, etc.), and for this reason, the approach may provide many technological breakthroughs.

Table 4 compares the two approaches by outlining and summarizing the pros and cons for both.

Finally, the review shows the efficiency of L-aa as a reductant agent. In the future, this green reduction method of GO may provide fascinating results in terms of graphene quality, size, and also production.

The aim of this review was to collect all the literature papers concerning the most important approach for the GO reduction based on the use of L-aa. The achieved r-GO quality allows for the technological exploitation of this nanostructure in a variety of forms (coating, self-supported, embedded in a polymer, etc.). These factional materials have potential applications for flexible electronics, sensing applications, optics applications, etc. In particular, the use of this green chemical method is an emerging technology of a fundamental importance for the production of large-area, lightweight, low cost, and mechanically stable devices. The knowledge of all aspects related to the synthesis and properties of r-GO obtained by the L-aa reduction technique is a critical point for bringing this process to mass production.

Table 4. Summary of principal advantages and disadvantages of liquid-phase and gel-phase approaches.

Liquid-Phase Reduction		*Gel-Phase Reduction*	
Advantage	**Disadvantage**	**Advantage**	**Disadvantage**
✓ Uniform degrees of reduction	• Separation of r-GO from solvent	✓ Treatment of a large area	• Need to control the diffusion of L-aa to have a uniform degree of reduction in the coating
✓ Control of morphology	• Re-dispersion of r-GO in selected matrix to possible applications	✓ Preparation of r-GO coating in one-step	
		✓ Use of thermally unstable substrates	

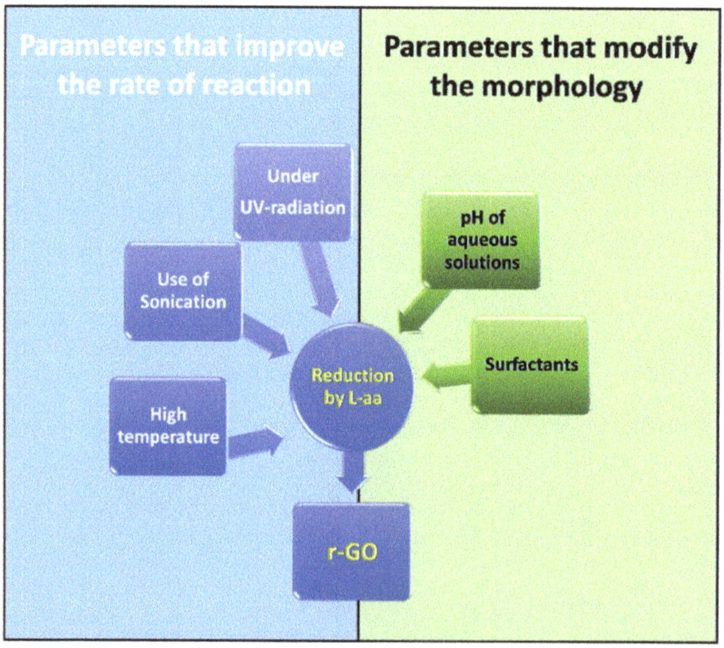

Figure 6. Parameters that act on the reduction mechanism.

Author Contributions: Conceptualization, A.L.; validation, A.L., G.C. and M.P.; formal analysis, A.L.; investigation, A.L.; resources, A.L.; data curation, A.L.; writing—original draft preparation, A.L.; writing—review and editing, A.L., M.P. and G.C. All authors have read and agreed to the published version of the manuscript.

Funding: This research received no external funding.

Acknowledgments: The authors are grateful to Maria Cristina Del Barone of LAMEST laboratory (IPCB-CNR) and to Maria Rosaria Marcedula of Thermo-Analysis Laboratory (IPCB-CNR) for experimental measurements.

Conflicts of Interest: The authors declare no conflict of interest.

References

1. Shafraniuk, S. *Graphene: Fundamentals, Devices, and Applications*, 1st ed.; Jenny Stanford Publishing: Dubai, United Arab Emirates, 2015; ISBN 9789814613477.
2. Geim, A.K. Graphene: Status and prospects. *Science* **2009**, *324*, 1530–1534. [CrossRef] [PubMed]
3. Geim, A.K.; Novoselov, K.S. The rise of graphene. *Nat. Mater.* **2007**, *6*, 183–191. [CrossRef] [PubMed]
4. Novoselov, K.S.; Jiang, Z.; Zhang, Y.; Morozov, S.V.; Stormer, H.L.; Zeitler, U.; Maang, J.C.; Boebingerp, G.S.; Geim, A.K. Room-temperature quantum hall effect in graphene. *Science* **2007**, *315*, 1379. [CrossRef] [PubMed]
5. Novoselov, K.S.; Geim, A.K.; Morozov, S.V.; Jiang, D.; Zhang, Y.S.; Dubonos, V.; Grigorieva, I.V.; Firsov, A.A. Electric field effect in atomically thin carbon films. *Science* **2004**, *306*, 666–669. [CrossRef] [PubMed]
6. Tiwari, S.K.; Sahoo, S.; Wang, N.; Huczko, A. Graphene research and their outputs: Status and prospect. *J. Sci.-Adv. Mater. Dev.* **2020**, *5*, 10–29. [CrossRef]
7. Rao, C.N.R.; Sood, A.K.; Subrahmanyam, K.S.; Govindaraj, A. Graphene: The New Two-Dimensional Nanomaterial. *Angew. Chem.* **2009**, *48*, 7752–7777. [CrossRef]
8. Liu, Y.; Zhang, S.; He, J.; Wang, Z.M.; Liu, Z. Recent Progress in the Fabrication, Properties, and Devices of Heterostructures Based on 2D Materials. *Nano-Micro Lett.* **2019**, *11*, 13. [CrossRef]
9. Nurazzi, N.M.; Abdullah, N.; Demon, S.Z.N.; Halim, N.A.; Azmi, A.F.M.; Knight, V.F.; Mohamad, I.S. The frontiers of functionalized graphene-based nanocomposites as chemical sensors. *Nanotechnol. Rev.* **2021**, *10*, 330–369. [CrossRef]
10. Kavitha, M.K.; Jaiswal, M. Graphene: A review of optical properties and photonic applications. *Asian J. Phys.* **2016**, *25*, 809–831.
11. Zhu, Y.; Ji, H.; Cheng, H.M.; Ruoff, R.S. Mass production and industrial applications of graphene materials. *Natl. Sci. Rev.* **2018**, *5*, 90–101. [CrossRef]
12. Yan, Y.; Nashath, F.Z.; Chen, S.; Manickam, S.; Lim, S.S.; Zhao, H.; Lester, E.; Wu, T.; Pang, C.H. Synthesis of graphene: Potential carbon precursors and approaches. *Nanotechnol. Rev.* **2020**, *9*, 1284–1314. [CrossRef]
13. Bianco, A.; Cheng, H.-M.; Enoki, T.; Gotosi, Y.; Hurt, R.H.; Koratkar, N.; Kyotani, T.; Monthioux, M.; Park, C.R.; Tascon, J.M.D.; et al. All in the graphene family—A recommended nomenclature for two-dimensional carbon materials. *Carbon* **2013**, *65*, 1–6. [CrossRef]
14. Pei, S.; Cheng, H.-M. The reduction of graphene. *Carbon* **2012**, *50*, 3210–3228. [CrossRef]
15. Lerf, A.; He, H.; Forster, M.; Klinowski, J. Structure of graphite oxide revisited. *J. Phys. Chem. B* **1998**, *102*, 4477–4482. [CrossRef]
16. He, H.; Klinowski, J.; Forster, M.; Lerf, A. A new structural model for graphite oxide. *Chem Phys Lett.* **1998**, *287*, 53–56. [CrossRef]
17. Smith, A.T.; LaChance, A.M.; Zeng, S.; Liu, B.; Sun, L. Synthesis, properties, and applications of graphene oxide/reduced grapheneoxide and their nanocomposites. *Nano Mater. Sci.* **2019**, *1*, 31–47. [CrossRef]
18. Roweley-Neale, S.J.; Randviir, E.P.; Abo Dena, A.S.; BanKs, C.E. An overview of recent applications of reduced graphene oxide as a basis of electroanalytical sensing platforms. *Appl. Mater. Today* **2018**, *10*, 218–226. [CrossRef]
19. Ray, S. *Applications of Graphene and Graphene-Oxide Based Nanomaterials Micro and Nano Technologies*; William Andrew—Elsevier: Oxford, UK, 2015.
20. Zhu, Y.; Murali, S.; Cai, W.; Li, X.; Suk, J.W.; Potts, J.R.; Ruoff, R.S. Graphene and Graphene Oxide: Synthesis, Properties, and Applications. *Adv. Mater.* **2010**, *22*, 3906–3924. [CrossRef]
21. Chen, D.; Feng, H.; Li, J. Graphene Oxide: Preparation, Functionalization, and Electrochemical Applications. *Chem. Rev.* **2012**, *112*, 6027–6053. [CrossRef]
22. Wan, X.; Huang, Y.; Chen, Y. Focusing on Energy and Optoelectronic Applications: A Journey for Graphene and Graphene Oxide at Large Scale. *Acc. Chem. Res.* **2012**, *45*, 598–607. [CrossRef]
23. Schniepp, H.C.; Li, J.-L.; McAllister, M.J.; Sai, H.; Herrera-Alonso, M.; Adamson, D.H.; Prud'homme, R.K.; Car, R.; Saville, D.A.; Aksay, I.A. Functionalized single graphene sheets derived from splitting graphite oxide. *J. Phys. Chem. B* **2006**, *110*, 8535–8549. [CrossRef] [PubMed]
24. Mkhoyan, K.A.; Contryman, A.W.; Silcox, J.; Stewart, D.A.; Eda, G.; Mattevi, C.; Miller, S.; Chhowalla, M. Atomic and electronic structure of graphene-oxide. *Nano Lett.* **2009**, *9*, 1058–1063. [CrossRef] [PubMed]
25. Dreyer, D.R.; Park, S.; Bielawski, C.W.; Ruoff, R.S. The chemistry of graphene oxide. *Chem. Soc. Rev.* **2009**, *39*, 228–240. [CrossRef]
26. Pandey, D.; Reifenberger, R.; Piner, R. Scanning probe microscopy study of exfoliated oxidized graphene sheets. *Surf. Sci.* **2008**, *602*, 1607–1613. [CrossRef]
27. Simsikova, M.; Sikola, T. Interaction of Graphene Oxide with Proteins and Applications of their Conjugates. *J. Nanomed. Res.* **2017**, *5*, 00109. [CrossRef]
28. Guex, L.G.; Sacchi, B.; Peuvot, K.F.; Andersson, R.L.; Pourrahimi, A.M.; Ström, V.; Farris, S.; Olsson, R.T. Experimental review: Chemical reduction of graphene oxide (GO) to reduced graphene oxide (rGO) by aqueous chemistry. *Nanoscale* **2017**, *9*, 9562–9571. [CrossRef] [PubMed]
29. Chua, C.K.; Pumera, M. Chemical reduction of graphene oxide: A synthetic chemistry viewpoint. *Chem. Soc. Rev.* **2014**, *43*, 291–312. [CrossRef]
30. Furst, A.; Berlo, R.C.; Hooton, S. Hydrazine as a Reducing Agent for Organic Compounds (Catalytic Hydrazine Reductions). *Chem. Rev.* **1965**, *65*, 51–68.
31. Chua, C.K.; Pumera, M. The reduction of graphene oxide with hydrazine: Elucidating its reductive capability based on a reaction-model approach. *Chem. Commun.* **2016**, *52*, 72–75. [CrossRef]

32. Park, S.; An, J.; Potts, J.R.; Velamakanni, A.; Murali, S.; Ruoff, R.S. Hydrazine-reduction of graphite- and graphene oxide. *Carbon* **2011**, *49*, 3019–3023. [CrossRef]
33. Seung, H.H. *Thermal Reduction of Graphene Oxide—Physics and Applications of Graphene—Experiments*; Mikhailov, S., Ed.; InTech: Rijeka, Croatia, 2011. [CrossRef]
34. Longo, A.; Carotenuto, G. Graphene oxide reduction by microwave heating. *AIP Conf. Proc.* **2016**, *1736*, 020164–020168.
35. Jakhar, R.; Yap, J.E.; Joshib, R. Microwave reduction of graphene oxide. *Carbon* **2020**, *170*, 277–293. [CrossRef]
36. Chaban, V.V.; Prezhdo, O.V. Microwave reduction of graphene oxide rationalized by reactive molecular dynamics. *Nanoscale* **2017**, *9*, 4024–4033. [CrossRef]
37. Qiu, Y.; Guo, F.; Hurt, R.; Külaots, I. Explosive thermal reduction of graphene oxide-based materials: Mechanism and safety implications. *Carbon* **2014**, *72*, 215–223. [CrossRef]
38. Slobodian, O.M.; Lytvyn, P.M.; Nikolenko, A.S.; Naseka, V.M.; Khyzhun, O.Y.; Vasin, A.V.; Sevostianov, S.V.; Nazarov, A.N. Low-Temperature Reduction of Graphene Oxide: Electrical Conductance and Scanning Kelvin Probe Force Microscopy. *Nanoscale Res. Lett.* **2018**, *13*, 139. [CrossRef] [PubMed]
39. Abdelsayed, V.; Moussa, S.; Hassan, H.; Aluri, H.S.; Collinson, M.M.; El-Shall, M.S. Photothermal deoxygenation of graphite oxide with laser excitation in solution and graphene-aided increase in water temperature. *J. Phys. Chem. Lett.* **2010**, *1*, 2804–2809. [CrossRef]
40. Gao, W.; Singh, N.; Song, L.; Liu, Z.; Reddy, A.L.; Ci, L.; Vajtai, R.; Zhang, Q.; Wei, B.; Ajayan, P.M. Direct laser writing of micro-supercapacitors on hydrated graphite oxide films. *Nat. Nanotechnol.* **2011**, *6*, 496–500. [CrossRef] [PubMed]
41. Zhou, Y.; Bao, Q.; Varghese, B.; Tang, L.A.; Tan, C.K.; Sow, C.H.; Loh, K.P. Microstructuring of graphene oxide nanosheets using direct laser writing. *Adv. Mater.* **2010**, *22*, 67–71. [CrossRef] [PubMed]
42. Zhang, Y.; Guo, L.; Wei, S.; He, Y.; Xia, H.; Chen, Q.; Sun, H.-B.; Xiao, F.-S. Direct imprinting of microcircuits on graphene oxides film by femtosecond laser reduction. *Nano Today* **2010**, *5*, 15–20. [CrossRef]
43. Sokolov, D.A.; Shepperd, K.R.; Orlando, T.M. Formation of graphene features from direct laser-induced reduction of graphite oxide. *J. Phys. Chem. Lett.* **2010**, *1*, 2633–2636. [CrossRef]
44. Sokolov, D.A.; Rouleau, C.M.; Geohegan, D.B.; Orlando, T.M. Excimer laser reduction and patterning of graphite oxide. *Carbon* **2013**, *53*, 81–89. [CrossRef]
45. Longo, A.; Verrucchi, R.; Aversa, L.; Tatti, R.; Ambrosio, A.; Orabona, E.; Coscia, U.; Carotenuto, G.; Maddalena, P. Graphene oxide prepared by graphene nanoplatelets and reduced by laser treatment. *Nanotecnology* **2017**, *28*, 224002–224008. [CrossRef] [PubMed]
46. Orabona, E.; Ambrosio, A.; Longo, A.; Carotenuto, G.; Nicolais, L.; Maddalena, P. Holographic patterning of graphene-oxide films by light-driven reduction. *Opt. Lett.* **2014**, *39*, 4263–4266. [CrossRef] [PubMed]
47. Wang, G.; Qian, F.; Saltikov, C.W.; Jiao, Y.; Yat, L. Microbial Reduction of Graphene Oxide by Shewanella. *Nano Res.* **2011**, *4*, 563–570. [CrossRef]
48. Utkan, G.; Öztürk, T.; Duygulu, Ö.; Tahtasakal, E.; Denizci, A.A. Microbial Reduction of Graphene Oxide by Lactobacillus Plantarum. *Int. J. Nanosci. Nanotechnol.* **2019**, *15*, 127–136.
49. Salas, E.C.; Sun, Z.; Lüttge, A.; Tour, J.M. Reduction of Graphene Oxide via Bacterial Respiration. *ACS Nano* **2010**, *4*, 4852–4856. [CrossRef]
50. Serrano-Luján, L.; Víctor-Román, S.; Toledo, C.; Sanahuja-Parejo, O.; Mansour, A.E.; Abad, J.; Amassian, A.; Benito, A.M.; Maser, W.K.; Urbina, A. Environmental impact of the production of graphene oxide and reduced graphene oxide. *SN Appl. Sci.* **2019**, *1*, 179. [CrossRef]
51. Zhang, J.; Yang, H.; Sheng, G.; Cheng, P.; Zang, J.; Guo, S. Reduction of graphene oxide vial-ascorbic acid. *Chem. Commun.* **2010**, *46*, 1112–1114. [CrossRef]
52. Liu, J.; Liu, L.; Wu, X.; Zhang, X.; Li, T. Environmentally friendly synthesis of graphene–silver composites with surface-enhanced Raman scattering and antibacterial activity via reduction with l-ascorbic acid/water vapor. *New J. Chem.* **2015**, *39*, 5272–5281. [CrossRef]
53. Habte, A.T.; Ayele, D.W. Synthesis and Characterization of Reduced Graphene Oxide (rGO) Started from Graphene Oxide (GO) Using the Tour Method with Different Parameters. *Adv. Mater. Sci. Eng.* **2019**, *15*, 5058163. [CrossRef]
54. Fernández-Merino, M.J.; Guardia, L.; Paredes, J.L.; Villar-Rodil, S.; Solís-Fernández, P.; Martínez-Alonso, A.; Tascón, J.M.D. Vitamin C is an ideal substitute for hydrazine in the reduction of graphene oxide suspensions. *J. Phys. Chem. C* **2010**, *214*, 6426–6432. [CrossRef]
55. Chen, D.; Li, L.; Guo, L. An Environment-Friendly Preparation of Reduced Graphene Oxide Nanosheets via Amino Acid. *Nanotecnology* **2011**, *22*, 325601. [CrossRef] [PubMed]
56. Muralikrishna, S.; Sureshkumar, K.; Varley, T.S.; Nagaraju, D.H.; Ramakrishnappa, T. In situ reduction and functionalization of graphene oxide with l-cysteine for simultaneous electrochemical determination of cadmium(ii), lead(ii), copper(ii), and mercury(ii) ions. *Anal. Methods* **2014**, *6*, 8698–8705. [CrossRef]
57. Abdelhalim, A.O.E.; Sharoyko, V.V.; Meshcheriakov, A.A.; Martynova, S.D.; Ageev, S.V.; Iurev, G.O.; Mulla, H.A.; Petrov, A.V.; Solovstova, I.L.; Vasina, L.V.; et al. Reduction and functionalisation of graphene oxide with L-cysteine: Synthesis, characterization and biocompatibility. *Nanomed. Nanotechnol. Biol. Med.* **2020**, *29*, 102284. [CrossRef] [PubMed]

58. Kumar, A.; Khandelwal, M. Amino acid mediated functionalization and reduction of graphene oxide—Synthesis and the formation mechanism of nitrogen-doped graphene. *New J. Chem.* **2014**, *38*, 3457–3467. [CrossRef]
59. Bose, S.; Kuila, T.; Mishra, A.K.; Kim, N.H.; Lee, J.H. Dual role of glycine as a chemical functionalize and a reducing agent in the preparation of graphene: An environmentally friendly method. *J. Mater. Chem.* **2012**, *22*, 9696–9703. [CrossRef]
60. Wang, Y.; Shi, Z.; Yin, J. Facile Synthesis of Soluble Graphene via a Green Reduction of Graphene Oxide in Tea Solution and Its Biocomposites. *ACS Appl. Mater. Interfaces* **2011**, *3*, 1127–1133. [CrossRef]
61. Moosa, A.A.; Jaafar, N. Green Reduction of Graphene Oxide Using Tea Leaves Extract with Applications to Lead Ions Removal from Water. *Nanosci. Nanotechnol.* **2017**, *7*, 38–47.
62. Vatandosta, E.; Azade Ghorbani-Hasan, S.; Chekin, F.; Raeisi, S.N.; Shahidi, S.-A. Green tea extract assisted green synthesis of reduced graphene oxide: Application for highly sensitive electrochemical detection of sunset yellow in food products. *Food Chem. X* **2020**, *6*, 100085. [CrossRef]
63. Kurt, B.Z.; Durmus, Z.; Sevgi, E. In situ reduction of graphene oxide by different plant extracts as a green catalyst for selective hydrogenation of nitroarenes. *Int. J. Hydrog. Energy* **2019**, *44*, 26322–26337. [CrossRef]
64. Zulhelmi, I. Green reduction of graphene oxide by plant extracts: A short review. *Ceram. Int.* **2019**, *45A*, 23857–23868.
65. Lee, G.; Kim, B.S. Biological reduction of graphene oxide using plant leaf extracts. *Biotechnol. Prog.* **2014**, *30*, 463–469. [CrossRef] [PubMed]
66. Gao, J.; Liu, F.; Liu, Y.; Ma, N.; Wang, Z.; Zang, X. Environment-Friendly Method to Produce Graphene That Employs Vitamin C and Amino Acid. *Chem. Mater.* **2010**, *22*, 2213–2218. [CrossRef]
67. Longo, A.; Palomba, M.; Carotenuto, G. Green Solid-State Chemical Reduction of Graphene Oxide Supported on a Paper Substrate. *Coatings* **2020**, *10*, 693. [CrossRef]
68. Palomba, M.; Longo, A.; Carotenuto, G. Gel-Phase Reduction of Graphene Oxide Coatings by L-Ascorbic Acid. *Mater. Proc.* **2021**, *4*, 33.
69. Tas, M.; Altin, Y.; Bedeloglu, A.C. Reduction of graphene oxide thin films using a stepwise thermal annealing assisted by L-ascorbic acid. *Diam. Relat. Mater.* **2019**, *92*, 242–247. [CrossRef]
70. Li, J.-C.; Weng, C.-H.; Tsai, F.-C.; Shih, W.-P.; Chang, P.-Z. Porous reduced graphene oxide membrane with enhanced gauge factor. *Appl. Phys. Lett.* **2016**, *108*, 013108. [CrossRef]
71. Sui, Z.; Zhang, X.; Lei, Y.; Luo, Y. Easy and green synthesis of reduced graphite oxide-based hydrogels. *Carbon* **2011**, *43*, 4314–4321. [CrossRef]
72. Ha, T.; Kim, S.K.; Choi, J.-W.; Chang, H.; Jang, H.D. pH-controlled synthesis of porous graphene sphere and application to supercapacitors. *Adv. Powder Technol.* **2019**, *30*, 18–22. [CrossRef]
73. Li, D.; Müller, M.B.; Gilje, S.; Kaner, R.B.; Wallace, G.G. Processable aqueous dispersions of graphene nanosheets. *Nat. Nanotechnol.* **2008**, *3*, 101. [CrossRef]
74. Everett, D.H. *Basic Principles of Colloid Science*; The Royal Society of Chemistry: Cambridge, UK, 1988.
75. Kondratowicz, I.; Żelechowska, K.; Nadolska, M.; Jażdżewska, A.; Gazda, M. Comprehensive study on graphene hydrogels and aerogels synthesis and their ability of gold nanoparticles adsorption. *Colloids Surf. A* **2017**, *528*, 65–73. [CrossRef]
76. Wang, F.; Jia, Z.; Su, W.; Shang, Y.; Wang, Z.-L. Adsorption of phenanthrene and 1-naphthol to graphene oxide and L-ascorbic-acid-reduced graphene oxide: Effects of pH and surfactants. *Environ. Sci. Pollut. Res.* **2019**, *26*, 11062–11073. [CrossRef] [PubMed]
77. De Silva, K.K.H.; Huang, H.-H.; Yoshimura, M. Progress of reduction of graphene oxide by ascorbic acid. *App. Surf. Sci.* **2018**, *447*, 338–346. [CrossRef]
78. Abulizi, A.; Okitsu, K.; Zhu, J.-J. Ultrasound assisted reduction of graphene oxide to graphene in L-ascorbic acid aqueous solutions: Kinetics and effects of various factors on the rate of graphene formation. *Ultrason. Sonochem.* **2014**, *21*, 1174–1181. [CrossRef]
79. Go, S.-H.; Kim, H.; Yu, J.; You, N.-H.; Ku, B.-C.; Kim, Y.-K. Synergistic effect of UV and L-ascorbic acid on the reduction of graphene oxide: Reduction kinetics and quantum chemical simulations. *Solid State Sci.* **2018**, *84*, 120–125. [CrossRef]
80. Tikekar, R.V.; Anantheswaran, R.C.; Elias, R.J.; LaBorde, L.F. Ultraviolet-induced oxidation of ascorbic acid in a model juice system: Identification of degradation products. *J. Agric. Food Chem.* **2011**, *59*, 8244–8248. [CrossRef]
81. Ding, H.; Zhang, S.; Chen, J.-T.; Hu, X.-P.; Du, Z.-F.; Qiu, Y.-X.; Zao, D.-L. Reduction of graphene oxide at room temperature with vitamin C for RGO–TiO_2 photoanodes in dye-sensitized solar cell. *Thin Solid Films* **2015**, *584*, 29–36. [CrossRef]
82. Dan, L.; Pope, M.A.; Elias, A.L. Solution-Processed Conductive Biocomposites Based on Polyhydroxybutyrate and Reduced Graphene Oxide. *J. Phys. Chem. C* **2018**, *122*, 17490–17500. [CrossRef]
83. Xiong, K.; Wu, T.; Fan, Q.; Chen, L.; Yan, M. Novel Reduced Graphene Oxide/Zinc Silicate/Calcium Silicate Electroconductive Biocomposite for Stimulating Osteoporotic Bone Regeneration. *Appl. Mater. Interfaces* **2017**, *9*, 44356–44368. [CrossRef]
84. Chen, Y.; Pötschke, P.; Pionteck, J.; Voit, B.; Qi, H. Aerogels Based on Reduced Graphene Oxide/Cellulose Composites: Preparation and Vapour Sensing Abilities. *Nanomaterials* **2020**, *10*, 1729. [CrossRef]
85. Guerrero-Contreras, J.; Caballero-Briones, F. Graphene oxide powders with different oxidation degree, prepared by synthesis variations of the Hummers method. *Mater. Chem. Phys.* **2015**, *153*, 209–220. [CrossRef]
86. Kondratowicz, I.; Nadolska, M.; Żelechowska, K. Reduced Graphene Oxide Joins Graphene Oxide to Teach Undergraduate Students Core Chemistry and Nanotechnology Concepts. *J. Chem. Educ.* **2018**, *95*, 1012–1017. [CrossRef]

87. Neklyudov, V.V.; Khafizov, N.R.; Sedov, I.A.; Dimiev, A.M. New insights into the solubility of graphene oxide in water and alcohols. *Phys. Chem. Chem. Phys.* **2017**, *19*, 17000–17008. [CrossRef] [PubMed]
88. Cho, Y.H.; Kim, H.W.; Lee, H.D.; Shin, J.E.; Yoo, B.M.; Park, H.B. Water and ion sorption, diffusion, and transport in graphene oxide membranes revisited. *J. Membr. Sci.* **2017**, *544*, 425–435. [CrossRef]

materials

MDPI

Review

Graphene-Based Fiber Materials for Gas Sensing Applications: State of the Art Review

Susanna Vu [1,2], Mohamed Siaj [2] and Ricardo Izquierdo [1,*]

1 Department of Electrical Engineering, École de Technologie Supérieure, 1100 Rue Notre-Dame Ouest, Montréal, QC H3C 1K3, Canada; susanna.vu.1@ens.etsmtl.ca
2 Department of Chemical Engineering and Biotechnological Engineering, Université de Sherbrooke, 2500 Boulevard de l'Université, Sherbrooke, QC J1K 2R1, Canada; mohamed.siaj@usherbrooke.ca
* Correspondence: ricardo.izquierdo@etsmtl.ca

Abstract: The importance of gas sensors is apparent as the detection of gases and pollutants is crucial for environmental monitoring and human safety. Gas sensing devices also hold the potential for medical applications as health monitoring and disease diagnostic tools. Gas sensors fabricated from graphene-based fibers present a promising advancement in the field of sensing technology due to their enhanced sensitivity and selectivity. The diverse chemical and mechanical properties of graphene-based fibers—such as high surface area, flexibility, and structural stability—establish them as ideal gas-sensing materials. Most significantly, graphene fibers can be readily tuned to detect a wide range of gases, making them highly versatile in gas-sensing technologies. This review focuses on graphene-based composite fibers for gas sensors, with an emphasis on the preparation processes used to achieve these fibers and the gas sensing mechanisms involved in their sensors. Graphene fiber gas sensors are presented based on the chemical composition of their target gases, with detailed discussions on their sensitivity and performance. This review reveals that graphene-based fibers can be prepared through various methods and can be effectively integrated into gas-sensing devices for a diverse range of applications. By presenting an overview of developments in this field over the past decade, this review highlights the potential of graphene-based fiber sensors and their prospective integration into future technologies.

Keywords: gas sensor; fiber; nanofiber; fiber composite; graphene oxide; reduced graphene oxide; electrospinning; wet-spinning; gas sensing mechanism; flexible electronics

1. Introduction

Gas sensors, devices designed to detect gases in a chosen environment, play a critical role in applications ranging from environmental monitoring to industrial systems monitoring and healthcare diagnostics [1–5]. Gas detection must be highly efficient, reliable, and accurate, especially when concerning the safety of both the individual and the environment [6]. For example, an occurrence such as a gas leak in an industrial setting can quickly become catastrophic and require immediate action [7], while the rapid detection and assessment of biomarkers must be precise for proper medical treatment [8]. Gas sensing of pollutants is also crucial in maintaining human health and safety, as well as ecosystem preservation and enabling effective regulation of harmful emissions [9–12]. The need for gas sensors in these settings has resulted in increasing sophistication in both gas sensor design and nanofabrication approaches [13]. In addition, as automated and remotely operated equipment becomes more widely adopted, gas sensors have been proposed to act as an electronic "nose" for these systems [14,15], underscoring the importance of the continued development of these sensors and their place in modern technologies [16].

Gas sensors can be categorized as physical and chemical sensors or have elements of both. Physical sensors rely on measurements of physical quantities, such as light or heat, while chemical sensors rely on measurement of chemical interactions at the gas–sensor interface [17]. Although the change in resistance is a physical measurement, the mechanism driving this effect may be physical absorption or chemical reaction, depending on the interactions between the gas and the sensing material [18,19]. Gas sensor performance can be influenced by the dimensionality of the sensor and the loading of the sensing materials [20,21]. A higher specific surface area of a sensing material leads to higher sensitivity, and several different fabrication techniques have been employed with the aim of producing sensors with intricate and fine structures to maximize surface area [22,23]. Many factors affect the gas sensing mechanism of a sensor, including the reactivity and atomicity of the gas, and the conductivity and morphology of the sensing material. Gas molecules can influence the response of a gas sensor due to the rate of diffusion of gas and the kinetics of collision between gas and surface [24,25]. Larger gas molecules may exhibit higher reactivity due to their greater collision diameters, enhancing interactions with the sensing material and generating a stronger response [26–28]. Conversely, smaller gas molecules can more easily penetrate the pores of the sensing material and diffuse more rapidly, resulting in faster response times [25,27,29]. These effects have led to efforts to tailor sensing materials by modifying surface properties [30–32] and tuning pore sizes to target specific gases based on molecular size [25,33–35]. As other areas of technology evolve, there is great interest in developing sensors with efficient sensing materials that are suited to these new technological landscapes [36,37]. A prominent example of this is the development of sensors that follow the design tenets of flexible and wearable electronics [38].

Challenges in developing flexible gas sensing platforms often involve selecting suitable sensing materials that can maintain their structure and functionality when strained [39]. Research focuses on optimizing these sensors to achieve comparable sensitivity to rigid substrates, along with high selectivity, fast response and recovery, and durability, ensuring consistent performance under repeated bending or stretching without damage [40]. Carbon-based materials offer mechanical stability and electronic properties and have been demonstrated to be promising sensitive materials for sensing applications [41]. Among these materials, graphene, defined as a single layer of sp^2 carbon atoms, exhibits desirable characteristics and has been thoroughly investigated for use in sensors [42–44]. Many of the properties of graphene, including high conductivity, excellent chemical stability at ambient temperature, flexibility, intrinsic high surface area, and low fabrication costs make it an ideal candidate in gas sensing applications [45]. The large surface-to-volume ratio of graphene offers numerous active sites for gas adsorption, while its exceptional electrical conductivity enables fast response times, contributing to the sensitivity of the sensor. The sensitivity of graphene is such that it has been shown to be able to sense a single gas molecule [46].

Graphene materials, including pristine graphene, graphene oxide (GO), and reduced graphene oxide (RGO) [47], also demonstrate distinct gas detection capabilities that can be exploited in sensor fabrication [43]. GO can be produced at a low cost by chemically exfoliating graphite in the presence of a suitable oxidant [48]. GO has semiconducting properties superior to raw graphite and can be enhanced significantly by reduction to RGO using chemical, thermal, or UV reduction processes [43,48]. RGO is particularly advantageous as it possesses a defined band gap and has accessible functional groups capable of selectively binding gas molecules [48]. The implementation of graphene materials in composite fibers for gas sensing has garnered the most significant interest, as graphene and its derivatives have been proven to be robust nanofillers [49].

Graphene-based composite fibers can be fabricated with a wide range of materials including polymers, metal oxide nanoparticles, and even other 2D materials and synthetic fibers [17,50]. Combining graphene with other semiconducting materials, such as conductive polymers and metal oxides, often enhances sensing capabilities by increasing the surface available for gas interaction [3,51]. Graphene is a competitive gas sensing material due to its tolerance to humidity, unlike many polymers, and its broader operating temperature range, compared to some metal oxides [3]. Graphene-based fibers are advantageous functional materials as they benefit from the high sensitivity and stability of graphene, along with their good electrical and mechanical properties [3,43,52]. While harnessing the beneficial characteristics of graphene, graphene fibers provide increased flexibility and structural integrity, making them well-suited for sensor fabrication. With the growing interest in micro and nano gas sensors, graphene-based fibers are strong contenders owing to their conductivity, compatibility, conformability, and ease of integration into lightweight, flexible devices [53,54].

A popular method to fabricate nanofiber gas sensors is electrospinning, where an electrostatic force is employed to draw threads from a composite solution to form nanofibers [55]. Electrospinning offers an advantage over other nanofiber fabrication methods, such as chemical vapor deposition, sol-gel methods, and template-assisted fabrication, due to its simplicity, cost-effectiveness, and versatility relative to these competing nanofabrication techniques [17]. Another emerging technique for the fabrication of fibers is wet-spinning, whereby a composite material is extruded through a spinneret into a coagulation bath composed of a non-solvent [56,57]. In the coagulation bath, the material undergoes rapid drawing, resulting in fiber formation through phase inversion [57,58]. Electrospinning allows for more precise control of fiber diameter while wet-spinning can be more easily scalable [59,60]. Fibers produced through spinning methods are distinguished by their exceptionally high aspect ratio and surface area, as well as the ease with which graphene materials can be uniformly integrated throughout the fiber, enhancing the performance of the resulting sensor [61].

This review aims to provide an overview of the state-of-the-art graphene-based fiber materials for gas detection sensors, focusing on advancements from the past decade. This review will discuss both gas sensing systems designed to detect specific gases and those capable of detecting multiple gases. It will cover the synthesis and processing methods for these fiber sensors and delve into the sensing mechanisms, including response dynamics and modes of detection. To demonstrate the scope and selectivity of these gas sensors, the application of graphene-based fibers in sensing gases with varying atomicity (diatomic, triatomic, and polyatomic gases) is described to illustrate the various modes to target these different gases. Additionally, sensors designed to detect volatile organic compounds (VOCs), at times termed gas vapors, are discussed. In addition to sensors fabricated to sense one gas, detection systems for two or more gases are reviewed. These gases represent an extensive range of gases with respect to their molecular size, atomic composition, thermal conductivity, and oxidizing or reducing potential. This review distinguishes gas sensors designed for various target gases, to spotlight the range of sensing capabilities of graphene fibers and highlight their applicability in diverse settings and applications.

This review will focus on sensing systems that incorporate graphene and gas sensors with a graphene fiber component, examining how graphene was integrated and how it enhanced the overall performance of each system. This work will emphasize sensing systems that utilize graphene fibers; however, in some cases, gas sensing performance relies on the combination of graphene with other materials, such as metal oxides and polymers. In these instances, we will focus on the role of graphene, while briefly discussing its interaction with these materials in relation to sensor sensitivity. Throughout this review, research trends in this field will be revealed and breakthrough findings will be highlighted. By examining recent literature, it aims to showcase the potential of graphene-based composite fibers in gas sensing systems and illustrate how these advances set a foundation for their integration into diverse applications, including medical diagnosis, health management, environmental monitoring, and wearable electronics.

2. Graphene-Based Fiber Sensors for Diatomic Gases

It is imperative to detect diatomic gases, such as hydrogen (H_2) and carbon monoxide (CO), for a plethora of reasons related to safety concerns that endanger human health and the environment. H_2 is a non-toxic, colorless, odorless gas; however, it is primarily produced by fossil fuels and is highly flammable [17,62]. This is due to its low ignition energy, where even an H_2 volume fraction of 4% in the air can trigger explosions [63,64].

H_2 gas is also susceptible to leakage into the atmosphere, making its safe transportation and storage very challenging, especially in industrial settings [64]. Similarly to H_2, CO is a colorless, odorless, and flammable gas generated from fossil fuels and industrial processes, as well as vehicle exhaust emissions and wildfires [17,65]. However, CO exposure is toxic to humans and can be lethal in high concentrations, while lower concentrations exposure can lead to adverse symptoms, including headache, nausea, and dizziness [65]. This occurs as CO binds to hemoglobin in blood with high affinity, competing with oxygen and reducing its capacity to carry oxygen by displacing it [66]. In addition to being colorless and odorless, both H_2 and CO gas are less dense than air, which enables their accumulation in enclosed spaces; therefore, efficient sensors for these gases are essential [64,66].

2.1. H_2 Gas Sensor

While graphene alone has demonstrated effective gas-sensing properties, numerous studies are focused on optimizing these capabilities by combining graphene with other sensing materials [46,52,67]. Extensive research has focused on enhancing the sensitivity and selectivity of gas sensors by incorporating graphene and its derivatives with metal oxides, forming nanocomposite materials [50,68]. Kim et al. employed reduced graphene oxide to enhance the gas sensing capabilities of zinc oxide (ZnO) nanofibers for the selective detection of hydrogen gas [69]. The electrospun nanofibers were constructed from RGO-loaded ZnO, produced by incorporating RGO nanosheets with zinc acetate, and had an average diameter of 190 nm. The sensor exhibited the highest response of 2542 (R_a/R_g, where R_a is the resistance in the absence and R_g is the resistance in the presence of hydrogen) to 10 ppm H_2 gas at 400 °C (Table 1). At the lowest concentration of H_2 of 100 ppb, a response of 866 is shown by the sensor. The study suggests that there are electrical potential barriers at the interfaces of RGO/ZnO, RGO/Zn, and Zn/ZnO at equilibrium (Figure 1a). When introducing H_2 gas to RGO-ZnO nanofibers, ZnO became n-type and was reduced to metallic Zn at the surface of the nanofiber, as hydrogen atoms reacted with surface oxygen ions of bulk ZnO. An energy potential barrier at the RGO/Zn interface prevented the flow of electrons into RGO, thus the addition and removal of H_2 is a resistance modulation and generates a sensing signal. The high sensitivity of RGO-loaded ZnO nanofibers stands in contrast to SnO_2 nanofibers fabricated in the same study, which are less responsive than the ZnO counterparts. Following similar methodologies, the same research group also fabricated RGO-ZnO nanofibers for sensing various other gases and reported their findings in a separate paper [70].

2.2. CO Gas Sensor

Incorporating graphene into a sensing system can amplify the overall electrical properties of the sensor, increasing its conductivity due to the high electron mobility of graphene [20,71]. Additionally, doping a graphene-based system with other semiconducting materials, or doping graphene itself, can further elevate the sensitivity and response time of the sensor, enabling more accurate and rapid detection of target gases [52,72]. Shams et al. electrospun cadmium-doped tin oxide and reduced graphene oxide composite nanofibers to function as a carbon monoxide gas sensor [73]. They reported the best-performing sensor, containing 1.6% cadmium, had a band gap of 2.80 eV and a diameter of 200.57 nm. In the presence of CO gas, the Cd-doped RGO-SnO_2 nanofibers were able to respond after 25 s at 100 °C, while the nanofibers consisting of SnO_2 alone showed a delayed response at 35 s (Table 1). The sensing mechanism of this system is enhanced by RGO due to its energy level, which lies between the LUMO orbital of CO and the conduction band of SnO_2, and it exhibits p-type behavior relative to SnO_2 (Figure 1b). This facilitates the transfer of electrons and decreases the resistance. In addition, the authors note that doping with Cd allowed for more sites for oxygen adsorption, which promoted the oxidation of Cd and generated more electrons back to the nanofiber. Therefore, a combination of RGO with Cd dopant further optimized the adsorption and desorption kinetics of the gas sensing system. To promote personal safety by minimizing potential hazards, this study presents a method for the rapid detection of toxic gases using sensors based on graphene-enhanced composites.

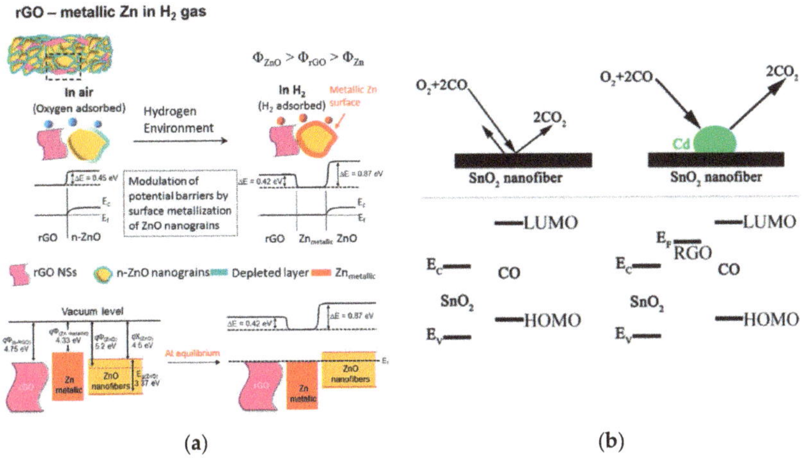

Figure 1. Examples of graphene-based fiber sensors for diatomic gases: (**a**) Schematic illustration of RGO-ZnO sensing mechanism for H_2 gas [69]; (**b**) Schematic diagram of Cd/RGO/SnO_2 sensing mechanism to CO gas [73].

Table 1. Summary of graphene-based fiber gas sensors for H_2 and CO gas.

Gas	Conc.	Material	Response	Temp.	Ref.
Hydrogen (H_2)	100 ppm	RGO-ZnO	2542	400 °C	[69]
Carbon Monoxide (CO)	100 ppm	Cd/SnO_2/RGO	25 s	100 °C	[73]

3. Graphene-Based Fiber Sensors for Triatomic Gases

Triatomic gases like carbon dioxide (CO_2), hydrogen sulfide (H_2S), and nitrogen dioxide (NO_2) are gases that can come from anthropogenic sources, and high concentrations of each gas in the atmosphere are considered undesirable [74,75]. CO_2 is a colourless and odourless gas and is the most prevalent greenhouse gas in our atmosphere [17]. The absorption of infrared radiation from the sun by atmospheric CO_2 is understood to be the primary driver of climate change [76]. On a smaller scale, control of CO_2 concentration in systems, enabled by quick and accurate measurements, is important in air quality, food preservation, and early fire detection [77]. Atmospheric CO_2 levels remained at approximately 250 ppm from human evolution until the Industrial Revolution but doubled between the years 1813 and 2019 [78]. Prolonged exposure to CO_2 levels up to 1000 ppm poses health risks to humans [79], making the detection and management of CO_2 essential both indoors and outdoors. H_2S is a highly toxic and flammable gas, largely generated from petroleum refineries and oil and gas drilling operations [17,80–82]. Exposure to H_2S at low concentrations of 10 ppm should not exceed more than 10 min, while exposure to 100 ppm can cause instantons death [83]. Additionally, the distinct "rotten egg" odor of H_2S is unpleasant; however, anosmia, or olfactory desensitization, can sometimes prevent the human olfactory system from detecting this gas. [84–86]. NO_2 is a particularly important gas to detect as it is one of the primary emissions from the manufacturing and automobile industries [17]. Global average NO_2 levels are on the rise, with motor vehicle exhaust contributing up to 80% of NO_2 emissions in certain cities [87,88]. NO_2 is associated with smog, and while not a greenhouse gas, it is toxic to humans, with the lethal dose being only 174 ppm [89]. NO_2 gas detection is thus critical, especially in densely populated urban areas [77]. Given these factors, it is therefore essential to rapidly detect any present in the environment for the protection of individuals and the Earth. The presence of hazardous gases in the atmosphere, such as CO_2, NO_2, and H_2S, necessitates constant environmental monitoring and emphasizes the need for high-performing gas sensors [90].

3.1. CO_2 Gas Sensor

Al-Thani et al. reported CO_2 sensors composed of polyaniline (PANI)-coated and RGO-PANI-coated electrospun polystyrene (PS) nanofibers, respectively [91]. The PS nanofibers underwent a plasma treatment, followed by coating with GO and PANI, whereby PANI was directly polymerized onto the fibers (Figure 2). The sensors containing GO were subjected to hydrogen reduction to convert to RGO with PANI-coated PS nanofibers. Although both sets of sensors showed sensitivity towards CO_2 gas, the sensors containing RGO showed higher sensitivity, with a more distinct change in resistance, when exposed to 60 ppm of CO_2 gas at room temperature. This behavior is credited to the broad electrochemical potential window and fast electron transfer rate of graphene [92]. The repeatability of the nanofiber sensors was examined, where the sensors exhibited a CO_2 gas sensing response and recovery time of 65 s (Table 2). Furthermore, the selectivity of the nanofibers was investigated, where the RGO/PANI/PS sensor exhibited a high response of 0.8 ($(R_g - R_a)/R_a$, where R_g and R_a are resistance in the presence of an analyte gas and N_2, respectively) to CO_2 and lower responses towards methanol, ethanol, and ammonia.

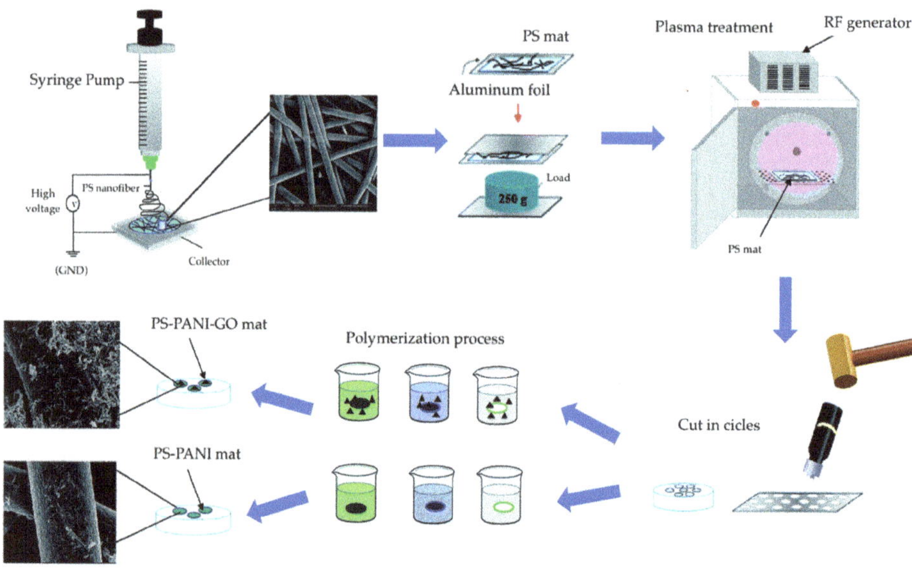

Figure 2. Schematic illustration of graphene/PANI/PS nanofibers preparation steps [91].

3.2. H_2S Gas Sensor

Kim et al. fabricated H_2S sensors using electrospun RGO-CuO nanofibers [93]. By testing RGO loadings ranging from 0.05 wt% to 1.5 wt%, they identified 0.5 wt% RGO-CuO as the optimal composition, which exhibited the highest sensitivity of 1.95 (R_g/R_a, where R_a is resistance in air and R_g is resistance in the presence of target gas) to 10 ppm of H_2S at 300 °C. The RGO-CuO nanofibers also demonstrated selectivity for H_2S when tested against other gases, including CO, C_6H_6, and C_7H_8, where the interfering gases showed minimal activity. The gas sensing efficiency was considerably influenced by the morphology of these nanofibers, which consist of nanoscale grains [94]. These nanograins play a critical role in determining the gas-sensing mechanism and enlarging the surface area of the fiber sensor. In the case of RGO-CuO, the varying RGO loadings alter the surface structure of the nanograins, leading to a distinct sensing response (Figure 3).

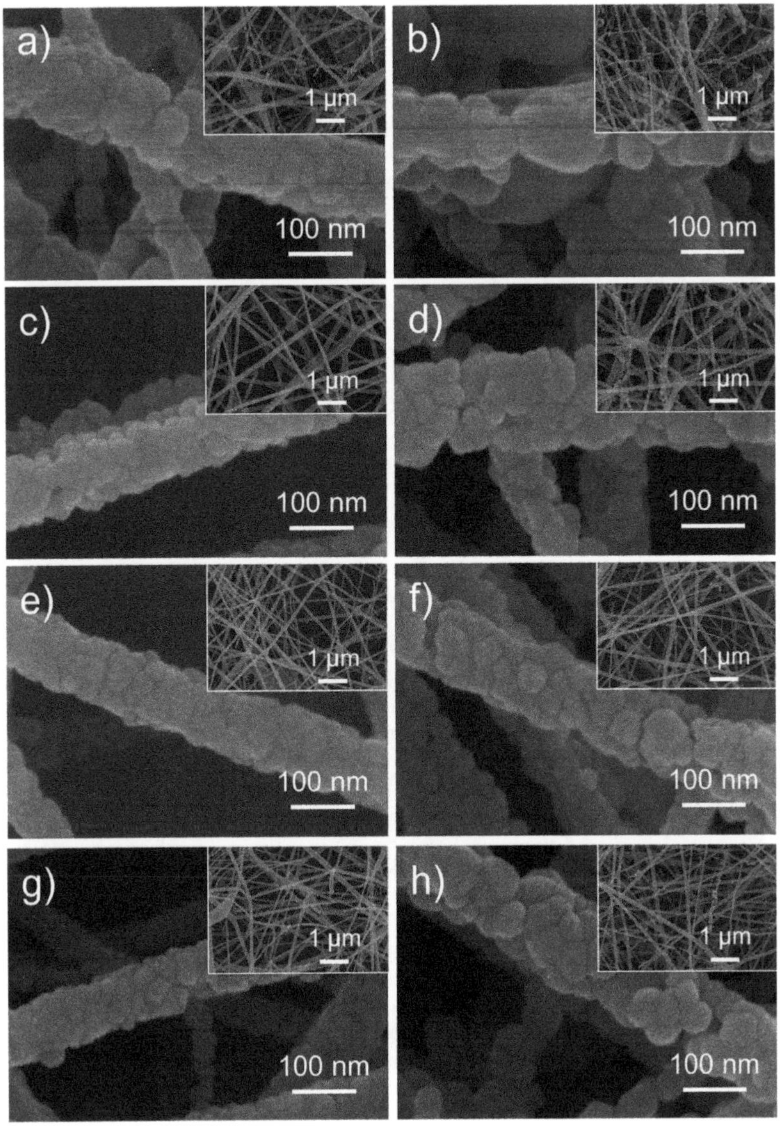

Figure 3. Field-emission scanning electron microscopy images of: (**a**) Pristine CuO nanofibers, and RGO-CuO nanofibers with different amounts of RGO; (**b**) 0.05 wt% RGO, (**c**) 0.1 wt% RGO, (**d**) 0.2 wt% RGO, (**e**) 0.3 wt% RGO, (**f**) 0.5 wt% RGO, (**g**) 1 wt% RGO, and (**h**) 1.5 wt RGO% [93].

Kim et al. reported nanofibers for H_2S detection based on non-oxidized graphene (NOGR), whereby pore size and distribution were controlled by a polymeric templating approach [95]. In the fabrication of their sensors, a composite containing colloidal polystyrene with tungsten precursor was electrospun, and the resulting nanofibers underwent a calcination procedure (Figure 4). In the process, the W precursor was oxidized, and PS colloids were decomposed. By tuning the size of the PS colloids, the size and distribution of pores can be controlled within the nanofibers as the PS colloids act as sacrificial templates and become void domains following the thermal treatment. In parallel, NOGR flakes were obtained through the chemical exfoliation of graphite intercalation compounds and subsequently combined with the PS-WO_3 nanofibers. The resulting PS/WO_3/NOGR nanofibers exhibited a sensitivity of 65.6 (R_{air}/R_{gas}) to 5 ppm of H_2S gas at 300 °C (Table 2), and this was achieved with only 0.1 wt% loading of NOGR flakes. The selectivity of sensors was also examined by exposing the nanofibers to various gases, including acetone, NO, toluene, ethanol, NH_3, CO, and pentane, where the sensor showed the highest response towards H_2S. The conductivity of NOGR contributes to the sensing performance as it facilitates the transport of charge carriers, as well as the pores on the nanofibers that allow for higher surface area and gas penetration [96,97].

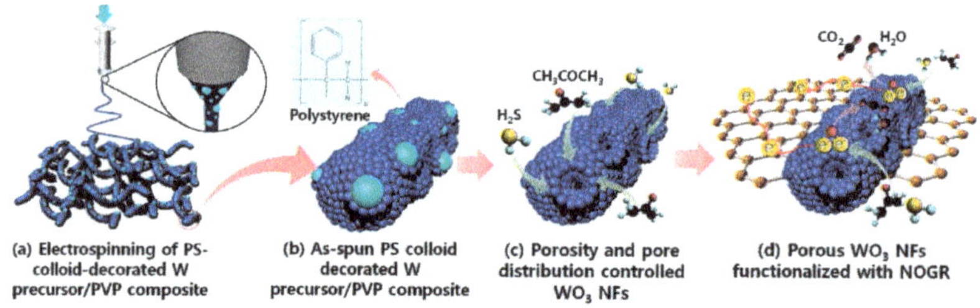

Figure 4. Schematic illustration of the fabrication process for PS-WO_3/NOGR nanofibers, whereby controlled pore distribution on the nanofiber is achieved [95].

Hieu et al. developed RGO/$ZnFe_2O_4$ nanofiber sensors for H_2S gas detection. The preparation of the sensors involved an on-chip electrospinning technique, wherein nanofibers were directly collected and assembled onto a microelectrode chip equipped with interdigitated electrodes [86]. The nanofiber sensors achieved a response of 147 (R_a/R_g, where R_a and R_g were the resistances of the sensors in the air and H_2S, respectively) to 1 ppm of H_2S at 350 °C (Table 2) [98]. The gas detection mechanism was attributed to the movement of electrons from RGO to $ZnFe_2O_4$ and the multi-porous structure of the sensor. The nanofibers were composed of nanograins, which induce the formation of depletion regions and potential barriers: one at the heterojunction between RGO and $ZnFe_2O_4$, and another at the boundaries between $ZnFe_2O_4$ nanograins (Figure 5). In the presence of air, oxygen molecules adsorb onto the surface of the nanofiber, capturing electrons from the conduction band to form oxygen ions. Upon exposure to H_2S, the gas molecules react with these oxygen ions, generating the electrons back into the conduction band. This interaction reduces the heterojunction and grain boundary barriers, leading to a decrease in the resistance of the sensor. The response of the sensor occurs owing to the heterojunction between RGO and $ZnFe_2O_4$ within the nanofiber.

Hieu et al. extended their work on H_2S gas sensors by fabricating RGO/α-Fe_2O_3 nanofiber sensors employing their previous on-chip electrospinning method [99]. The RGO/α-Fe_2O_3 nanocomposite was synthesized using poly(vinyl alcohol), a ferric salt precursor, and RGO reduced from GO. The nanofiber morphology was revealed to be significantly affected by changes in precursor concentration and annealing temperature while being independent of changes to the graphene content. The optimal sensor configuration, which yielded the highest response of 9.2 to 1 ppm H_2S gas at 350 °C (Table 2), consisted of nanofibers containing 1.0 wt% RGO, 11 wt% PVA, and 4 wt% $Fe(NO_3)_3{\cdot}9H_2O$, and was annealed at 600 °C. The sensitivity of the sensor was attributed to the morphology of the RGO/α-Fe_2O_3 nanofibers and the presence of nanograins, along with the large surface-to-volume ratio provided by the RGO. It was noted that the sensing mechanism involved potential barriers at both heterojunctions and homojunctions, consistent with the mechanisms described in their previous work [98]. This study presents a straightforward approach for the detection of toxic gases and environmental monitoring.

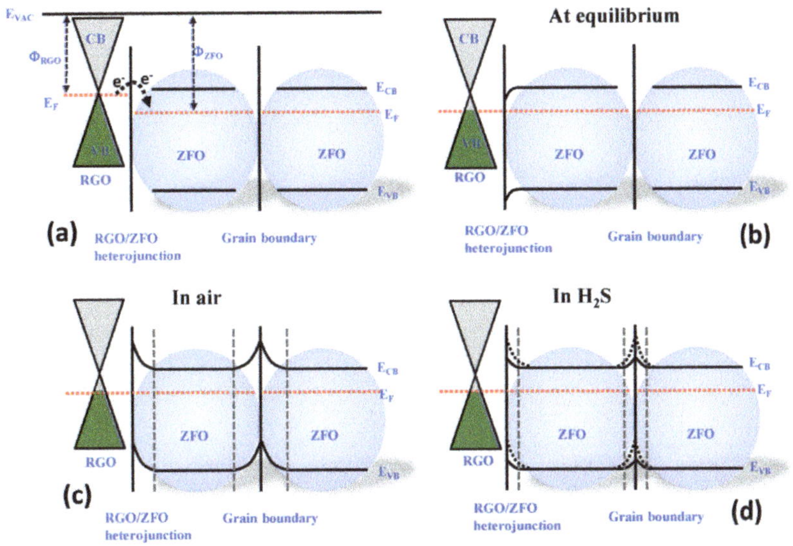

Figure 5. Schematic depicting the gas sensing mechanism of RGO-$ZnFe_2O_4$ nanofibers: (**a**) Band diagram of RGO and ZFO; (**b**) At equilibrium; (**c**) In air; (**d**) H_2S gas exposure [98].

Given rising concerns about air quality in our environment, accurate gas detection is critical in a multitude of settings. However, advancements must also focus on designing environmentally conscious sensing systems that can monitor gas in the environment without causing further harm to it. The environmental impact of graphene production itself must, therefore, also be evaluated. There are ongoing efforts to explore the production of graphite from biomass waste and the recycling of graphite from batteries to meet the demand for graphene [100,101]. With this, the lifespan of graphene-based sensors must be evaluated in future studies to ensure that they not only maintain optimal performance over their intended use but also that their materials can be reused to reduce the need for new production at the end of their operational life.

3.3. NO_2 Gas Sensor

Promising improvements in sensor performance have been made by extending the investigation of polymeric substrates and polymer composites to electrospun nanofiber-based gas sensors [102,103]. Shi et al. reported reduced graphene oxide and polymer composite nanofibers for the fabrication of nitrogen dioxide gas sensors [104]. The electrospun nanofibers were composed of a poly(vinyl alcohol) (PVA) and poly(ether imide) (PEI) polymer mixture and deposited onto an interdigitated electrode. The nanofibers were then dip-coated in a GO nanosheet solution, enabling the self-assembly of GO onto the nanofibers and were subsequently reduced to form RGO-polymer nanofiber gas sensors. The sensor showed repeatability over multiple cycles with exposure to NO_2 gas and N_2, where it reached 90% of the maximum response ($\Delta G/G_0$, the ratio of conductance change of sensor in target gas to N_2) upon exposure to 500 ppb of NO_2. At the highest NO_2 concentration of 5 ppm, the conductance increased by 159.4%, demonstrating that a higher NO_2 concentration resulted in a greater sensing response. This trend was primarily attributed to the accessibility of the RGO surface to NO_2 gas molecules [105,106].

Lee et al. also employed polymeric nanocomposites for the development of stretchable devices for the detection of NO_2 gas [107]. They described sensors fabricated from RGO, where GO was chemically reduced with hydrazine, layered onto electrospun polyurethane (PU) nanofibers, and assembled on polydimethylsiloxane (PDMS). During the electrospinning process, the collected fibers were rotated to form nanofibers in orthogonal directions with varied electrospinning times and number of fiber layers. The mechanical stability of the sensors was tested by stretching at 50% elongation up to 10,000 cycles. When considering both stretchability and gas sensitivity, the overall best-performing sensor exhibited a response of 176% ($\Delta I/I$, where I is the dynamic current intensity measured under stretching tests) to 5 ppm of NO_2 gas at room temperature (Table 2) and was comprised of five layers that were electrospun for 8, 3, 3, 3, and 1 min. This study illustrated an approach toward high-performing wearable gas sensors that maintain sensing capabilities under high strains.

As previously mentioned, a widely popular approach to obtaining gas-sensing materials is through the formation of nanocomposites with graphene derivatives and metal oxides [108–110]. Wang et al. demonstrated this with the fabrication of RGO-In_2O_3 nanofiber gas sensors for NO_2 detection [111]. These nanofibers were produced via electrospinning, incorporating In_2O_3 with RGO to form a composite material. The sensors containing 2.2 wt% RGO exhibited an enhanced gas response, with a sensitivity of 42 (R_g/R_a, where R_g is the resistance of the sensor in NO_2 and the R_a is the sensor resistance in the air) to 5 ppm NO_2 at 50 °C. The sensing mechanism of RGO-In_2O_3 nanofibers was influenced by the high surface area, structural defects, and functional groups of RGO, which provided ample adsorption sites for NO_2 gas molecules [112,113]. Additionally, RGO enhances the resistance modulation of the sensor through the formation of RGO-In_2O_3 heterojunctions (Figure 6). When the sensor is in the open air, oxygen molecules are adsorbed at the surface and between the juncture of adjacent In_2O_3 nanoparticles at the nanograin boundaries. Potential barriers and depletion layers are formed at the nanograin boundaries between In_2O_3 nanoparticles, as well as between In_2O_3 and RGO heterojunctions when oxygen species are generated. When the nanofiber is exposed to NO_2, the gas reacts with the adsorbed oxygen and expends electrons from the conduction band. This results in the expansion of the depletion layer, thereby altering the resistance of the sensor and producing a sensing signal.

The research group of Kim et al. are recognized for their work on RGO-loaded metal oxide electrospun nanofibers for gas sensing applications [69,70,93,114]. In two separate studies, the authors examined the gas sensing properties of electrospun RGO-SnO_2 and RGO-ZnO nanofibers, evaluating their response to various oxidizing gases (NO_2, SO_2, O_2) and reducing gases (CO, C_6H_6, C_2H_5OH) [70,114]. Although both nanofibers demonstrated sensitivity to a range of gases, the studies primarily focused on NO_2 due to the notably high response observed. The high response was attributed to the inherent high reactivity of NO_2 molecules as opposed to the selectivity of the nanofibers for NO_2 [70,114].

In their investigations, nanofibers with RGO concentrations ranging from 0.04 to 1.04 wt% were evaluated. It was found that nanofibers with 0.44 wt% RGO exhibited the optimal sensing performance for both SnO_2 and ZnO [70,114]. The authors suggest that increasing RGO concentrations beyond this optimal level reduced gas sensing performance due to percolation effects, wherein RGO forms conducting networks that interfere with the sensor. The optimal RGO concentration in the preparation of RGO-SnO_2 nanofibers resulted in the most pronounced resistance modulation and a response of approximately 100 (R_g/R_a, where R_g is the resistance of the sensor in NO_2 and R_a is the resistance in the air) when exposed to 5 ppm NO_2 at 200 °C (Figure 7a, Table 2) [114]. Similarly, in the RGO-ZnO study, nanofibers with the same RGO loading exhibited the highest response, with a maximum response of 150 when exposed to 5 ppm NO_2 at 400 °C (Figure 7b, Table 2) [70].

Scaling up the production of graphene-based fibers presents several challenges, particularly in transitioning spinning techniques to industrial-scale manufacturing. The studies discussed thus far all fabricated nanofibers using electrospinning methods, underlining the popularity of this technique for producing graphene fibers [49,115,116]. Despite the successful employment of this technique in the literature, it has been less widely adopted at an industrial scale due to its relatively slow production rates and challenges with maintaining consistency [117,118]. Wet-spinning fabrication techniques are currently regarded as a promising approach for the scalable production of microscale graphene composite fibers [57,119]. Although wet-spinning has been employed in the production of textiles such as viscose rayon fibers [120], adapting this technique for more complex functional materials remains an area of active research [60,121]. Further efforts are needed to optimize spinning techniques, enabling large-scale production while achieving precise control over fiber morphology and ensuring the functional performance of the fibers.

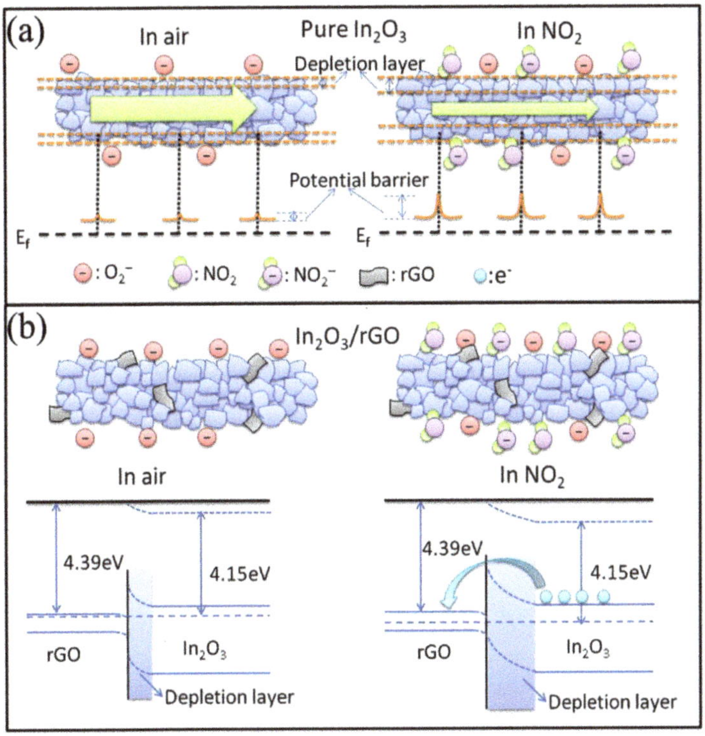

Figure 6. Schematic illustration of the sensing mechanism of (**a**) In_2O_3 compared to (**b**) RGO-In_2O_3 towards NO_2 gas [111].

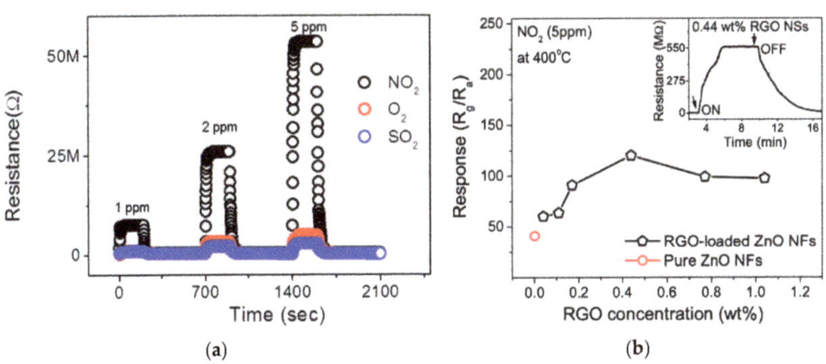

Figure 7. (**a**) Response of RGO-SnO_2 nanofibers to NO_2, O_2, and SO_2 gases, where the concentration was set to of 1, 2, and 5 ppm, respectively [114]; (**b**) Response of RGO-ZnO nanofibers, with varying RGO concentrations, to 5 ppm of NO_2 gas [70].

Han et al. used a continuous wet-spinning technique to synthesize Cu-Cu_2O and Ni-NiO graphene fibers [122]. In this process, a GO dispersion was extruded into a coagulation bath containing a catalytic solution with Cu or Ni ions, followed by thermal treatment of the fibers (Figure 8). The metal cations aid in binding the GO into fiber assemblies, making the wet-spinning technique suitable due to the even dispersion of cations in the coagulation bath. In addition, this preparation method allowed for sensors with flexibility and compatibility, enabling the fibers to be integrated into other fabrics. The resulting Cu/Cu_2O/RGO and Ni/NiO/RGO fiber sensors demonstrated sensitivities of 18.90% and 0.82% ($(R_{air} - R_{gas}/R_{air}) \times 100$) respectively, to exposure of 5 ppm of NO_2 gas at 150 °C (Table 2). Although the Ni/NiO/RGO fibers exhibited lower sensitivity compared to the Cu/Cu_2O/RGO fibers, they outperformed other Ni-containing graphene fibers, showing double the response of NiO-graphene fibers. The gas sensing mechanism of the fibers involves the spillover effect, whereby adsorbed gas molecules are dissociated by metal into more reactive species and subsequently dispersed onto the adjacent surface [123,124].

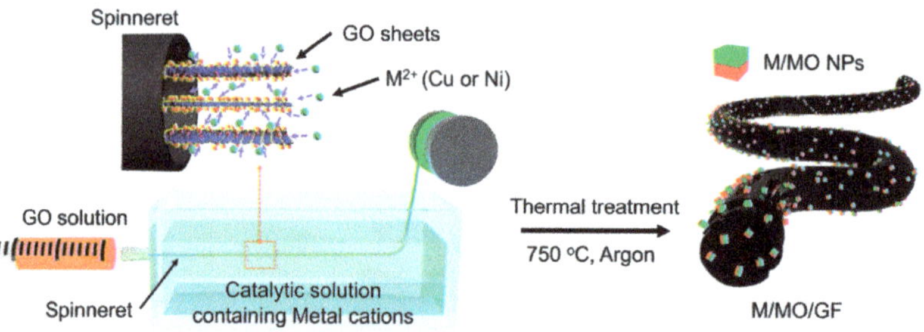

Figure 8. Schematic illustration of Cu/Cu_2O/graphene and Ni/NiO/graphene fiber (M/MO/GF) fabrication, involving wet-spinning and thermal treatment of fibers [122].

Kim et al. also utilized the wet-spinning technique to fabricate graphene fibers, incorporating tunicate cellulose nanofiber (TCNF) with GO to create TCNF-GO fibers [125]. These wet-spun fibers were then treated with a tungsten (W) precursor and subjected to thermal processing, resulting in reduction and calcination to produce porous RGO/WO_3/TCNF fibers. The inclusion of TCNF facilitated the formation of mesopores and created a wrinkled surface morphology, which increased the surface area of the fiber. The maximum response observed by the sensor was a sensitivity of 9.67% ($(R_{air} - R_{gas}/R_{air}) \times 100$) towards 5 ppm of NO_2 at 100 °C (Table 2), although they remained functional at room temperature. To further demonstrate the practical application of these fibers, the authors integrated them into wearable devices, showcasing the potential of the fibers in wearable sensing systems (Figure 9).

Graphene can be utilized to elevate existing commercially available fiber materials, presenting straightforward approaches to high performance composites [126,127]. These graphene fiber composites not only offer cost-effective and widely accessible fabrication methodologies, but also have the potential to support large-scale production of wearable electronics [128,129]. Ren et al. developed RGO-enhanced mesoporous ZnO nanosheet hybrid fibers using cotton and elastic thread and evaluated their sensing response to NO_2 gas [128]. The synthetic process involved treating the cotton and elastic treads with an adhesive and annealing treatment, followed by immersion in GO, a chemical reduction reaction, and subsequent coating with ZnO to produce RGO/ZnO/thread sensors (Figure 10a). These hybrid fibers demonstrated effective gas sensing capabilities, exhibiting a 44% response (R (%) = $(R_g - R_a)/R_a \times 100$, where R_a is the initial resistance value in air and R_g is the resistance value in NO_2) to 15 ppm of NO_2 at room temperature, with response and recovery times of 140 and 630 s, respectively (Table 2). To explore practical applications, the RGO/ZnO sensors were integrated into fabric, forming a wearable multi-sensor array network, which was successfully tested for NO_2 detection (Figure 10b).

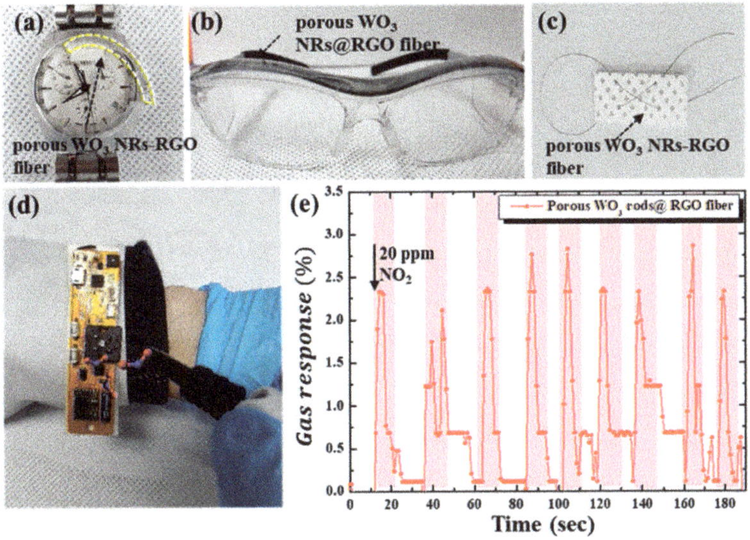

Figure 9. Images of RGO/WO_3/TCNF fibers integrated into various objects: (**a**) A wristwatch (the area highlighted in yellow is where the fiber was integrated); (**b**) A pair of safety goggles; (**c**) Sown onto Kimtech paper; (**d**) A wearable sensing module. NO_2 gas monitoring from the portable sensing device (**d**) is depicted in (**e**) [125].

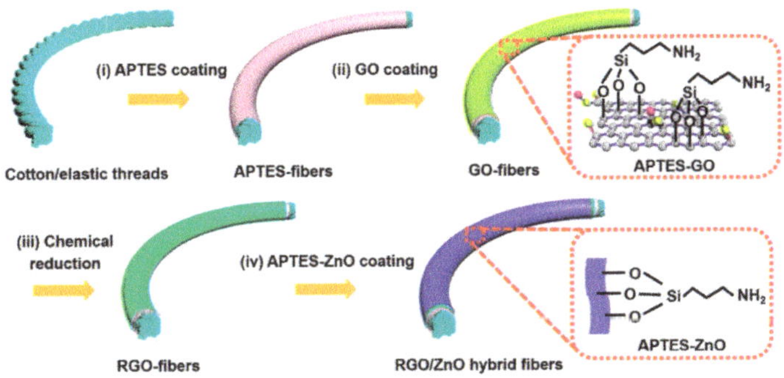

(**a**)

(**b**)

Figure 10. (**a**) Schematic illustration depicting the multi-step fabrication of RGO/ZnO/thread sensors; (**b**) Image of sensor array network of RGO/ZnO/thread and its integration onto wearable fabric [128].

The group of Lee et al. carried out extensive studies using commercially available fiber materials and coating them with graphene for various applications [129–131]. They reported the development of RGO-decorated cotton and polyester yarns [129]. The yarns were dip-coated in GO, which self-assembled onto the fibers. Subsequently, the GO-coated yarns were reduced to RGO through a low-temperature chemical reduction process. Utilizing these RGO fibers, they constructed devices capable of selectively detecting NO_2 gas at concentrations as low as 0.25 ppm at room temperature. The RGO-cotton yarn sensors exhibited a response of -7.0% (R (%) = $(R_g - R_a)/R_a \times 100$, where R_g and R_a denote the electrical resistance upon exposure of NO_2 and air, respectively), whereas RGO-polyester yarn sensors yielded a response -6.0% (Figure 11a, Table 2). When RGO-yarn sensors are exposed to NO_2, the resistance of the sensors decreases. This decline was attributed to an increase in hole concentrations, resulting in the observed negative response.

Following their initial report, they investigated the use of cotton yarn coated with RGO and MoS_2, utilizing similar processing methods [130]. This study revealed that incorporating MoS_2 into RGO-containing fibers increased their sensitivity to NO_2 by a factor of four compared to fibers containing only RGO. When exposed to 0.45 ppm of NO_2, RGO/MoS_2/yarn had a response of 28% ($\Delta R/R_0$ (%) = $(R_g - R_0)/R_0 \times 100$, where R_0 and R_g are resistances of the yarn sensor before and after exposure to NO_2, respectively), while fibers without MoS_2 only exhibited a response of 6% (Figure 11b, Table 2). This improved sensitivity was ascribed to the large surface area of the RGO-MoS_2 composite and the synergistic interaction between RGO and MoS_2 [132,133].

Building on their previous reporting, Lee et al. further examined the use of elastic yarn coated with RGO, employing techniques consistent with earlier studies (Figure 11c) [131]. In this investigation, the sensors demonstrated a response of 50–55% (R (%) = $(R_g - R_a)/R_a \times 100$, where R_g and R_a denote the electrical resistance upon exposure to NO_2 and air, respectively) to 5 ppm of NO_2 even under 200% strain (Table 2). Leveraging this performance, they fabricated wearable gas-sensing wristbands, thereby highlighting the potential of these RGO-coated fibers for integration into wearable electronics.

In another study led by Yun et al., nylon-6, a widely used industrial synthetic polymer, was fabricated into a mesh fabric through electrospinning [134]. This technique resulted in the fabrication of nanofibers, which were subsequently functionalized with GO using a self-assembly dip-coating method. Following this coating process, a chemical reduction was applied, converting the GO to RGO, thereby creating RGO/nylon-6 nanofibers (Figure 11d). The resulting nanofibers demonstrated sensitivity to NO_2 gas, detecting concentrations at 1 ppm and exhibiting a response of 13.6% ($|R_g - R_0|/R_0$, where R_0 and R_g are resistances of the gas sensor before and after exposure to NO_2, respectively) at room temperature (Table 2). This response was attributed to the swelling of the hydrophilic and porous polymer, along with the high surface area of the nanofiber [135,136]. The bendability of the nanofibers was also examined, where a negligible change in response was observed for the sensors in flat and bent positions. These findings set a stage for the use of RGO-containing nanofibers in flexible electronics and electronic textiles applications.

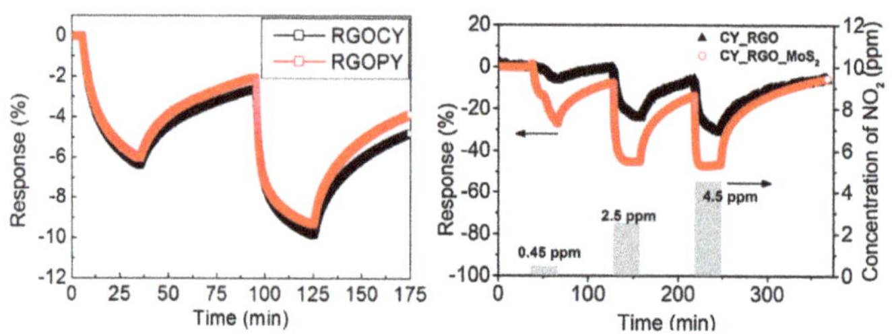

(a) (b)

Figure 11. *Cont.*

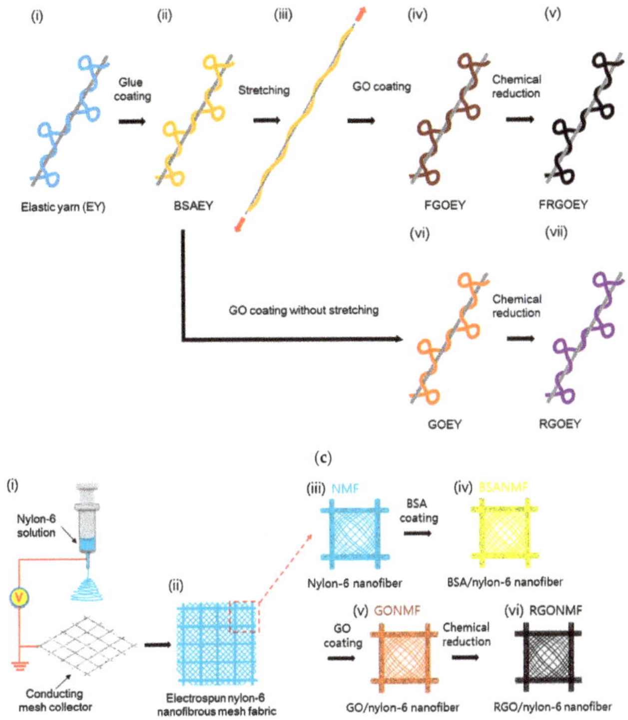

(**d**)

Figure 11. Examples of gas sensors using commercially available fibers: (**a**) Gas sensing performance of RGO-cotton yarn and RGO-polyester yarn exposed to 0.25 ppm and 1.25 ppm of NO_2 at room temperature [129]; (**b**) Gas sensing performance of RGO-cotton yarn and RGO-cotton yarn with MoS_2 exposed to 0.45 ppm, 2.5 ppm, and 4.5 ppm NO_2 gas at room temperature [130]; (**c**) Schematic illustration of the fabrication process for RGO-elastic yarn [131]; (**d**) Schematic illustration of the fabrication process of RGO/nylon-6 [134].

Table 2. Summary of graphene-based fiber gas sensors for CO_2, NO_2, and H_2S gas.

Gas	Conc.	Material	Response	Temp.	Ref.
Carbon dioxide (CO_2)	60 ppm	RGO/PANI/PS	0.8	RT	[91]
Hydrogen Sulfide (H_2S)	10 ppm	RGO-CuO	1.95	300 °C	[93]
	5 ppm	PS/WO_3/NOGR	65.5	300 °C	[95]
	1 ppm	RGO-$ZnFe_2O_4$	147	350 °C	[98]
	1 ppm	RGO/α-Fe_2O_3	9.2	350 °C	[99]
Nitrogen Dioxide (NO_2)	5 ppm	RGO/PVA/PEI	159.4%	RT	[104]
	5 ppm	RGO-PU	176%	RT	[107]
	5 ppm	Cu/Cu_2O/RGO	18.90%	150 °C	[122]
	5 ppm	Ni/NiO/RGO	0.82%	150 °C	[122]
	5 ppm	RGO/WO_3/TCNF	9.67%	100 °C	[125]
	5 ppm	RGO-In_2O_3	42	50 °C	[111]
	5 ppm	RGO-SnO_2	100	200 °C	[114]
	5 ppm	RGO-ZnO	150	400 °C	[70]
	15 ppm	RGO/ZnO/thread	44%	RT	[128]
	0.25 ppm	RGO-cotton yarn	−7.0%	RT	[129]
	0.25 ppm	RGO-polyester yarn	−6.0%	RT	[129]
	0.45 ppm	RGO/MoS_2/yarn	28%	RT	[130]
	5 ppm	RGO-elastic yarn	55%	RT	[131]
	1 ppm	RGO/nylon-6	13.6%	RT	[134]

4. Graphene-Based Fiber Sensors for Polyatomic Gases

The polyatomic gas, ammonia (NH_3), is an important feedstock for fertilizer, energy, and fine chemicals [137]. NH_3 is toxic and exposure to the gas should be restricted to 35 ppm for 10 min [138]. Other polyatomic gases, such as methane (CH_4) and propyl radical (C_3H_7), which are components of natural gas, have thermal conductivities that differ significantly from that of air, making calorimetric gas sensors highly effective for detecting these gases [139,140]. However, the thermal conductivity of NH_3 is similar to that of air, and detection of this gas by measuring thermal conductivity is difficult [139]. Substantial efforts have therefore been dedicated to the gas sensing of NH_3 gas using alternative methods [139,141]. In addition, as the normal concentration of NH_3 in a healthy person ranges from 0.5 to 2 ppm, NH_3 in human breath has been explored as a biomarker, showing considerable potential for liver and kidney disease screening [141,142].

NH_3 Gas Sensor

Gaskov et al. developed NH_3 gas sensors by encapsulating Co_3O_4 nanocrystals within a matrix of RGO [143]. The nanofibers were synthesized by combining cobalt, GO, and polyvinyl pyrrolidone (PVP). During the electrospinning process, GO enveloped cobalt ions, while PVP formed the nanofiber structure (Figure 12). Subsequent calcination resulted in the reduction of GO to RGO, carbonization of PVP into amorphous carbon, and the aggregation of cobalt oxide into larger crystals, ultimately forming RGO-Co_3O_4 nanofibers. The sensor demonstrated a sensitivity of 53.6% ($(R_g - R_a)/R_g$) to 50 ppm of NH_3 at room temperature (Table 3). The strong affinity of RGO for NH_3 contributed to the response of the sensor [144,145]. When NH_3 is adsorbed onto the nanofiber surface, it donates electron density in the relatively high energy lone pair orbital to the sp^2 carbon of graphene, which increases resistance by reducing the number of electron holes.

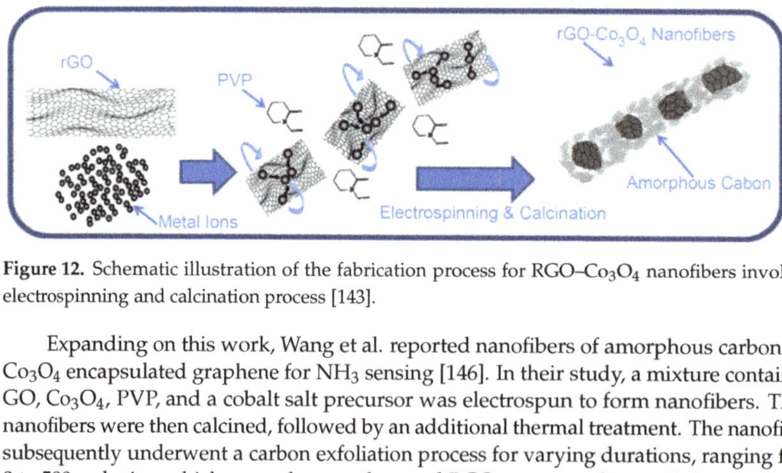

Figure 12. Schematic illustration of the fabrication process for RGO–Co_3O_4 nanofibers involving electrospinning and calcination process [143].

Expanding on this work, Wang et al. reported nanofibers of amorphous carbon and Co_3O_4 encapsulated graphene for NH_3 sensing [146]. In their study, a mixture containing GO, Co_3O_4, PVP, and a cobalt salt precursor was electrospun to form nanofibers. These nanofibers were then calcined, followed by an additional thermal treatment. The nanofibers subsequently underwent a carbon exfoliation process for varying durations, ranging from 0 to 580 s, during which amorphous carbon and RGO aggregated around the cobalt ions, forming carbon/RGO/Co_3O_4 nanofibers. It was observed that the sensor thermally etched for 250 s exhibited the highest response, with a 123% ($(R_g - R_a)/R_a$) sensitivity to 50 ppm of NH_3 at room temperature (Table 3). Compared to the earlier work by Gaskov et al., the carbon/RGO/Co_3O_4 sensors demonstrated a significant improvement in performance, with over a 50% increase in sensitivity achieved through optimized carbon exfoliation.

Wu et al. developed nanofiber sensors for NH_3 gas based on polyaniline, nitrogen-doped graphene quantum dots (N-GQD), and In_2O_3 [147]. The N-GQDs were synthesized through a hydrothermal process, while hollow In_2O_3 nanofibers were fabricated via electrospinning (Figure 13). The N-GQDs were subsequently combined with In_2O_3 nanofibers through electrostatic interaction, and PANI/N-GQD/In_2O_3 nanofibers were prepared through in-situ chemical oxidative polymerization. The assembly of the nanofiber sensors was completed by depositing them onto gold-interdigitated electrodes. In assessing the effect of N-GQD loading, it was found that sensors with 20 wt% N-GQD exhibited the highest response, achieving a sensitivity of 15.2 (R_g/R_a) to 1 ppm NH_3 at room temperature (Table 3). This enhanced sensitivity was attributed to the increased surface area provided by the N-GQDs and hollow In_2O_3 nanofibers, which facilitate greater interaction with PANI and offer numerous adsorption sites for NH_3 gas. Notably, the sensor demonstrated effective NH_3 gas detection at room temperature concentrations ranging from 0.6 ppm to 2.0 ppm, the range in which kidney or liver diseases can be identified in human breath [148,149].

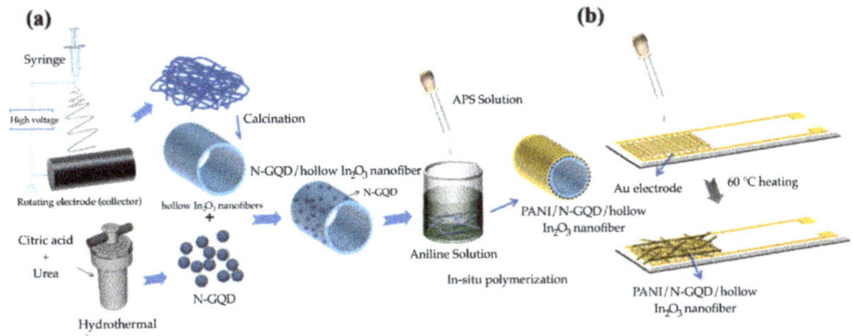

Figure 13. Schematic illustration of (**a**) The preparation of hollow In_2O_3 nanofibers, N-GQDs, and PANI/N-GQD/In_2O_3 nanofibers, and (**b**) nanofiber sensor fabrication [147].

Correa et al. also employed In_2O_3 in their preparation of sensors for NH_3 detection [150]. In their study, In_2O_3 nanofibers were fabricated via electrospinning, followed by a calcination procedure, while RGO was synthesized by partial chemical reduction of GO using sodium citrate. RGO was combined with In_2O_3 via ultrasonication to obtain RGO-In_2O_3 nanofibers, which were then cast onto gold interdigitated electrodes to form gas sensing devices. The sensors displayed a sensitivity of 95% ($[(R_a - R_g)/R_g] \times 100$, where R_a is the sensor resistance in air while R_g is the sensor resistance after being exposed to the gas) in response to 15 ppm of NH_3 gas at room temperature (Table 3). The sensors demonstrated selectivity for NH_3, showing a higher response to it compared to other gases such as acetone, ethanol, methanol, triethylamine, trimethylamine, and monomethylamine. The gas sensing performance is attributed to the formation of a depletion layer and p-n heterojunction between RGO and In_2O_3, where oxygen molecules from the air are adsorbed onto the surface of the nanofiber and electronics flow from n-type In_2O_3 to p-type RGO until equilibrium is reached (Figure 14). When exposed to NH_3, gas molecules react with the oxygen species, which generate electrons and eject them back into the nanofiber, thereby decreasing the resistance of the sensor and providing a sensing signal. RGO contributes to this response as it inherently provides the sensor with active sites at the surface and allows for more gas adsorption as it creates an interconnected structure with In_2O_3 [151,152].

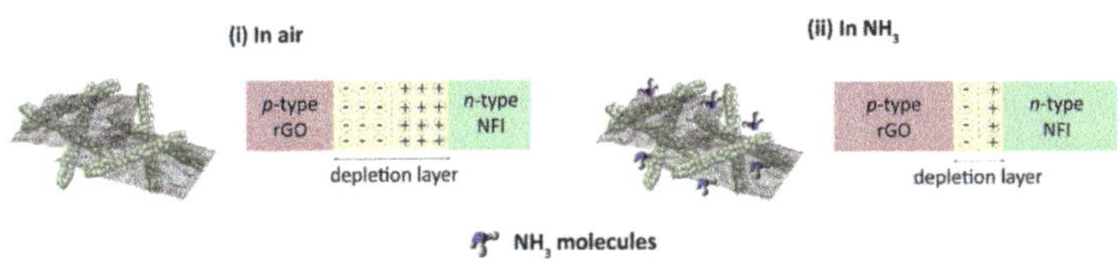

Figure 14. Schematic representation of RGO-In_2O_3 gas sensing mechanism, depicting depletion layer in air (**i**) and in NH_3 gas (**ii**) [150].

Han et al. also contributed advancements in NH_3 gas sensors through their work on wet-spun RGO/$Ti_3C_2T_x$ MXene hybrid fibers [153]. In addition to previously mentioned metal oxides, $Ti_3C_2T_x$ MXene, an emerging class of 2D material, is also recognized for its facilitation of gas adsorption [154,155]. The $Ti_3C_2T_x$ was synthesized by etching Ti_3AlC_2 and combined with GO, then the resulting composite was wet-spun into fibers and subjected to thermal reduction (Figure 15a). MXene and GO undergo galvanic displacement, where oxygen atoms from GO transfer to the MXene surface, while electrons of MXene reduce GO, driven by the difference in their relative potentials [156,157]. The sensors demonstrated a sensitivity of 6.77% ($\Delta R/R_0$) in response to 50 ppm of NH_3 at room temperature (Table 3). When tested for selectivity, the fibers demonstrated a notably higher response to NH_3, while sensitivity to other gases remained low at approximately 1%. (Figure 15b). The potential of these fibers as wearable sensors was further explored by integrating them into a lab coat. The woven RGO/$Ti_3C_2T_x$ MXene sensor demonstrated a response of 7.21% when exposed to 100 ppm of NH_3 gas, underscoring the promise of these fibers for use in wearable and flexible sensing devices.

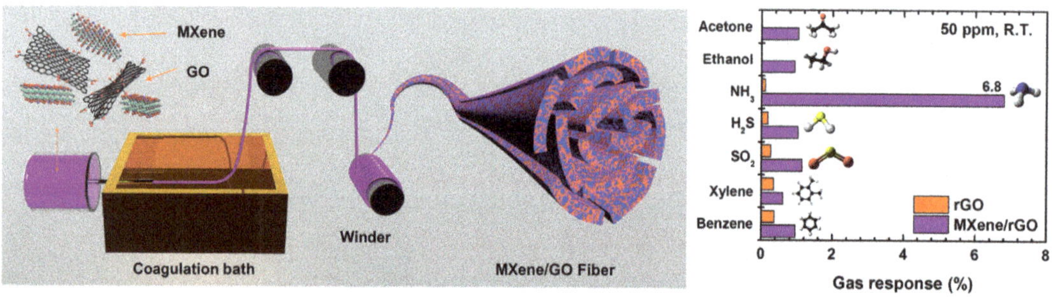

(**a**) (**b**)

Figure 15. (**a**) Schematic illustration of MXene/GO via wet-spinning; (**b**) Selectivity of RGO/$Ti_3C_2T_x$ MXene to NH_3 in comparison to other gases [153].

Dong et al. introduced an innovative approach to fabricating coatings for a quartz crystal microbalance (QCM), a technique used to determine the mass of an analyte absorbed by measuring the frequency changes related to adsorption activity on a quartz crystal [158,159]. They developed an NH_3 gas sensor by utilizing electrospun nanofibers made from polystyrene doped with carboxyl graphene (G-COOH) as the QCM coating [158]. G-COOH was specifically chosen for its high surface area and porosity, which serve to enhance the mass loading of NH_3 molecules onto the QCM [160,161]. The G-COOH and PS composite was electrospun and directly deposited onto the QCM, forming a G-COOH/PS/QCM sensor. When tested with NH_3 concentrations ranging from 1 to 40 ppm at room temperature, the sensor exhibited a decrease in frequency as ammonia concentration increased (Figure 16a). The sensor, exhibiting an inherent frequency of 5 MHz, achieved a sensitivity of 17.67 ng Hz^{-1}.

This work was further expanded upon by Li et al., as they proposed that improved dispersibility of graphene could enhance the response of the sensor [162]. They described their approach as an electrostatic layer-by-layer self-assembly technique, whereby negatively-charged electrospun cellulose acetate (CA) nanofibers were encased with a layer of positively-charged poly(ether imide) and a layer of negatively-charged GO. The resulting nanofiber membrane was utilized as sensing coatings for QCM to form CA/PEI/GO/QCM NH_3 gas sensors. With this modified method, the inherent frequency of the sensor was 5 MHz and the sensitivity of the sensor increased to 53.01 ng Hz^{-1} (Figure 16b). When exposed to 1 ppm of NH_3, the sensor observed a higher response of 0.9 Hz, compared to a response of 0.3 Hz by the previously reported G-COOH/PS QCM sensor (Table 3) [158]. This effect was attributed to the improved uniformity of GO, which was more evenly distributed on the nanofiber as a coating, rather than being mixed into a spinning solution. The authors also attributed this enhanced sensitivity to the 3D structure of the layered fiber, with the CA nanofibers providing permeable space for gaseous NH_3 molecules.

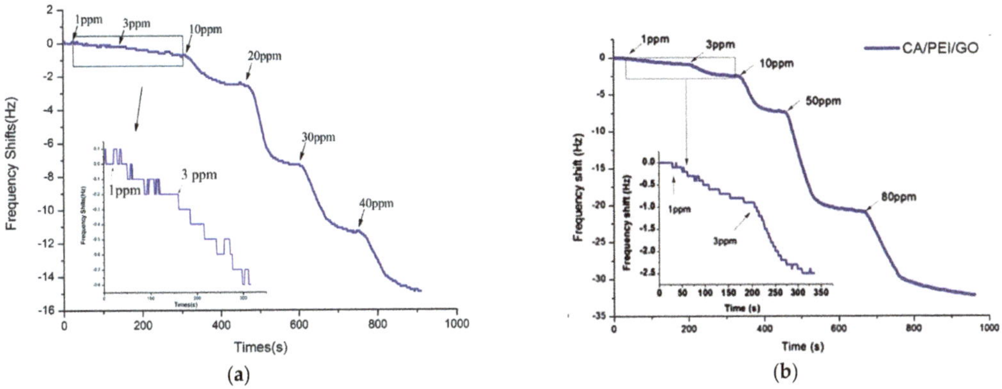

Figure 16. (**a**) Frequency shifts of G-COOH/PS QCM sensors upon exposure to increasing NH_3 concentrations [158]; (**b**) Frequency shifts of CA/PEI/GO QCM sensors upon exposure to increasing NH_3 concentrations [162].

Table 3. Summary of graphene-based fiber gas sensors for NH_3 gas.

Gas	Conc.	Material	Response	Temp.	Ref.
Ammonia (NH_3)	50 ppm	RGO-Co_3O_4	53.6%	RT	[143]
	50 ppm	Carbon/RGO/Co_3O_4	123%	RT	[146]
	1 ppm	PANI/N-GQD/In_2O_3	15.2	RT	[147]
	15 ppm	RGO-In_2O_3	95%	RT	[150]
	50 ppm	$Ti_3C_2T_x$ MXene/RGO	6.77%	RT	[153]
	1 ppm	G-COOH/PS/QCM	0.3 Hz	RT	[158]
	1 ppm	CA/PEI/GO/QCM	0.9 Hz	RT	[162]

5. Graphene-Based Fiber Sensors for Volatile Organic Compounds

Volatile organic compounds (VOCs) are a class of organic compounds that vaporize and aerosolize readily due to a relatively high vapor pressure at standard temperature and pressure [163]. VOCs are emitted from both natural and anthropogenic sources, with several being acutely toxic to humans [164]. Their release into the atmosphere can also lead to the formation of harmful secondary pollutants [165], thus there is a need to monitor levels of these compounds to assess indoor and outdoor air quality. Formaldehyde is a toxic VOC found in common products like paint and preservatives, while chlorobenzene, although less prevalent, is associated with carcinogenic effects [166,167]. Additionally, the emerging interest in using certain gases as biomarkers for human health highlights the importance of detecting and quantifying these gases [168–171]. Analyzing breath samples for the presence or absence of VOCs, such as acetone and ethanol, has shown that they are useful indicators of disease and adverse health conditions [172].

5.1. Acetone

Wang et al. realized porous GO-WO_3 electrospun nanofibers for the gas sensing of acetone [173]. In this study, various volumes of GO ranging from 0 to 1.5 mL were added to a tungsten precursor solution and processed using electrospinning. The resulting nanofibers were calcined to obtain CO-WO_3 nanofiber sensors. The sensor exhibiting the highest response contained nanofibers fabricated from 1 mL of GO-WO_3, with a sensitivity of 35.9 (R_a/R_g) to 100 ppm of acetone vapor at 375 °C (Table 4). The enhanced sensing performance of GO-WO_3 nanofibers is primarily due to the formation of ohmic contact between conductive GO nanosheets and WO_3 nanograins, which facilitates electron migration and resistance modulation [174]. The morphology of the nanofibers, including high surface area and porosity, enables better adsorption and faster diffusion of acetone molecules, leading to quicker response and recovery times [175].

Ghafarinia et al. developed RGO-ZnO nanofibers for the detection of acetone gas via electrospinning [176]. The nanofibers were prepared with different ratios of zinc acetate and GO, facilitated by PVA, followed by a calcination treatment (Figure 17). The sensors were fabricated by depositing the nanofibers onto a silicon wafer. It was determined that the sensor containing a zinc acetate concentration of 4 weight fractions and a GO concentration of 0.07 weight fractions performed the best. The sensor exhibited a sensitivity of 4 (R_{air}/R_{gas}) to 200 ppm of acetone at 200 °C (Table 4). Interestingly, a study of the sensors containing ZnO without RGO revealed that the addition of graphene decreased the optimal operating temperature from 400 °C to 200 °C. This improvement was attributed to the efficient charge transfer capabilities of RGO that refine electrical conductivity [177].

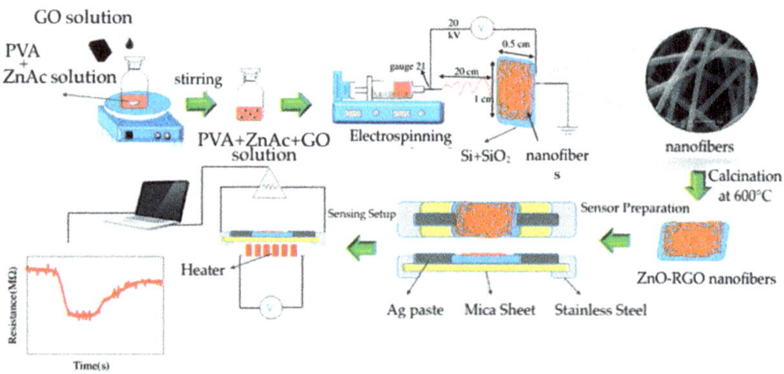

Figure 17. Schematic illustration of the preparation of RGO-ZnO nanofiber sensors [176].

Lu et al. reported RGO/α-Fe_2O_3 nanofibers for acetone gas detection [178]. The nanofibers were fabricated with different loadings of RGO via electrospinning. The optimal sensor, containing 1 wt% of RGO/α-Fe_2O_3 nanofibers, reached a maximum of 8.9 (R_a/R_g) to 100 ppm acetone at 375 °C (Table 4), which was 4.5 times higher than the sensors without RGO. The formation of RGO and Fe_2O_3 heterojunctions generated ohmic contacts, enhancing the sensing signal, while defects and functional groups in RGO provided strong adsorption sites for gas molecules. The sensing mechanism is also influenced by the catalytic effect of RGO in adsorption, where the pores between layers of RGO nanosheets are efficient gas diffusion channels, offering active sites for acetone gas molecules [51].

Shen et al. recognized the potential of RGO-poly(vinylidene fluoride) (PVDF) nanofibers for sensing and energy storage applications [179]. The nanofibers were fabricated by electrospinning a composite of PVDF and GO, followed by the reduction of GO to RGO using hydrazine, resulting in RGO-PVDF nanofibers. These nanofibers were employed in the fabrication of three sensor types, including pressure, photodetector, and gas sensors, as well as three micro-supercapacitors. For each sensor, nickel film electrodes were placed on two ends of the nanofibers at different distances between the electrodes, depending on the type of sensor. All device types were integrated onto a single PDMS substrate with thermally evaporated Ni and Ag tape electrodes serving as electrical interconnections. The entire structure was then encapsulated with an additional PDMS layer, exposing only the sensing materials to air, to create a self-powered multifunctional electronic skin system. The gas sensing function of the electronic skin (e-skin) demonstrated a response of 0.25 ($S = \Delta I/I_0$, where ΔI is the difference current between in the air and in the target gas, I_0 is the current in the air) to 500 ppm of acetone at room temperature (Table 4), with rapid response and recovery times of 5.5 and 30 s, respectively (Figure 18). This work establishes a promising platform for advancing e-skin technologies and integrating sensors into wearable devices.

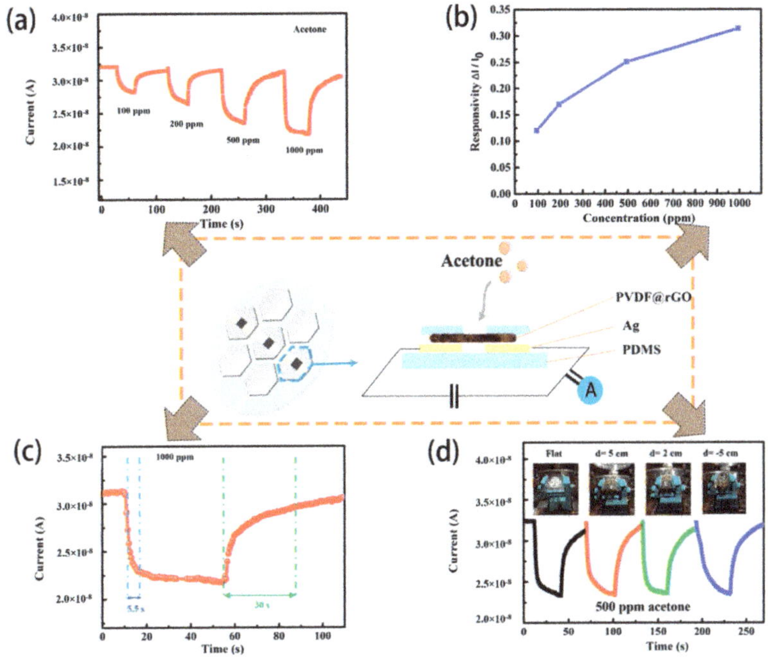

Figure 18. Gas sensor on e-skin device: (**a**) Response of gas sensor to different concentrations of acetone vapor; (**b**) Response of gas sensor to increasing acetone concentrations; (**c**) Response and recovery time of gas sensor; (**d**) sensing stability of device under different bending states in 500 ppm of acetone vapor (each color on the plot represents the response for different bending states) [179].

5.2. Chlorobenzene

Park et al. described chlorobenzene gas sensors made from RGO fibers embedded with copper iodide and metallic copper, developed using an innovative wet spinning method [180]. They began with a GO liquid crystal dispersion, extruding it into a $CuCl_2$-ethylene glycol coagulation solution where the GO aligned due to shear forces and formed fibers with Cu cations through ionic and van der Waals interactions (Figure 19). A portion of the Cu cations is converted to copper hydroxide, which was subsequently reduced to metallic Cu via a redox reaction with hydrogen iodide and acetic acid. This process led to the formation of CuI alloys, with any residual iodine rinsed away. The resulting RGO-Cu fibers served dual purposes as both gas and temperature sensors, where the response of the sensors was measured by the change in conductance (ΔG) upon varying rates of chlorobenzene evaporation. As gas sensors, the fibers showed increased conductance in response to rising chlorobenzene evaporation rates, with sensors containing higher Cu concentrations displaying greater conductance change at 20 °C. The conductance of the sensor reached 12.5×10^{-6} G when exposed to chlorobenzene vapor, with response and recovery times of approximately 70 s (Table 4). The gas sensitivity of RGO-Cu fibers is likely due to the activation of surface oxygen ions in the encapsulated oxidized Cu particles. The change in conductance results from gas molecules absorbed on the surface of Cu, which increase hole conductivity by accepting electrons from the oxygen ions. Additionally, these electrical properties and large accessible surface area for gas activity are promoted by RGO in the fiber sensor [181].

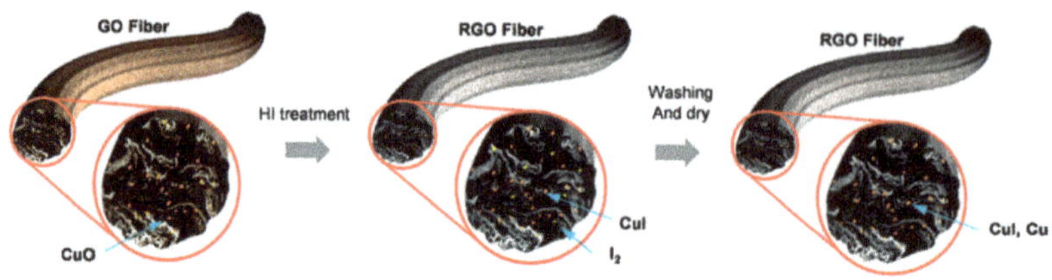

Figure 19. Schematic depicting RGO-Cu fiber preparation illustrating the content of Cu in the fiber [180].

5.3. Ethanol

In et al. reported the fabrication of GO-SnO_2 nanofibers for the detection of ethanol gas [182]. The process of preparing GO-SnO_2 nanofibers involved electrospinning SnO_2 nanofibers, followed by GO dip-coating and thermal annealing. The optimal operating temperature of the sensors was determined to be 300 °C, with a response of 85.3 (R_a/R_g) to 100 ppm of ethanol vapor (Table 4). Additionally, the functionality of the sensors persisted under high relative humidity conditions of 96%, showing a response of 51.75 to 100 ppm ethanol gas (Figure 20a). This impact from humidity decreases the response of the sensor by affecting conductivity, as water molecules compete with the target gas for adsorption sites [183]. The selectivity of the sensor for ethanol vapor was also demonstrated by testing against other gases such as ammonia, acetone, methanol, and ammonia acetate (Figure 20b). The interaction between graphene and SnO_2 was proposed to contribute to the gas sensing mechanism, as electron transfer from SnO_2 to graphene increases the number of active sites that are available for ethanol molecule adsorption [184,185].

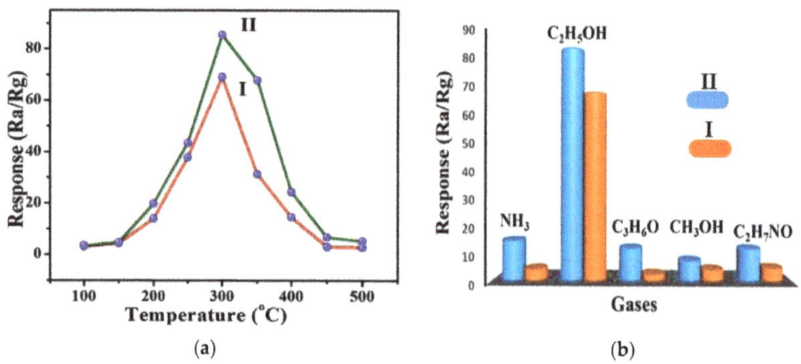

Figure 20. (**a**) Response of GO-SnO_2 nanofibers (II) and SnO_2 without GO (I); (**b**) Selectivity of GO-SnO_2 nanofibers (II) and SnO_2 without GO (I) to ethanol gas compared to other gases [182].

5.4. Formaldehyde

Yang et al. investigated electrospun hollow SnO_2 nanofibers with carbon materials, including graphene, carbon nanotubes, and graphene oxide, and the resulting nanofibers were examined as gas sensors for formaldehyde [186]. The study found that the GO-SnO_2 sensors demonstrated superior gas response and selectivity to formaldehyde vapor, compared to SnO_2 nanofiber sensors and SnO_2 with other nanocarbons (Figure 21a). The optimal operating temperature was 120 °C, where the sensors containing 1 wt% GO achieved a response value of 32 (R_a/R_g) to 100 ppm of formaldehyde (Table 4), which is four times higher than that of nanofibers without GO (Figure 21b). The sensing mechanism of GO-SnO_2 involves the interaction of gas molecules with the surface of metal oxides, leading to changes in electrical conductivity [187]. Hollow SnO_2 nanofibers with porous morphology allow gas molecules to permeate, facilitating gas adsorption and electron transfer. With the addition of GO, the selectivity and sensitivity of the sensor are enhanced by lowering the energy barrier for electron transfer and providing active sites for oxygen species generation [188].

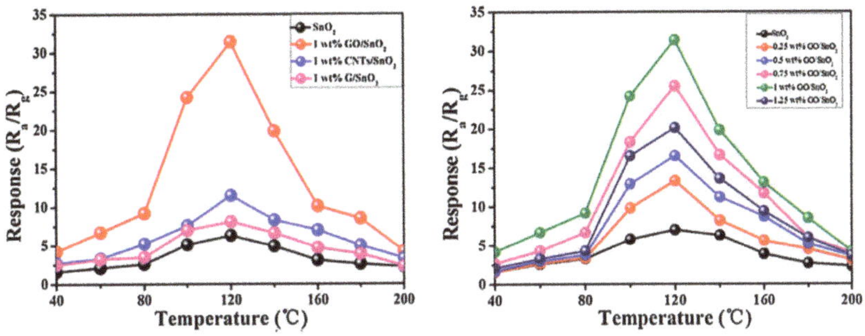

(a) (b)

Figure 21. (**a**) Response of RGO-SnO_2 sensor compared to RGO with other nanocarbons to 100 ppm of formaldehyde as a function of temperature; (**b**) Response of RGO-SnO_2 sensor containing different RGO concentrations as a function of temperature [186].

Table 4. Summary of graphene-based fiber gas sensors for VOCs.

Gas	Conc.	Material	Response	Temp.	Ref.
Acetone (C_3H_6O)	100 ppm	GO-WO_3	35.9	375 °C	[173]
	200 ppm	RGO-ZnO	4	200 °C	[176]
	100 ppm	RGO/α-Fe_2O_3	8.9	375 °C	[178]
	500 ppm	RGO-PVDF	0.25	RT	[179]
Chlorobenzene (C_6H_5Cl)	4.72 μg/s	RGO-Cu	12.5×10^{-6} G	20 °C	[180]
Ethanol (C_2H_6O)	100 ppm	GO-SnO_2	85.3	300 °C	[182]
Formaldehyde (CH_2O)	100 ppm	GO-SnO_2	32	120 °C	[186]

6. Graphene-Based Fiber Sensors for Detection of Multiple Gases

Thus far in this review, the focus has been placed on gas sensors specifically designed for the detection of individual gases. However, gas sensing systems capable of detecting two or more gasses are also extremely advantageous in real-world, complex environments. Similarly to single gas detection systems, sensing systems for multiple gases can be useful for different scenarios where it is desirable to detect many gases, such as environmental monitoring for air pollutants [189]. Furthermore, multi-gas sensors have been increasingly used in human health diagnoses based on breath, which can contain many trace gases, several of which are biomarkers for disease and adverse health conditions [190–192].

Kim et al. reported a gas-sensing system capable of detecting H_2S and acetone with RGO-SnO_2 nanofibers [193]. SnO_2 nanofibers were obtained via electrospinning and subsequently calcined via thermal treatment. The resulting nanofibers were combined with GO, followed by a thermal reduction to achieve an RGO-SnO_2 nanofiber composite (Figure 22). Sensors containing 0.01 wt% of RGO exhibited a response of 34 (R_{air}/R_{gas}) to 5 ppm of H_2S at 200 °C, while sensors containing 5 wt% RGO displayed a response of 10 to 5 ppm of acetone at 350 °C (Table 5). It was found that at low RGO loading concentrations, the sensing properties were predominately influenced by the SnO_2 component, whereas the RGO component determined the electrical transport at higher RGO concentrations. Notably, the sensors were examined in a humid atmosphere at the respective optimal operating temperatures to investigate the interference of water vapor. As the sensors were able to detect the gases in humid air, this shows the potential of these sensors for breath analyzers that use acetone and H_2S as biomarkers for the diagnosis of diabetes and halitosis [194,195]. This gas sensing system is notable as the loading of RGO in the nanofiber composite results in greater sensitivity to one gas, rather than the other. In addition, it underscores the challenges of gas sensing in the presence of interfering substances.

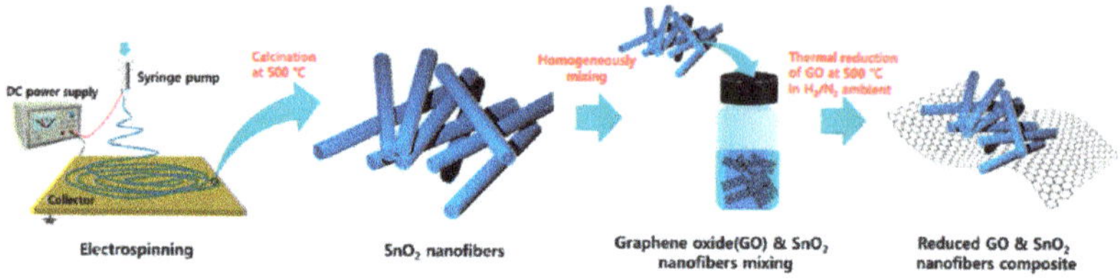

Figure 22. Schematic illustration of the preparation of RGO-SnO_2 nanofiber composite [193].

Fabricating gas sensors often encounter problems related to susceptibility to environmental changes, such as humidity and temperature [196]. A primary limitation of gas sensors is achieving high selectivity to detect a specific gas amid interfering substances, as cross-sensitivity remains a persistent challenge in both single-gas and multi-gas sensors, affecting accuracy in real-world conditions [197]. Addressing this problem involves controlling the morphology of sensing materials to enhance the detection of particular gases. It is necessary to optimize the sensing material by fine-tuning its surface properties and nanostructure to achieve better interaction with gases while taking into account the chemical composition, molecular size, and reactivity of the target gas molecule. It is significant to note that ongoing research involving gas sensors, including electronic nose technology, is exploring artificial intelligence and machine learning algorithms in sensing systems to enhance the accuracy and precision of gas detection and differentiation [15,198]. Further research is required to mitigate cross-sensitivity issues and facilitate precise discrimination between gases in multi-gas sensing applications.

Ren et al. also fabricated electrospun RGO-SnO_2 nanofibers and studied their sensing behavior to NO_2 and sulfur dioxide (SO_2) under different intensities of UV light illumination [199]. RGO-SnO_2 sensors were prepared through electrospinning, following calcination, and ultrasonic treatments, where different concentrations of SnO_2 were combined with RGO, and the resulting sensors were investigated from dark to UV light irradiation with different light intensities. The sensor containing a mass ratio of RGO:SnO_2 at 1:40 showed relatively similar responses of 23% and 22% ($(R_g - R_a)/R_a \times 100\%$, where R_g and R_a are the resistance values in the gas and air, respectively) to 3 ppm of NO_2 and 30 ppm SO_2, respectively, in a dark environment at room temperature (Table 5). Interestingly, under 97 mW/cm^2 of UV illumination, the sensor exhibited the highest response of 102% to NO_2 but also the lowest response of 11% to SO_2. In the presence of UV light, SnO_2 absorbs UV light and collects photo-electrons, whereas RGO accepts these photo-electrons and facilitates charge transport. The enhanced selectivity is likely due to photocatalytic oxidation and photochemical desorption effects, leading to varied responses depending on the gas [200]. The findings of this study suggest that sensor selectivity can be improved by optimizing the intensity of excitation light, presenting the role of UV light in improving gas detection for specific gases.

Kim et al. presented nanofibers composed of RGO and SnO_2, loaded with platinum (Pt) or palladium (Pd), for the selective detection of benzene and toluene, respectively [201]. Unlike the previously discussed study that modulates graphene to target different gases, this approach optimized the gas-sensing properties of the sensor by varying the type of metal used. The sensors were fabricated by incorporating either Pt or Pd nanoparticles, grown via UV irradiation, into a SnO_2 and RGO composite, which was then processed into nanofibers using electrospinning. The RGO/Pd/SnO_2 sensors exhibited the highest sensitivity to benzene, with a response of 12.3 ($R = R_a/R_g$, where R_a and R_g are the resistances in the presence of air and target gas) at 5 ppm at 200 °C. In contrast, the RGO/Pt/SnO_2 nanofibers demonstrated the strongest response to toluene, achieving a sensitivity of 16.0 at 5 ppm, which is 255% greater than their response to benzene (Table 5). This behavior can be attributed to toluene generating more hydrogen molecules than benzene, enabling Pt to dissociate toluene more efficiently, resulting in a stronger response in Pt-loaded sensors [202,203]. In the case of Pd-loaded sensors, Pd nanoparticles demonstrate higher catalytic activity for benzene decomposition owing to their lower adsorption energy [204,205]. Conversely, the adsorption of toluene by Pt is electronically favorable, while benzene dehydrogenation is thermodynamically unfavorable [205]. RGO also contributes to the gas sensing mechanism by absorbing electrons from adjacent SnO_2, which increases the resistivity within the nanofibers, thereby reducing their conductivity and intensifying resistance modulation [124]. Electron flow through the connected SnO_2 nanograins and p-n junctions at the SnO_2 and RGO interfaces alters resistance as the depletion region contracts upon gas exposure (Figure 23).

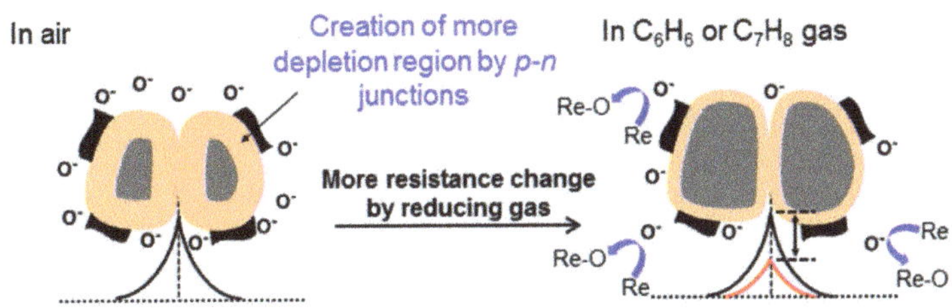

Figure 23. Schematic illustration of gas sensing mechanism of RGO/(Pt or Pd)/SnO_2 (the change in the potential barrier is presented by the black and red curves) [201].

The same research group also reported RGO and ZnO nanofiber sensors following similar methodologies for the gas sensing of CO and benzene [206]. The preparation of the sensors involved the synthesis of Au and Pd nanoparticles through UV radiation and incorporation with RGO and ZnO (Figure 24). The resulting composite solution formed nanofibers via electrospinning, and the as-spun nanofibers were subjected to a calcination treatment. The nanofibers containing Au exhibited a higher response to CO, whereas the sensors with Pd demonstrated a greater sensitivity to benzene. The RGO/Au/ZnO sensor showed a response of 23.5 to 1 ppm of CO, while RGO/Pd/ZnO had a response of 11.8 to 1 ppm of benzene at 400 °C (Table 5). This sensitivity is due in part to the high catalytic efficiency of Au nanoparticles for CO oxidation by lowering the oxidation barrier and the small kinetic diameter of CO, which allows molecules to permeate into the sensor and result in an amplified response. This gas sensing system is notable for its ability to achieve a stronger response to one gas over another by adjusting the type of metal nanoparticle employed.

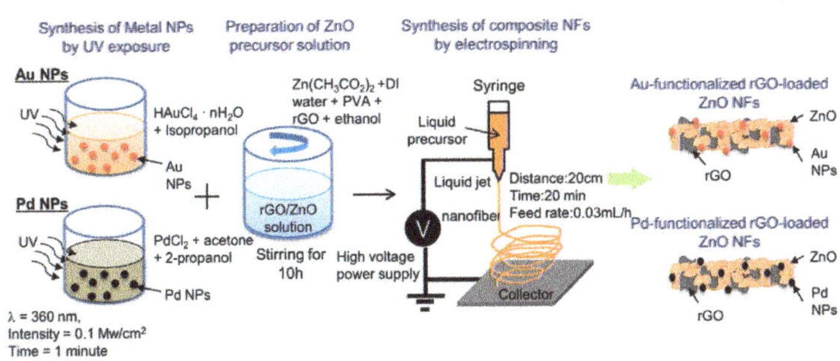

Figure 24. Schematic illustration of the synthesis of RGO/(Au or Pd)/ZnO [206].

Ruiz-Valdepeñas et al. also realized the sensing properties of graphene-doped SnO_2 nanofibers, utilizing graphene synthesized through a liquid phase exfoliation process, whereby direct exfoliation of graphite was achieved in water-based solutions without the use of stabilizing agents [207,208]. The sensors consisted of electrospun nanofibers with a diameter of around 50 nm and nanoribbons approximately 1 μm in diameter deposited onto interdigitated electrodes. When exposed to various gases, the sensor exhibited responses exceeding 35% ($(R_a - R) \times 100/R$, where R_a and R are sensor resistance under exposure to air and selected gas, respectively) for acetone and ethanol gases at temperatures ranging from 25–300 °C, with peak responses of approximately 85% to 4 ppm of acetone and 90% to 2 ppm of ethanol at 300 °C (Table 5). The sensor demonstrated negligible responses for CO and NO gases, thereby displaying a preference for acetone and ethanol. This behavior does not indicate selectivity for specific gases but rather shows a stronger sensitivity to certain gases compared to others. Although sensitivity is optimal at higher temperatures, varying the temperature allows for different sensor responses, facilitating their use in multi-sensor systems. It was proposed that the presence of graphene increases both the detection of the sensors at low temperatures and the response to gases. This enhancement was attributed to the existence of n-p heterojunctions that form potential barriers influenced by gas adsorption, while the porous structure of the nanofibers and nanoribbons improves gas penetration, thereby increasing sensitivity [209].

Cheng et al. investigated the effects of varying types of RGO on gas sensitivity performance by fabricating nanofibers from polyaniline, camphorsulfonic acid (HCSA), polyethylene oxide (PEO), and different RGO forms, including thermally reduced (trGO), chemically reduced (crGO), chemically reduced for 6 h (crGO-6), and chemically reduced for 24 h (crGO-24) [210]. These electrospun nanofibers were deposited onto interdigitated microelectrodes to fabricate sensors, which were then tested for responses to aliphatic alcohol vapors: methanol, ethanol, and 1-propanol. The sensor with crGO-6 exhibited the highest resistance modulation, showing the strongest response to 1-propanol, followed by methanol and ethanol. It exhibited responses of 22.6, 7.9, and 2.1 ($\Delta R/R_0$, where R_0 is baseline resistance and ΔR is change in resistance upon exposure to analyte vapor) to 200 ppm of 1-propanol, methanol, and ethanol, respectively, at room temperature (Table 5), outperforming the other sensors containing differing RGO variants. Upon adsorption of vapor molecules, the nanofiber swells, increasing the separation between PANI chains, widening the electron transport gap, and increasing the resistance, with larger analytes amplifying this effect. The enhanced response of crGO-6 compared to crGO-24 suggests that hydrogen bonding between vapor molecules and RGO plays a role in resistance modulation and contributes to the overall sensing mechanism [211]. This study not only examines the impact of various reduction methods but also points to the effects of gas molecule size on the gas-sensing response.

He et al. developed an RGO-MoS_2 composite fiber with NO_2 and NH_3 gas-sensing properties [212]. The synthetic approach involved wet-spinning a composite containing GO and sodium molybdate, followed by treatment with L-cysteine, hydrothermal process, then thermal annealing. The resulting RGO-MoS_2 fiber consists of MoS_2 domains anchored onto the surface of graphene. It is observed that the sensor displayed a sensitivity of −85% (S (%) = $100 \times \Delta R/R_0 = 100 \times R_g - R_0/R_0$, where R_g is the resistance under target gas exposure and R_0 is the initial resistance under N_2 exposure) to 100 ppm of NO_2 and 100% to 100 ppm of NH_3 gas (Table 5). RGO-MoS_2 conjugates facilitate rapid charge transfer, leading to fast resistance fluctuations. When the sensor is exposed to NO_2 gas, the p-type dopant accepts electrons from MoS_2, resulting in a decrease in the resistance of the sensor (Figure 25a). When exposed to NH_3 gas, an n-type dopant, electrons are donated to MoS_2, resulting in an increase in the resistance (Figure 25b). This study demonstrates a gas-sensing system achieved by a singular fiber material that exhibits an inverse response to either NO_2 or NH_3. It should be noted that while selectivity between these two gases is high, the detection of either of these gases from the ambient atmosphere, where cross-selectivity poses challenges, was not tested in this study. However, these findings, particularly the sensitivity measured by resistance changes, warrant further study to identify potential response patterns that could enable differentiation between gases from the sensor. As demonstrated in this study, different gases interact uniquely with the sensing material, resulting in characteristic resistance changes. This distinction can be employed in sensors to improve pattern recognition and enhance the selectivity of sensing systems [213,214].

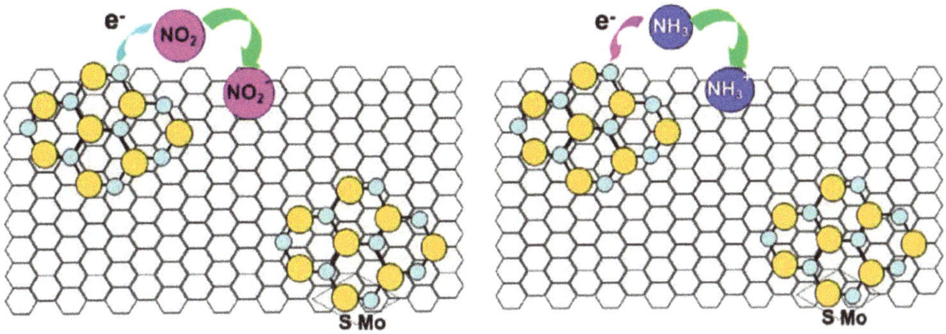

Figure 25. (**a**) Schematic illustrating the gas sensing mechanism of RGO-MoS_2 composite fiber to (**a**) NO_2 and (**b**) NH_3 gas [212].

Yoo et al. reported fiber sensors composed of RGO and ZnO composites with the ability to sense NO_2 and H_2S gas [215]. The preparation of the fibers involved the wet-spinning of GO suspension into a coagulation bath containing calcium chloride, followed by hydrothermal treatment with zinc nitrate to obtain RGO-ZnO fibers (Figure 26a). The sensors were constructed by depositing the fibers onto interdigitated gold electrodes and adhering them with tape. The fiber sensor exhibited a sensitivity of 1.65 $((R_a - R_g)/R_g] \times 100$ (%), where R_a and R_g are the resistances of the sensor material when air and gas were injected into the sensor, respectively) when exposed to 4 ppm of NO_2 and 2.68 to 20 ppm of H_2S at room temperature, respectively (Table 5). When exposed to CO_2 and H_2 gases, the sensor showed low sensitivity, thus demonstrating greater responses towards NO_2 and H_2S. The stability of the sensor was evaluated by subjecting the sensor to continuous NO_2 and H_2S exposure for 10 days, where the sensor performance was observed to be constant for both gases. The fiber sensor possesses p type semiconductor characteristics, which causes the formation of a hole accumulation layer in the open air (Figure 26b). Electron transfer from the fiber increases hole concentration upon exposure to NO_2. In contrast, exposure to H_2S causes electrons to be transferred to the fiber and reduces the hole concentration, thereby inducing n-type behavior in the fiber [216,217]. The sensing mechanism is further influenced by the morphology of ZnO and RGO, which possess a high surface area of catalytic sites for gas adsorption and desorption. This work presents a promising platform for real-time human health monitoring, leveraging H_2S as a biomarker, as well as in environmental monitoring, as NO_2 is considered a hazardous gas. Crucially, the stability of these nanocomposites is demonstrated, marking a significant advancement towards the development of lightweight and robust sensors with potential applications in wearable electronics and automated portable devices.

Although graphene itself is known for its chemical stability and mechanical durability, the longevity of graphene-based fibers is influenced by interfacial properties between graphene and fiber-forming composite materials [218,219]. Ensuring long-term operational stability under various conditions is crucial for sensor performance reliability. While graphene itself demonstrates mechanical and chemical stability, the long-term stability of graphene-based composite fibers often depends on the non-graphene components, such as metal oxides or polymers [113,218]. In the case of polymer-containing composites, the durability of these fibers is influenced by the interaction between graphene sheets and the polymer matrix, which serves as the fiber-forming framework [218]. Therefore, the strain on this matrix plays a critical role in determining the overall stability and performance of the fiber. However, in composites consisting of metal oxides, graphene can enhance the structural integrity and improve the stability of the overall network owing to synergistic effects [220]. Continued research into these composite materials will aid in developing graphene fibers with optimal stability to achieve the best performance.

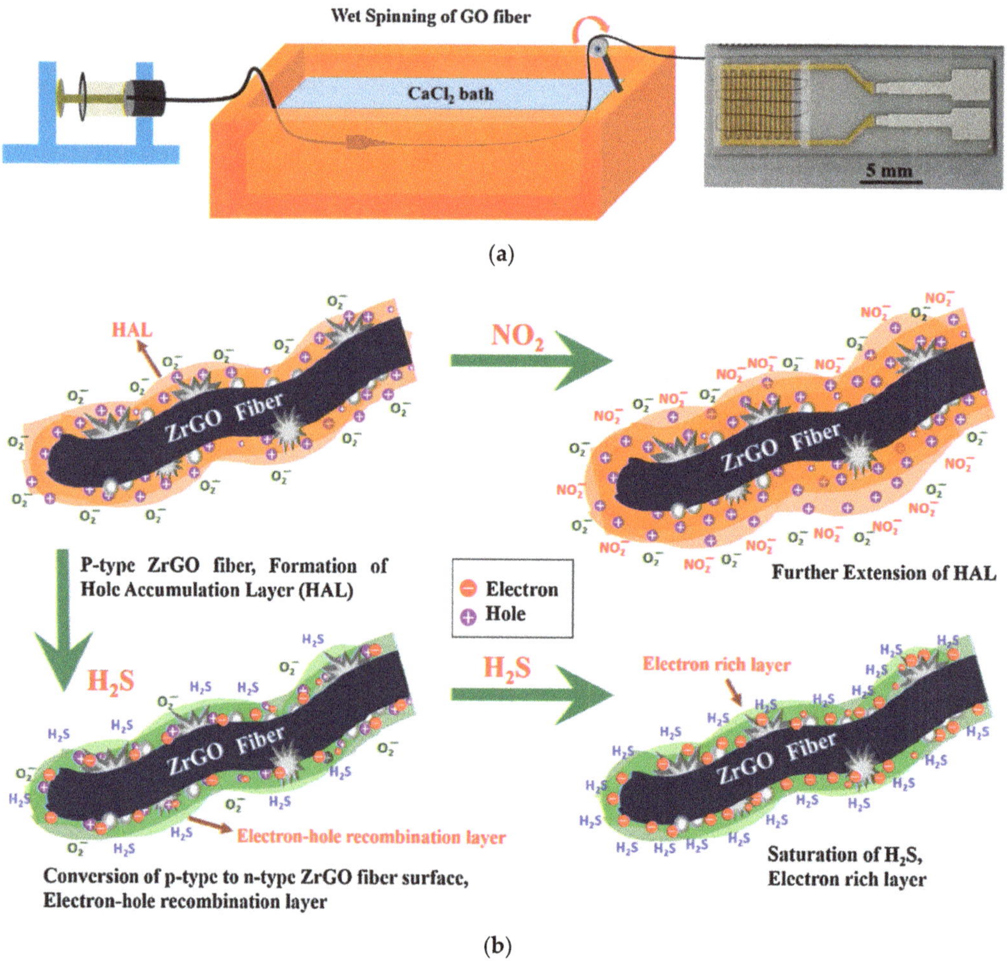

Figure 26. (**a**) Schematic illustration of wet-spinning of GO fiber and fiber sensor; (**b**) Schematic illustrating the sensing mechanism of RGO-ZnO fiber [215].

It is evident that sensor systems for multiple gases demonstrate significant potential across various applications, owing to their ability to target a wide range of gases. As mentioned, this versatility enables their use in diverse fields such as environmental monitoring, industrial safety, and medical diagnostics, where the detection of multiple gases simultaneously is crucial for accurate assessments and timely responses. The examples presented in this review showcase the current framework and obstacles to overcome for the expansion of more sophisticated systems with high sensitivity and selectivity.

Table 5. Summary of gas sensing systems for multiple gases.

Material	Gas	Conc.	Response	Temp.	Ref.
RGO-SnO_2	H_2S Acetone	5 ppm	34 10	200 °C 350 °C	[193]
RGO-SnO_2 I UV light	NO_2 SO_2	3 ppm 20 ppm	102% 11%	RT	[199]
RGO/Pd/SnO_2 RGO/Pt/SnO_2	Benzene Toluene	5 ppm	12.3 16.0	200 °C	[201]
RGO/Au/ZnO RGO/Pd/ZnO	CO Benzene	5 ppm	35.8 22.8	400 °C	[206]
Graphene-SnO_2	Acetone Ethanol	4 ppm 2 ppm	85% 90%	300 °C	[207]
RGO/PANI/HCSA/PEO	1-propanol Methanol Ethanol	200 ppm	22.6 7.9 2.1	RT	[210]
RGO-MoS_2	NO_2 NH_3	100 ppm	−85% 100%	RT	[212]
RGO-ZnO	NO_2 H_2S	8 ppm	1.86 0.87	RT	[215]

7. Conclusions and Perspectives

This review highlights a decade of advancements in graphene-based composite fibers for gas sensing applications. It explores their preparation, fabrication into sensors, and gas sensing mechanisms, emphasizing graphene's ability to enhance sensitivity and selectivity through its high surface area, electrical conductivity, and chemical tunability. These fibers also offer flexibility and mechanical strength, enabling integration into wearable and flexible electronics. The performance of these fibers in detecting various gases, including diatomic (H_2, CO), triatomic (CO_2, NO_2, H_2S), polyatomic (NH_3), and VOCs (acetone, ethanol, formaldehyde) is detailed, with multi-gas sensing systems summarized for broader applications. As graphene-based fiber sensors continue to evolve, their seamless incorporation into everyday objects as wearable and portable devices holds great potential to enable real-time monitoring in fields such as industrial safety, environmental monitoring, and medical diagnostics.

Despite their promise, challenges persist. Environmental factors like temperature and humidity, as well as cross-sensitivity, impact real-world accuracy. Optimization of material morphology and surface properties is critical to enhancing selectivity and performance. Key hurdles in practical deployment include ensuring long-term stability, scalability for industrial production, and sustainable manufacturing. Stability depends on the interactions between graphene and other components within the composite, such as polymers or metal oxides. Scaling up production requires refining techniques like wet-spinning and electrospinning for consistent, high-quality output. Sustainability efforts must focus on renewable graphene production and recycling to minimize environmental impact.

Continued research is vital to overcoming challenges and driving innovation, paving the way for environmentally conscious, next-generation gas sensing technologies. This review has presented the versatility of graphene-based fibers and the significant potential these materials hold for gas sensing systems, highlighting the foundation established for the next breakthrough in the field.

Author Contributions: Writing—original draft preparation, review, and editing, S.V.; Supervision, writing—review and editing, M.S. and R.I. All authors have read and agreed to the published version of the manuscript.

Funding: This research received no external funding.

Institutional Review Board Statement: Not applicable.

Informed Consent Statement: Not applicable.

Data Availability Statement: Not applicable.

Conflicts of Interest: The authors declare no conflicts of interest.

References

1. Jiménez-Cadena, G.; Riu, J.; Rius, F.X. Gas Sensors Based on Nanostructured Materials. *Analyst* **2007**, *132*, 1083. [CrossRef]
2. Dhall, S.; Mehta, B.R.; Tyagi, A.K.; Sood, K. A Review on Environmental Gas Sensors: Materials and Technologies. *Sens. Int.* **2021**, *2*, 100116. [CrossRef]
3. Nikolic, M.V.; Milovanovic, V.; Vasiljevic, Z.Z.; Stamenkovic, Z. Semiconductor Gas Sensors: Materials, Technology, Design, and Application. *Sensors* **2020**, *20*, 6694. [CrossRef]
4. Shaalan, N.M.; Ahmed, F.; Saber, O.; Kumar, S. Gases in Food Production and Monitoring: Recent Advances in Target Chemiresistive Gas Sensors. *Chemosensors* **2022**, *10*, 338. [CrossRef]
5. Essl, C.; Seifert, L.; Rabe, M.; Fuchs, A. Early Detection of Failing Automotive Batteries Using Gas Sensors. *Batteries* **2021**, *7*, 25. [CrossRef]
6. Kumar, P.; Skouloudis, A.N.; Bell, M.; Viana, M.; Carotta, M.C.; Biskos, G.; Morawska, L. Real-Time Sensors for Indoor Air Monitoring and Challenges Ahead in Deploying Them to Urban Buildings. *Sci. Total Environ.* **2016**, *560–561*, 150–159. [CrossRef] [PubMed]
7. Gomes, J.B.A.; Rodrigues, J.J.P.C.; Rabêlo, R.A.L.; Kumar, N.; Kozlov, S. IoT-Enabled Gas Sensors: Technologies, Applications, and Opportunities. *J. Sens. Actuator Netw.* **2019**, *8*, 57. [CrossRef]
8. Nasiri, N.; Clarke, C. Nanostructured Gas Sensors for Medical and Health Applications: Low to High Dimensional Materials. *Biosensors* **2019**, *9*, 43. [CrossRef]
9. Tsujita, W.; Yoshino, A.; Ishida, H.; Moriizumi, T. Gas Sensor Network for Air-Pollution Monitoring. *Sens. Actuators B Chem.* **2005**, *110*, 304–311. [CrossRef]
10. Calderón-Garcidueñas, L.; Azzarelli, B.; Acuna, H.; Garcia, R.; Gambling, T.M.; Osnaya, N.; Monroy, S.; Del Rosario Tizapantzi, M.; Carson, J.L.; Villarreal-Calderon, A.; et al. Air Pollution and Brain Damage. *Toxicol. Pathol.* **2002**, *30*, 373–389. [CrossRef]
11. Bernstein, J.A.; Alexis, N.; Barnes, C.; Bernstein, I.L.; Nel, A.; Peden, D.; Diaz-Sanchez, D.; Tarlo, S.M.; Williams, P.B.; Bernstein, J.A. Health Effects of Air Pollution. *J. Allergy Clin. Immunol.* **2004**, *114*, 1116–1123. [CrossRef] [PubMed]
12. Meixner, H.; Gerblinger, J.; Fleischer, M. Sensors for Monitoring Environmental Pollution. *Sens. Actuators B Chem.* **1993**, *15*, 45–54. [CrossRef]
13. Tai, H.; Wang, S.; Duan, Z.; Jiang, Y. Evolution of Breath Analysis Based on Humidity and Gas Sensors: Potential and Challenges. *Sens. Actuators B Chem.* **2020**, *318*, 128104. [CrossRef]
14. James, D.; Scott, S.M.; Ali, Z.; O'Hare, W.T. Chemical Sensors for Electronic Nose Systems. *Microchim. Acta* **2005**, *149*, 1–17. [CrossRef]
15. Röck, F.; Barsan, N.; Weimar, U. Electronic Nose: Current Status and Future Trends. *Chem. Rev.* **2008**, *108*, 705–725. [CrossRef]
16. Kakani, V.; Nguyen, V.H.; Kumar, B.P.; Kim, H.; Pasupuleti, V.R. A Critical Review on Computer Vision and Artificial Intelligence in Food Industry. *J. Agric. Food Res.* **2020**, *2*, 100033. [CrossRef]
17. Lang, K.; Liu, T.; Padilla, D.J.; Nelson, M.; Landorf, C.W.; Patel, R.J.; Ballentine, M.L.; Kennedy, A.J.; Shih, W.-S.; Scotch, A.; et al. Nanofibers Enabled Advanced Gas Sensors: A Review. *Adv. Sens. Energy Mater.* **2024**, *3*, 100093. [CrossRef]
18. Jian, Y.; Hu, W.; Zhao, Z.; Cheng, P.; Haick, H.; Yao, M.; Wu, W. Gas Sensors Based on Chemi-Resistive Hybrid Functional Nanomaterials. *Nano-Micro Lett.* **2020**, *12*, 71. [CrossRef]
19. Sharma, A.; Eadi, S.B.; Noothalapati, H.; Otyepka, M.; Lee, H.-D.; Jayaramulu, K. Porous Materials as Effective Chemiresistive Gas Sensors. *Chem. Soc. Rev.* **2024**, *53*, 2530–2577. [CrossRef]
20. Chakraborthy, A.; Nuthalapati, S.; Nag, A.; Afsarimanesh, N.; Alahi, M.E.E.; Altinsoy, M.E. A Critical Review of the Use of Graphene-Based Gas Sensors. *Chemosensors* **2022**, *10*, 355. [CrossRef]
21. Sabri, S.S.; Guillemette, J.; Guermoune, A.; Siaj, M.; Szkopek, T. Enhancing Gas Induced Charge Doping in Graphene Field Effect Transistors by Non-Covalent Functionalization with Polyethyleneimine. *Appl. Phys. Lett.* **2012**, *100*, 113106. [CrossRef]
22. Ding, B.; Wang, M.; Yu, J.; Sun, G. Gas Sensors Based on Electrospun Nanofibers. *Sensors* **2009**, *9*, 1609–1624. [CrossRef]
23. Fakih, I.; Mahvash, F.; Siaj, M.; Szkopek, T. Sensitive Precise pH Measurement with Large-Area Graphene Field-Effect Transistors at the Quantum-Capacitance Limit. *Phys. Rev. Appl.* **2017**, *8*, 044022. [CrossRef]
24. Kunze, S.; Groll, R.; Besser, B.; Thöming, J. Molecular Diameters of Rarefied Gases. *Sci. Rep.* **2022**, *12*, 2057. [CrossRef] [PubMed]
25. Yuan, C.; Ma, J.; Zou, Y.; Li, G.; Xu, H.; Sysoev, V.V.; Cheng, X.; Deng, Y. Modeling Interfacial Interaction between Gas Molecules and Semiconductor Metal Oxides: A New View Angle on Gas Sensing. *Adv. Sci.* **2022**, *9*, 2203594. [CrossRef]
26. Marcus, Y. The Sizes of Molecules—Revisited. *J. Phys. Org. Chem.* **2003**, *16*, 398–408. [CrossRef]
27. Jo, Y.K.; Jeong, S.-Y.; Moon, Y.K.; Jo, Y.-M.; Yoon, J.-W.; Lee, J.-H. Exclusive and Ultrasensitive Detection of Formaldehyde at Room Temperature Using a Flexible and Monolithic Chemiresistive Sensor. *Nat. Commun.* **2021**, *12*, 4955. [CrossRef]

28. Nogueira, J.J.; Hase, W.L.; Martínez-Núñez, E. Understanding Energy Transfer in Gas–Surface Collisions from Gas-Phase Models. *J. Phys. Chem. C* **2014**, *118*, 2609–2621. [CrossRef]
29. Wang, C.; Yin, L.; Zhang, L.; Xiang, D.; Gao, R. Metal Oxide Gas Sensors: Sensitivity and Influencing Factors. *Sensors* **2010**, *10*, 2088–2106. [CrossRef]
30. Wusiman, M.; Taghipour, F. Methods and Mechanisms of Gas Sensor Selectivity. *Crit. Rev. Solid State Mater. Sci.* **2022**, *47*, 416–435. [CrossRef]
31. Song, R.; Zhou, X.; Wang, Z.; Zhu, L.; Lu, J.; Xue, D.; Wang, Z.; Huang, L.; Chi, L. High Selective Gas Sensors Based on Surface Modified Polymer Transistor. *Org. Electron.* **2021**, *91*, 106083. [CrossRef]
32. Keshavaraja, A.; Jayashri, B.S.; Ramaswamy, A.V.; Vijayamohanan, K. Effect of Surface Modification Due to Superacid Species in Controlling the Sensitivity and Selectivity of SnO_2 Gas Sensors. *Sens. Actuators B Chem.* **1995**, *23*, 75–81. [CrossRef]
33. Kida, T.; Fujiyama, S.; Suematsu, K.; Yuasa, M.; Shimanoe, K. Pore and Particle Size Control of Gas Sensing Films Using SnO_2 Nanoparticles Synthesized by Seed-Mediated Growth: Design of Highly Sensitive Gas Sensors. *J. Phys. Chem. C* **2013**, *117*, 17574–17582. [CrossRef]
34. Zhang, R.; Song, W.; Wang, M.; Ji, H. Controlling the Size of a Zn-MOF through Ligand Exchange and Pore-Tailored ZnO Assemblies for Size-Selective Gas Sensing. *CrystEngComm* **2019**, *21*, 6414–6422. [CrossRef]
35. van den Broek, J.; Weber, I.C.; Güntner, A.T.; Pratsinis, S.E. Highly Selective Gas Sensing Enabled by Filters. *Mater. Horiz.* **2021**, *8*, 661–684. [CrossRef] [PubMed]
36. Jung, H.-T. The Present and Future of Gas Sensors. *ACS Sens.* **2022**, *7*, 912–913. [CrossRef]
37. Fakih, I.; Sabri, S.; Mahvash, F.; Nannini, M.; Siaj, M.; Szkopek, T. Large Area Graphene Ion Sensitive Field Effect Transistors with Tantalum Pentoxide Sensing Layers for pH Measurement at the Nernstian Limit. *Appl. Phys. Lett.* **2014**, *105*, 083101. [CrossRef]
38. Liu, E.; Cai, Z.; Ye, Y.; Zhou, M.; Liao, H.; Yi, Y. An Overview of Flexible Sensors: Development, Application, and Challenges. *Sensors* **2023**, *23*, 817. [CrossRef]
39. Bag, A.; Lee, N. Recent Advancements in Development of Wearable Gas Sensors. *Adv. Mater. Technol.* **2021**, *6*, 2000883. [CrossRef]
40. Alrammouz, R.; Podlecki, J.; Abboud, P.; Sorli, B.; Habchi, R. A Review on Flexible Gas Sensors: From Materials to Devices. *Sens. Actuators Phys.* **2018**, *284*, 209–231. [CrossRef]
41. Shooshtari, M. Ammonia Gas Sensors Based on Multi-Wall Carbon Nanofiber Field Effect Transistors by Using Gate Modulation. *Colloids Surf. Physicochem. Eng. Asp.* **2025**, *704*, 135563. [CrossRef]
42. Novoselov, K.S.; Geim, A.K.; Morozov, S.V.; Jiang, D.; Zhang, Y.; Dubonos, S.V.; Grigorieva, I.V.; Firsov, A.A. Electric Field Effect in Atomically Thin Carbon Films. *Science* **2004**, *306*, 666–669. [CrossRef] [PubMed]
43. Varghese, S.S.; Lonkar, S.; Singh, K.K.; Swaminathan, S.; Abdala, A. Recent Advances in Graphene Based Gas Sensors. *Sens. Actuators B Chem.* **2015**, *218*, 160–183. [CrossRef]
44. Eissa, S.; N'diaye, J.; Brisebois, P.; Izquierdo, R.; Tavares, A.C.; Siaj, M. Probing the Influence of Graphene Oxide Sheets Size on the Performance of Label-Free Electrochemical Biosensors. *Sci. Rep.* **2020**, *10*, 13612. [CrossRef]
45. Wang, T.; Huang, D.; Yang, Z.; Xu, S.; He, G.; Li, X.; Hu, N.; Yin, G.; He, D.; Zhang, L. A Review on Graphene-Based Gas/Vapor Sensors with Unique Properties and Potential Applications. *Nano-Micro Lett.* **2016**, *8*, 95–119. [CrossRef]
46. Schedin, F.; Geim, A.K.; Morozov, S.V.; Hill, E.W.; Blake, P.; Katsnelson, M.I.; Novoselov, K.S. Detection of Individual Gas Molecules Adsorbed on Graphene. *Nat. Mater.* **2007**, *6*, 652–655. [CrossRef]
47. Brisebois, P.P.; Siaj, M. Harvesting Graphene Oxide—Years 1859 to 2019: A Review of Its Structure, Synthesis, Properties and Exfoliation. *J. Mater. Chem. C* **2020**, *8*, 1517–1547. [CrossRef]
48. Toda, K.; Furue, R.; Hayami, S. Recent Progress in Applications of Graphene Oxide for Gas Sensing: A Review. *Anal. Chim. Acta* **2015**, *878*, 43–53. [CrossRef]
49. Al-Dhahebi, A.M.; Gopinath, S.C.B.; Saheed, M.S.M. Graphene Impregnated Electrospun Nanofiber Sensing Materials: A Comprehensive Overview on Bridging Laboratory Set-up to Industry. *Nano Converg.* **2020**, *7*, 27. [CrossRef]
50. Majhi, S.M.; Mirzaei, A.; Kim, H.W.; Kim, S.S. Reduced Graphene Oxide (rGO)-Loaded Metal-Oxide Nanofiber Gas Sensors: An Overview. *Sensors* **2021**, *21*, 1352. [CrossRef]
51. Meng, F.-L.; Guo, Z.; Huang, X.-J. Graphene-Based Hybrids for Chemiresistive Gas Sensors. *TrAC Trends Anal. Chem.* **2015**, *68*, 37–47. [CrossRef]
52. Bilisik, K.; Akter, M. Graphene Nanocomposites: A Review on Processes, Properties, and Applications. *J. Ind. Text.* **2022**, *51*, 3718S–3766S. [CrossRef]
53. Meng, F.; Lu, W.; Li, Q.; Byun, J.; Oh, Y.; Chou, T. Graphene-Based Fibers: A Review. *Adv. Mater.* **2015**, *27*, 5113–5131. [CrossRef] [PubMed]
54. Xu, T.; Zhang, Z.; Qu, L. Graphene-Based Fibers: Recent Advances in Preparation and Application. *Adv. Mater.* **2019**, 32, 1901979. [CrossRef] [PubMed]
55. Li, Y.; Zhu, J.; Cheng, H.; Li, G.; Cho, H.; Jiang, M.; Gao, Q.; Zhang, X. Developments of Advanced Electrospinning Techniques: A Critical Review. *Adv. Mater. Technol.* **2021**, *6*, 2100410. [CrossRef]
56. Nag, A.; Alahi, M.E.E.; Mukhopadhyay, S.C. Recent Progress in the Fabrication of Graphene Fibers and Their Composites for Applications of Monitoring Human Activities. *Appl. Mater. Today* **2021**, *22*, 100953. [CrossRef]
57. Cong, H.-P.; Ren, X.-C.; Wang, P.; Yu, S.-H. Wet-Spinning Assembly of Continuous, Neat and Macroscopic Graphene Fibers. *Sci. Rep.* **2012**, *2*, 613. [CrossRef]

58. Rohani Shirvan, A.; Nouri, A.; Sutti, A. A Perspective on the Wet Spinning Process and Its Advancements in Biomedical Sciences. *Eur. Polym. J.* **2022**, *181*, 111681. [CrossRef]
59. Xue, J.; Wu, T.; Dai, Y.; Xia, Y. Electrospinning and Electrospun Nanofibers: Methods, Materials, and Applications. *Chem. Rev.* **2019**, *119*, 5298–5415. [CrossRef]
60. Eom, W.; Shin, H.; Ambade, R.B.; Lee, S.H.; Lee, K.H.; Kang, D.J.; Han, T.H. Large-Scale Wet-Spinning of Highly Electroconductive MXene Fibers. *Nat. Commun.* **2020**, *11*, 2825. [CrossRef]
61. Zhang, Z.; Shao, C.; Li, X.; Zhang, L.; Xue, H.; Wang, C.; Liu, Y. Electrospun Nanofibers of $ZnO-SnO_2$ Heterojunction with High Photocatalytic Activity. *J. Phys. Chem. C* **2010**, *114*, 7920–7925. [CrossRef]
62. Ilnicka, A.; Lukaszewicz, J.P. Graphene-Based Hydrogen Gas Sensors: A Review. *Processes* **2020**, *8*, 633. [CrossRef]
63. Kayfeci, M.; Kecebaş, A.; Bayat, M. Hydrogen Production. In *Solar Hydrogen Production: Processes, Systems and Technologies*; Academic Press: London, UK, 2019; pp. 45–83.
64. Yang, F.; Wang, T.; Deng, X.; Dang, J.; Huang, Z.; Hu, S.; Li, Y.; Ouyang, M. Review on Hydrogen Safety Issues: Incident Statistics, Hydrogen Diffusion, and Detonation Process. *Int. J. Hydrogen Energy* **2021**, *46*, 31467–31488. [CrossRef]
65. You, S.; Li, G.; Fan, Z.; Li, X.; Fu, L.; Wu, W. Nanotechnology-Assisted Sensors for the Detection of Carbon Monoxide: A Review. *Int. J. Electrochem. Sci.* **2023**, *18*, 100314. [CrossRef]
66. Rose, J.J.; Wang, L.; Xu, Q.; McTiernan, C.F.; Shiva, S.; Tejero, J.; Gladwin, M.T. Carbon Monoxide Poisoning: Pathogenesis, Management, and Future Directions of Therapy. *Am. J. Respir. Crit. Care Med.* **2017**, *195*, 596–606. [CrossRef]
67. Kausar, A.; Ahmad, I.; Zhao, T.; Aldaghri, O.; Eisa, M.H. Graphene in Polymeric Nanocomposite Membranes—Current State and Progress. *Processes* **2023**, *11*, 927. [CrossRef]
68. Khan, M.; Tahir, M.N.; Adil, S.F.; Khan, H.U.; Siddiqui, M.R.H.; Al-warthan, A.A.; Tremel, W. Graphene Based Metal and Metal Oxide Nanocomposites: Synthesis, Properties and Their Applications. *J. Mater. Chem. A* **2015**, *3*, 18753–18808. [CrossRef]
69. Abideen, Z.U.; Kim, H.W.; Kim, S.S. An Ultra-Sensitive Hydrogen Gas Sensor Using Reduced Graphene Oxide-Loaded ZnO Nanofibers. *Chem. Commun.* **2015**, *51*, 15418–15421. [CrossRef]
70. Abideen, Z.U.; Katoch, A.; Kim, J.-H.; Kwon, Y.J.; Kim, H.W.; Kim, S.S. Excellent Gas Detection of ZnO Nanofibers by Loading with Reduced Graphene Oxide Nanosheets. *Sens. Actuators B Chem.* **2015**, *221*, 1499–1507. [CrossRef]
71. Nag, A.; Mitra, A.; Mukhopadhyay, S.C. Graphene and Its Sensor-Based Applications: A Review. *Sens. Actuators Phys.* **2018**, *270*, 177–194. [CrossRef]
72. Yoon, H.J.; Jun, D.H.; Yang, J.H.; Zhou, Z.; Yang, S.S.; Cheng, M.M.-C. Carbon Dioxide Gas Sensor Using a Graphene Sheet. *Sens. Actuators B Chem.* **2011**, *157*, 310–313. [CrossRef]
73. Mobasheri, A.; Parhoodeh, S.; Shams, G. Cd-Doped SnO_2-Reduced Graphene Oxide Composite Nanofibrous Mats as CO Gas Sensors. *Fibers Polym.* **2023**, *23*, 784–790. [CrossRef]
74. Mohammadi, M.D.; Abbas, F.; Louis, H.; Mathias, G.E.; Unimuke, T.O. Trapping of CO, CO_2, H_2S, NH_3, NO, NO_2, and SO_2 by Polyoxometalate Compound. *Comput. Theor. Chem.* **2022**, *1215*, 113826. [CrossRef]
75. Khan, M.A.H.; Rao, M.V.; Li, Q. Recent Advances in Electrochemical Sensors for Detecting Toxic Gases: NO_2, SO_2 and H_2S. *Sensors* **2019**, *19*, 905. [CrossRef] [PubMed]
76. Anagnostou, E.; John, E.H.; Edgar, K.M.; Foster, G.L.; Ridgwell, A.; Inglis, G.N.; Pancost, R.D.; Lunt, D.J.; Pearson, P.N. Changing Atmospheric CO_2 Concentration Was the Primary Driver of Early Cenozoic Climate. *Nature* **2016**, *533*, 380–384. [CrossRef]
77. Marsal, A.; Dezanneau, G.; Cornet, A.; Morante, J.R. A New CO_2 Gas Sensing Material. *Sens. Actuators B Chem.* **2003**, *95*, 266–270. [CrossRef]
78. Karnauskas, K.B.; Miller, S.L.; Schapiro, A.C. Fossil Fuel Combustion Is Driving Indoor CO_2 Toward Levels Harmful to Human Cognition. *GeoHealth* **2020**, *4*, e2019GH000237. [CrossRef]
79. Jacobson, T.A.; Kler, J.S.; Hernke, M.T.; Braun, R.K.; Meyer, K.C.; Funk, W.E. Direct Human Health Risks of Increased Atmospheric Carbon Dioxide. *Nat. Sustain.* **2019**, *2*, 691–701. [CrossRef]
80. Georgiadis, A.G.; Charisiou, N.D.; Goula, M.A. Removal of Hydrogen Sulfide From Various Industrial Gases: A Review of The Most Promising Adsorbing Materials. *Catalysts* **2020**, *10*, 521. [CrossRef]
81. Onaizi, S.A.; Gawish, M.A.; Murtaza, M.; Gomaa, I.; Tariq, Z.; Mahmoud, M. H_2S Scavenging Capacity and Rheological Properties of Water-Based Drilling Muds. *ACS Omega* **2020**, *5*, 30729–30739. [CrossRef]
82. Murtaza, M.; Alarifi, S.A.; Abozuhairah, A.; Mahmoud, M.; Onaizi, S.A.; Al-Ajmi, M. Optimum Selection of H_2S Scavenger in Light-Weight and Heavy-Weight Water-Based Drilling Fluids. *ACS Omega* **2021**, *6*, 24919–24930. [CrossRef] [PubMed]
83. Shaik, R.; Kampara, R.K.; Kumar, A.; Sharma, C.S.; Kumar, M. Metal Oxide Nanofibers Based Chemiresistive H_2S Gas Sensors. *Coord. Chem. Rev.* **2022**, *471*, 214752. [CrossRef]
84. Chen, Y.; Xu, P.; Xu, T.; Zheng, D.; Li, X. ZnO-Nanowire Size Effect Induced Ultra-High Sensing Response to *Ppb*-Level H_2S. *Sens. Actuators B Chem.* **2017**, *240*, 264–272. [CrossRef]
85. Kumar, A.; Samanta, S.; Singh, A.; Roy, M.; Singh, S.; Basu, S.; Chehimi, M.M.; Roy, K.; Ramgir, N.; Navaneethan, M.; et al. Fast Response and High Sensitivity of ZnO Nanowires—Cobalt Phthalocyanine Heterojunction Based H_2S Sensor. *ACS Appl. Mater. Interfaces* **2015**, *7*, 17713–17724. [CrossRef]
86. Van Hoang, N.; Hung, C.M.; Hoa, N.D.; Van Duy, N.; Van Hieu, N. Facile On-Chip Electrospinning of $ZnFe_2O_4$ Nanofiber Sensors with Excellent Sensing Performance to H_2S down Ppb Level. *J. Hazard. Mater.* **2018**, *360*, 6–16. [CrossRef]

87. Jamali, S.; Klingmyr, D.; Tagesson, T. Global-Scale Patterns and Trends in Tropospheric NO_2 Concentrations, 2005–2018. *Remote Sens.* **2020**, *12*, 3526. [CrossRef]
88. Lebrusán, I.; Toutouh, J. Car Restriction Policies for Better Urban Health: A Low Emission Zone in Madrid, Spain. *Air Qual. Atmos. Health* **2021**, *14*, 333–342. [CrossRef]
89. Elsayed, N.M. Toxicity of Nitrogen Dioxide: An Introduction. *Toxicology* **1994**, *89*, 161–174. [CrossRef]
90. Zegebreal, L.T.; Tegegne, N.A.; Hone, F.G. Recent Progress in Hybrid Conducting Polymers and Metal Oxide Nanocomposite for Room-Temperature Gas Sensor Applications: A Review. *Sens. Actuators Phys.* **2023**, *359*, 114472. [CrossRef]
91. Bhadra, J.; Popelka, A.; Abdulkareem, A.; Ahmad, Z.; Touati, F.; Al-Thani, N. Fabrication of Polyaniline–Graphene/Polystyrene Nanocomposites for Flexible Gas Sensors. *RSC Adv.* **2019**, *9*, 12496–12506. [CrossRef]
92. Bolotin, K.I.; Sikes, K.J.; Jiang, Z.; Klima, M.; Fudenberg, G.; Hone, J.; Kim, P.; Stormer, H.L. Ultrahigh Electron Mobility in Suspended Graphene. *Solid State Commun.* **2008**, *146*, 351–355. [CrossRef]
93. Kim, J.-H.; Mirzaei, A.; Zheng, Y.; Lee, J.-H.; Kim, J.-Y.; Kim, H.W.; Kim, S.S. Enhancement of H_2S Sensing Performance of P-CuO Nanofibers by Loading p-Reduced Graphene Oxide Nanosheets. *Sens. Actuators B Chem.* **2019**, *281*, 453–461. [CrossRef]
94. Katoch, A.; Choi, S.-W.; Kim, S.S. Nanograins in Electrospun Oxide Nanofibers. *Met. Mater. Int.* **2015**, *21*, 213–221. [CrossRef]
95. Choi, S.; Choi, C.; Kim, S.-J.; Cho, H.-J.; Hakim, M.; Jeon, S.; Kim, I. Highly Efficient Electronic Sensitization of Non-Oxidized Graphene Flakes on Controlled Pore-Loaded WO_3 Nanofibers for Selective Detection of H_2S Molecules. *Sci. Rep.* **2015**, *5*, 8067. [CrossRef]
96. Park, K.H.; Kim, B.H.; Song, S.H.; Kwon, J.; Kong, B.S.; Kang, K.; Jeon, S. Exfoliation of Non-Oxidized Graphene Flakes for Scalable Conductive Film. *Nano Lett.* **2012**, *12*, 2871–2876. [CrossRef]
97. Lee, S.; Lee, K.; Moon, G.D.; Won, Y.S.; Yoon, Y.-J.; Park, S.S.; Kim, Y.-R.; Jeong, U. Preparation of Macroporous Carbon Nanofibers with Macroscopic Openings in the Surfaces and Their Applications. *Nanotechnology* **2009**, *20*, 445702. [CrossRef]
98. Van Hoang, N.; Hung, C.M.; Hoa, N.D.; Van Duy, N.; Park, I.; Van Hieu, N. Excellent Detection of H_2S Gas at Ppb Concentrations Using $ZnFe_2O_4$ Nanofibers Loaded with Reduced Graphene Oxide. *Sens. Actuators B Chem.* **2019**, *282*, 876–884. [CrossRef]
99. Van Hoang, N.; Hung, C.M.; Hoa, N.D.; Van Duy, N.; Van Toan, N.; Hong, H.S.; Hong Van, P.T.; Sơn, N.T.; Yoon, S.-G.; Van Hieu, N. Enhanced H_2S Gas-Sensing Performance of α-Fe_2O_3 Nanofibers by Optimizing Process Conditions and Loading with Reduced Graphene Oxide. *J. Alloys Compd.* **2020**, *826*, 154169. [CrossRef]
100. Shi, Z.; Wang, S.; Jin, Y.; Zhao, L.; Chen, S.; Yang, H.; Cui, Y.; Svanberg, R.; Tang, C.; Jiang, J.; et al. Establishment of Green Graphite Industry: Graphite from Biomass and Its Various Applications. *SusMat* **2023**, *3*, 402–415. [CrossRef]
101. Abdollahifar, M.; Doose, S.; Cavers, H.; Kwade, A. Graphite Recycling from End-of-Life Lithium-Ion Batteries: Processes and Applications. *Adv. Mater. Technol.* **2023**, *8*, 2200368. [CrossRef]
102. Tian, W.; Liu, X.; Yu, W. Research Progress of Gas Sensor Based on Graphene and Its Derivatives: A Review. *Appl. Sci.* **2018**, *8*, 1118. [CrossRef]
103. Zamiri, G.; Haseeb, A.S.M.A. Recent Trends and Developments in Graphene/Conducting Polymer Nanocomposites Chemiresistive Sensors. *Materials* **2020**, *13*, 3311. [CrossRef] [PubMed]
104. Yuan, W.; Huang, L.; Zhou, Q.; Shi, G. Ultrasensitive and Selective Nitrogen Dioxide Sensor Based on Self-Assembled Graphene/Polymer Composite Nanofibers. *ACS Appl. Mater. Interfaces* **2014**, *6*, 17003–17008. [CrossRef] [PubMed]
105. Dan, Y.; Lu, Y.; Kybert, N.J.; Luo, Z.; Johnson, A.T.C. Intrinsic Response of Graphene Vapor Sensors. *Nano Lett.* **2009**, *9*, 1472–1475. [CrossRef]
106. Kong, L.; Enders, A.; Rahman, T.S.; Dowben, P.A. Molecular Adsorption on Graphene. *J. Phys. Condens. Matter* **2014**, *26*, 443001. [CrossRef]
107. Duy, L.T.; Trung, T.Q.; Hanif, A.; Siddiqui, S.; Roh, E.; Lee, W.; Lee, N.-E. A Stretchable and Highly Sensitive Chemical Sensor Using Multilayered Network of Polyurethane Nanofibres with Self-Assembled Reduced Graphene Oxide. *2D Mater.* **2017**, *4*, 025062. [CrossRef]
108. Leve, Z.D.; Iwuoha, E.I.; Ross, N. The Synergistic Properties and Gas Sensing Performance of Functionalized Graphene-Based Sensors. *Materials* **2022**, *15*, 1326. [CrossRef]
109. Gupta Chatterjee, S.; Chatterjee, S.; Ray, A.K.; Chakraborty, A.K. Graphene–Metal Oxide Nanohybrids for Toxic Gas Sensor: A Review. *Sens. Actuators B Chem.* **2015**, *221*, 1170–1181. [CrossRef]
110. Wang, C.; Wang, Y.; Yang, Z.; Hu, N. Review of Recent Progress on Graphene-Based Composite Gas Sensors. *Ceram. Int.* **2021**, *47*, 16367–16384. [CrossRef]
111. Yan, C.; Lu, H.; Gao, J.; Zhang, Y.; Guo, Q.; Ding, H.; Wang, Y.; Wei, F.; Zhu, G.; Yang, Z.; et al. Improved NO_2 Sensing Properties at Low Temperature Using Reduced Graphene Oxide Nanosheet–In_2O_3 Heterojunction Nanofibers. *J. Alloys Compd.* **2018**, *741*, 908–917. [CrossRef]
112. Cui, S.; Pu, H.; Mattson, E.C.; Wen, Z.; Chang, J.; Hou, Y.; Hirschmugl, C.J.; Chen, J. Ultrasensitive Chemical Sensing through Facile Tuning Defects and Functional Groups in Reduced Graphene Oxide. *Anal. Chem.* **2014**, *86*, 7516–7522. [CrossRef] [PubMed]
113. Yu, W.; Sisi, L.; Haiyan, Y.; Jie, L. Progress in the Functional Modification of Graphene/Graphene Oxide: A Review. *RSC Adv.* **2020**, *10*, 15328–15345. [CrossRef] [PubMed]
114. Lee, J.-H.; Katoch, A.; Choi, S.-W.; Kim, J.-H.; Kim, H.W.; Kim, S.S. Extraordinary Improvement of Gas-Sensing Performances in SnO_2 Nanofibers Due to Creation of Local p-n Heterojunctions by Loading Reduced Graphene Oxide Nanosheets. *ACS Appl. Mater. Interfaces* **2015**, *7*, 3101–3109. [CrossRef]

115. Han, Z.; Wang, J.; Liu, S.; Zhang, Q.; Liu, Y.; Tan, Y.; Luo, S.; Guo, F.; Ma, J.; Li, P.; et al. Electrospinning of Neat Graphene Nanofibers. *Adv. Fiber Mater.* **2022**, *4*, 268–279. [CrossRef]
116. Grant, J.J.; Pillai, S.C.; Hehir, S.; McAfee, M.; Breen, A. Biomedical Applications of Electrospun Graphene Oxide. *ACS Biomater. Sci. Eng.* **2021**, *7*, 1278–1301. [CrossRef]
117. Yu, M.; Dong, R.-H.; Yan, X.; Yu, G.-F.; You, M.-H.; Ning, X.; Long, Y.-Z. Recent Advances in Needleless Electrospinning of Ultrathin Fibers: From Academia to Industrial Production. *Macromol. Mater. Eng.* **2017**, *302*, 1700002. [CrossRef]
118. Wang, Y.; Lu, S.; Zheng, J.; Liang, L. Advances in Latest Application Status, Challenges, and Future Development Direction of Electrospinning Technology in the Biomedical. *J. Nanomater.* **2022**, *2022*, 3791908. [CrossRef]
119. Xu, Z.; Gao, C. Graphene in Macroscopic Order: Liquid Crystals and Wet-Spun Fibers. *Acc Chem Res* **2014**, *47*, 1267–1276. [CrossRef]
120. Gupta, B.S. Manufactured Textile Fibers. In *Handbook of Industrial Chemistry and Biotechnology*; Kent, J.A., Ed.; Springer US: Boston, MA, USA, 2012; pp. 419–473.
121. Panicker, P.S.; Kim, H.C.; Kim, J. An Integrated Wet-Spinning System for Continuous Fabrication of High-Strength Nanocellulose Long Filaments. *Sci. Rep.* **2023**, *13*, 13137. [CrossRef]
122. Eom, W.; Jang, J.-S.; Lee, S.H.; Lee, E.; Jeong, W.; Kim, I.-D.; Choi, S.-J.; Han, T.H. Effect of Metal/Metal Oxide Catalysts on Graphene Fiber for Improved NO_2 Sensing. *Sens. Actuators B Chem.* **2021**, *344*, 130231. [CrossRef]
123. Vedala, H.; Sorescu, D.C.; Kotchey, G.P.; Star, A. Chemical Sensitivity of Graphene Edges Decorated with Metal Nanoparticles. *Nano Lett.* **2011**, *11*, 2342–2347. [CrossRef] [PubMed]
124. Singhal, A.V.; Charaya, H.; Lahiri, I. Noble Metal Decorated Graphene-Based Gas Sensors and Their Fabrication: A Review. *Crit. Rev. Solid State Mater. Sci.* **2017**, *42*, 499–526. [CrossRef]
125. Jang, J.-S.; Yu, H.; Choi, S.-J.; Koo, W.-T.; Lee, J.; Kim, D.-H.; Kang, J.-Y.; Jeong, Y.J.; Jeong, H.; Kim, I.-D. Heterogeneous Metal Oxide–Graphene Thorn-Bush Single Fiber as a Freestanding Chemiresistor. *ACS Appl. Mater. Interfaces* **2019**, *11*, 10208–10217. [CrossRef] [PubMed]
126. Xu, Z.; Gao, C. Graphene Fiber: A New Trend in Carbon Fibers. *Mater. Today* **2015**, *18*, 480–492. [CrossRef]
127. da Luz, F.S.; Garcia Filho, F.d.C.; del-Río, M.T.G.; Nascimento, L.F.C.; Pinheiro, W.A.; Monteiro, S.N. Graphene-Incorporated Natural Fiber Polymer Composites: A First Overview. *Polymers* **2020**, *12*, 1601. [CrossRef]
128. Li, W.; Chen, R.; Qi, W.; Cai, L.; Sun, Y.; Sun, M.; Li, C.; Yang, X.; Xiang, L.; Xie, D.; et al. Reduced Graphene Oxide/Mesoporous ZnO NSs Hybrid Fibers for Flexible, Stretchable, Twisted, and Wearable NO_2 E-Textile Gas Sensor. *ACS Sens.* **2019**, *4*, 2809–2818. [CrossRef]
129. Ju Yun, Y.; Hong, W.G.; Choi, N.-J.; Hoon Kim, B.; Jun, Y.; Lee, H.-K. Ultrasensitive and Highly Selective Graphene-Based Single Yarn for Use in Wearable Gas Sensor. *Sci. Rep.* **2015**, *5*, 10904. [CrossRef]
130. Yun, Y.J.; Hong, W.G.; Kim, D.Y.; Kim, H.J.; Jun, Y.; Lee, H.-K. E-Textile Gas Sensors Composed of Molybdenum Disulfide and Reduced Graphene Oxide for High Response and Reliability. *Sens. Actuators B Chem.* **2017**, *248*, 829–835. [CrossRef]
131. Yun, Y.J.; Kim, D.Y.; Hong, W.G.; Ha, D.H.; Jun, Y.; Lee, H.-K. Highly Stretchable, Mechanically Stable, and Weavable Reduced Graphene Oxide Yarn with High NO_2 Sensitivity for Wearable Gas Sensors. *RSC Adv.* **2018**, *8*, 7615–7621. [CrossRef]
132. Yun, Y.J.; Hong, W.G.; Choi, N.-J.; Park, H.J.; Moon, S.E.; Kim, B.H.; Song, K.-B.; Jun, Y.; Lee, H.-K. A 3D Scaffold for Ultra-Sensitive Reduced Graphene Oxide Gas Sensors. *Nanoscale* **2014**, *6*, 6511–6514. [CrossRef]
133. Zhao, S.; Xue, J.; Kang, W. Gas Adsorption on MoS_2 Monolayer from First-Principles Calculations. *Chem. Phys. Lett.* **2014**, *595–596*, 35–42. [CrossRef]
134. Park, H.J.; Kim, W.-J.; Lee, H.-K.; Lee, D.-S.; Shin, J.-H.; Jun, Y.; Yun, Y.J. Highly Flexible, Mechanically Stable, and Sensitive NO_2 Gas Sensors Based on Reduced Graphene Oxide Nanofibrous Mesh Fabric for Flexible Electronics. *Sens. Actuators B Chem.* **2018**, *257*, 846–852. [CrossRef]
135. Inoue, K.; Hoshino, S. Swelling of Nylon 6 Film Due to Water Sorption. *J. Polym. Sci. Polym. Phys. Ed.* **1976**, *14*, 1513–1526. [CrossRef]
136. Yavari, F.; Chen, Z.; Thomas, A.V.; Ren, W.; Cheng, H.-M.; Koratkar, N. High Sensitivity Gas Detection Using a Macroscopic Three-Dimensional Graphene Foam Network. *Sci. Rep.* **2011**, *1*, 166. [CrossRef]
137. Ghavam, S.; Vahdati, M.; Wilson, I.A.G.; Styring, P. Sustainable Ammonia Production Processes. *Front. Energy Res.* **2021**, *9*, 580808. [CrossRef]
138. Bannov, A.G.; Popov, M.V.; Brester, A.E.; Kurmashov, P.B. Recent Advances in Ammonia Gas Sensors Based on Carbon Nanomaterials. *Micromachines* **2021**, *12*, 186. [CrossRef]
139. Demon, S.Z.N.; Kamisan, A.I.; Abdullah, N.; Noor, S.A.M.; Khim, O.K.; Kasim, N.A.M.; Yahya, M.Z.A.; Manaf, N.A.A.; Azmi, A.F.M.; Halim, N.A. Graphene-Based Materials in Gas Sensor Applications: A Review. *Sens. Mater.* **2020**, *32*, 759. [CrossRef]
140. He, J.; Li, X. Unveiling Methane Sensing Mechanisms of Graphene and Its Derivatives. *Sens. Actuators Phys.* **2024**, *376*, 115634. [CrossRef]
141. Kwak, D.; Lei, Y.; Maric, R. Ammonia Gas Sensors: A Comprehensive Review. *Talanta* **2019**, *204*, 713–730. [CrossRef]
142. Ricci, P.P.; Gregory, O.J. Sensors for the Detection of Ammonia as a Potential Biomarker for Health Screening. *Sci. Rep.* **2021**, *11*, 7185. [CrossRef]
143. Feng, Q.; Li, X.; Wang, J.; Gaskov, A.M. Reduced Graphene Oxide (rGO) Encapsulated Co_3O_4 Composite Nanofibers for Highly Selective Ammonia Sensors. *Sens. Actuators B Chem.* **2016**, *222*, 864–870. [CrossRef]

144. Lu, G.; Ocola, L.E.; Chen, J. Reduced Graphene Oxide for Room-Temperature Gas Sensors. *Nanotechnology* **2009**, *20*, 445502. [CrossRef]
145. Lu, G.; Yu, K.; Ocola, L.E.; Chen, J. Ultrafast Room Temperature NH_3 Sensing with Positively Gated Reduced Graphene Oxide Field-Effect Transistors. *Chem. Commun.* **2011**, *47*, 7761–7763. [CrossRef]
146. Feng, Q.; Zeng, Y.; Xu, P.; Lin, S.; Feng, C.; Li, X.; Wang, J. Tuning the Electrical Conductivity of Amorphous Carbon/Reduced Graphene Oxide Wrapped-Co_3O_4 Ternary Nanofibers for Highly Sensitive Chemical Sensors. *J. Mater. Chem. A* **2019**, *7*, 27522–27534. [CrossRef]
147. Hong, S.-Z.; Huang, Q.-Y.; Wu, T.-M. The Room Temperature Highly Sensitive Ammonia Gas Sensor Based on Polyaniline and Nitrogen-Doped Graphene Quantum Dot-Coated Hollow Indium Oxide Nanofiber Composite. *Polymers* **2021**, *13*, 3676. [CrossRef] [PubMed]
148. Haick, H.; Broza, Y.Y.; Mochalski, P.; Ruzsanyi, V.; Amann, A. Assessment, Origin, and Implementation of Breath Volatile Cancer Markers. *Chem. Soc. Rev.* **2014**, *43*, 1423–1449. [CrossRef] [PubMed]
149. Silva, L.G.; Bueno, S.C.E.; da Silva, M.G.; Mota, L.; Sthel, M.S.; de Castro, M.P.P.; Santiago Neto, R.M.; Kuba, V.M. Photoacoustic Detection of Ammonia Exhaled by Individuals with Chronic Kidney Disease. *Lasers Med. Sci.* **2022**, *37*, 983–991. [CrossRef]
150. Andre, R.S.; Mercante, L.A.; Facure, M.H.M.; Mattoso, L.H.C.; Correa, D.S. Enhanced and Selective Ammonia Detection Using In_2O_3/Reduced Graphene Oxide Hybrid Nanofibers. *Appl. Surf. Sci.* **2019**, *473*, 133–140. [CrossRef]
151. Poloju, M.; Jayababu, N.; Ramana Reddy, M.V. Improved Gas Sensing Performance of Al Doped ZnO/CuO Nanocomposite Based Ammonia Gas Sensor. *Mater. Sci. Eng. B* **2018**, *227*, 61–67. [CrossRef]
152. Zhang, D.; Liu, J.; Chang, H.; Liu, A.; Xia, B. Characterization of a Hybrid Composite of SnO_2 Nanocrystal-Decorated Reduced Graphene Oxide for Ppm-Level Ethanol Gas Sensing Application. *RSC Adv.* **2015**, *5*, 18666–18672. [CrossRef]
153. Lee, S.H.; Eom, W.; Shin, H.; Ambade, R.B.; Bang, J.H.; Kim, H.W.; Han, T.H. Room-Temperature, Highly Durable Ti3C2Tx MXene/Graphene Hybrid Fibers for NH_3 Gas Sensing. *ACS Appl. Mater. Interfaces* **2020**, *12*, 10434–10442. [CrossRef]
154. Bhardwaj, R.; Hazra, A. MXene-Based Gas Sensors. *J. Mater. Chem. C* **2021**, *9*, 15735–15754. [CrossRef]
155. Xia, Q.; Fan, Y.; Li, S.; Zhou, A.; Shinde, N.; Mane, R.S. MXene-Based Chemical Gas Sensors: Recent Developments and Challenges. *Diam. Relat. Mater.* **2023**, *131*, 109557. [CrossRef]
156. Cui, J.; Lai, Y.; Wang, W.; Li, H.; Ma, X.; Zhan, J. Galvanic Displacement Induced Reduction of Graphene Oxide. *Carbon* **2014**, *66*, 738–741. [CrossRef]
157. Fan, Z.-J.; Kai, W.; Yan, J.; Wei, T.; Zhi, L.-J.; Feng, J.; Ren, Y.; Song, L.-P.; Wei, F. Facile Synthesis of Graphene Nanosheets via Fe Reduction of Exfoliated Graphite Oxide. *ACS Nano* **2011**, *5*, 191–198. [CrossRef]
158. Jia, Y.; Chen, L.; Yu, H.; Zhang, Y.; Dong, F. Graphene Oxide/Polystyrene Composite Nanofibers on Quartz Crystal Microbalance Electrode for the Ammonia Detection. *RSC Adv.* **2015**, *5*, 40620–40627. [CrossRef]
159. Vashist, S.K.; Vashist, P. Recent Advances in Quartz Crystal Microbalance-Based Sensors. *J. Sens.* **2011**, *2011*, 571405. [CrossRef]
160. Masoudzadeh, M.; Karachi, N. Removal of Cadmium Ion from Waste Water Using Carboxylated Nanoporous Graphene (G-COOH). *Eurasian J. Anal. Chem.* **2018**, *13*, 1–10. [CrossRef]
161. Ionita, M.; Crica, L.E.; Vasile, E.; Dinescu, S.; Pandele, M.A.; Costache, M.; Haugen, H.J.; Iovu, H. Effect of Carboxylic Acid Functionalized Graphene on Physical-Chemical and Biological Performances of Polysulfone Porous Films. *Polymer* **2016**, *92*, 1–12. [CrossRef]
162. Jia, Y.; Yu, H.; Zhang, Y.; Dong, F.; Li, Z. Cellulose Acetate Nanofibers Coated Layer-by-Layer with Polyethylenimine and Graphene Oxide on a Quartz Crystal Microbalance for Use as a Highly Sensitive Ammonia Sensor. *Colloids Surf. B Biointerfaces* **2016**, *148*, 263–269. [CrossRef]
163. Khatib, M.; Haick, H. Sensors for Volatile Organic Compounds. *ACS Nano* **2022**, *16*, 7080–7115. [CrossRef] [PubMed]
164. He, C.; Cheng, J.; Zhang, X.; Douthwaite, M.; Pattisson, S.; Hao, Z. Recent Advances in the Catalytic Oxidation of Volatile Organic Compounds: A Review Based on Pollutant Sorts and Sources. *Chem. Rev.* **2019**, *119*, 4471–4568. [CrossRef] [PubMed]
165. Dumanoglu, Y.; Kara, M.; Altiok, H.; Odabasi, M.; Elbir, T.; Bayram, A. Spatial and Seasonal Variation and Source Apportionment of Volatile Organic Compounds (VOCs) in a Heavily Industrialized Region. *Atmos. Environ.* **2014**, *98*, 168–178. [CrossRef]
166. Mago, A.; Yang, Y.-S.; Shim, J.-H.; John, A.A. Wearable Device for Cumulative Chlorobenzene Detection and Accessible Mitigation Strategies. *Sensors* **2023**, *23*, 7904. [CrossRef] [PubMed]
167. Yuan, Z.; Yang, C.; Meng, F. Strategies for Improving the Sensing Performance of Semiconductor Gas Sensors for High-Performance Formaldehyde Detection: A Review. *Chemosensors* **2021**, *9*, 179. [CrossRef]
168. Pathak, A.K.; Swargiary, K.; Kongsawang, N.; Jitpratak, P.; Ajchareeyasoontorn, N.; Udomkittivorakul, J.; Viphavakit, C. Recent Advances in Sensing Materials Targeting Clinical Volatile Organic Compound (VOC) Biomarkers: A Review. *Biosensors* **2023**, *13*, 114. [CrossRef]
169. Wilson, A.D. Advances in Electronic-Nose Technologies for the Detection of Volatile Biomarker Metabolites in the Human Breath. *Metabolites* **2015**, *5*, 140–163. [CrossRef]
170. Gregis, G.; Sanchez, J.-B.; Bezverkhyy, I.; Guy, W.; Berger, F.; Fierro, V.; Bellat, J.-P.; Celzard, A. Detection and Quantification of Lung Cancer Biomarkers by a Micro-Analytical Device Using a Single Metal Oxide-Based Gas Sensor. *Sens. Actuators B Chem.* **2018**, *255*, 391–400. [CrossRef]

171. van Keulen, K.E.; Jansen, M.E.; Schrauwen, R.W.M.; Kolkman, J.J.; Siersema, P.D. Volatile Organic Compounds in Breath Can Serve as a Non-Invasive Diagnostic Biomarker for the Detection of Advanced Adenomas and Colorectal Cancer. *Aliment. Pharmacol. Ther.* **2020**, *51*, 334–346. [CrossRef]
172. Vishinkin, R.; Haick, H. Nanoscale Sensor Technologies for Disease Detection via Volatolomics. *Small* **2015**, *11*, 6142–6164. [CrossRef]
173. Zhang, J.; Lu, H.; Yan, C.; Yang, Z.; Zhu, G.; Gao, J.; Yin, F.; Wang, C. Fabrication of Conductive Graphene Oxide-WO_3 Composite Nanofibers by Electrospinning and Their Enhanced Acetone Gas Sensing Properties. *Sens. Actuators B Chem.* **2018**, *264*, 128–138. [CrossRef]
174. Hwang, J.D.; Lin, C.C. Gallium Nitridephotoconductive Ultraviolet Sensor with a Sputtered Transparent Indium–Tin–Oxide Ohmic Contact. *Thin Solid Films* **2005**, *491*, 276–279. [CrossRef]
175. Khalil, A.; Kim, J.J.; Tuller, H.L.; Rutledge, G.C.; Hashaikeh, R. Gas Sensing Behavior of Electrospun Nickel Oxide Nanofibers: Effect of Morphology and Microstructure. *Sens. Actuators B Chem.* **2016**, *227*, 54–64. [CrossRef]
176. Salehi, T.; Taherizadeh, A.; Bahrami, A.; Allafchian, A.; Ghafarinia, V. Toward a Highly Functional Hybrid ZnO Nanofiber–rGO Gas Sensor. *Adv. Eng. Mater.* **2020**, *22*, 2000005. [CrossRef]
177. Huang, X.; Yin, Z.; Wu, S.; Qi, X.; He, Q.; Zhang, Q.; Yan, Q.; Boey, F.; Zhang, H. Graphene-Based Materials: Synthesis, Characterization, Properties, and Applications. *Small* **2011**, *7*, 1876–1902. [CrossRef]
178. Guo, L.; Kou, X.; Ding, M.; Wang, C.; Dong, L.; Zhang, H.; Feng, C.; Sun, Y.; Gao, Y.; Sun, P.; et al. Reduced Graphene Oxide/α-Fe_2O_3 Composite Nanofibers for Application in Gas Sensors. *Sens. Actuators B Chem.* **2017**, *244*, 233–242. [CrossRef]
179. Ai, Y.; Lou, Z.; Chen, S.; Chen, D.; Wang, Z.M.; Jiang, K.; Shen, G. All rGO-on-PVDF-Nanofibers Based Self-Powered Electronic Skins. *Nano Energy* **2017**, *35*, 121–127. [CrossRef]
180. Yeo, C.S.; Kim, H.; Lim, T.; Kim, H.J.; Cho, S.; Cho, K.R.; Kim, Y.S.; Shin, M.K.; Yoo, J.; Ju, S.; et al. Copper-Embedded Reduced Graphene Oxide Fibers for Multi-Sensors. *J. Mater. Chem. C* **2017**, *5*, 12825–12832. [CrossRef]
181. Fowler, J.D.; Allen, M.J.; Tung, V.C.; Yang, Y.; Kaner, R.B.; Weiller, B.H. Practical Chemical Sensors from Chemically Derived Graphene. *ACS Nano* **2009**, *3*, 301–306. [CrossRef] [PubMed]
182. Reddy, C.S.; Murali, G.; Reddy, A.S.; Park, S.; In, I. GO Incorporated SnO_2 Nanotubes as Fast Response Sensors for Ethanol Vapor in Different Atmospheres. *J. Alloys Compd.* **2020**, *813*, 152251. [CrossRef]
183. Shooshtari, M.; Salehi, A.; Vollebregt, S. Effect of Humidity on Gas Sensing Performance of Carbon Nanotube Gas Sensors Operated at Room Temperature. *IEEE Sens. J.* **2021**, *21*, 5763–5770. [CrossRef]
184. Zhang, J.; Guo, J.; Xu, H.; Cao, B. Reactive-Template Fabrication of Porous SnO_2 Nanotubes and Their Remarkable Gas-Sensing Performance. *ACS Appl. Mater. Interfaces* **2013**, *5*, 7893–7898. [CrossRef]
185. Li, G.-J.; Zhang, X.-H.; Kawi, S. Relationships between Sensitivity, Catalytic Activity, and Surface Areas of SnO_2 Gas Sensors. *Sens. Actuators B Chem.* **1999**, *60*, 64–70. [CrossRef]
186. Wang, D.; Zhang, M.; Chen, Z.; Li, H.; Chen, A.; Wang, X.; Yang, J. Enhanced Formaldehyde Sensing Properties of Hollow SnO_2 Nanofibers by Graphene Oxide. *Sens. Actuators B Chem.* **2017**, *250*, 533–542. [CrossRef]
187. Kim, H.-J.; Lee, J.-H. Highly Sensitive and Selective Gas Sensors Using P-Type Oxide Semiconductors: Overview. *Sens. Actuators B Chem.* **2014**, *192*, 607–627. [CrossRef]
188. Chi, M.; Zhao, Y.-P. Adsorption of Formaldehyde Molecule on the Intrinsic and Al-Doped Graphene: A First Principle Study. *Comput. Mater. Sci.* **2009**, *46*, 1085–1090. [CrossRef]
189. Potyrailo, R.A. Toward High Value Sensing: Monolayer-Protected Metal Nanoparticles in Multivariable Gas and Vapor Sensors. *Chem. Soc. Rev.* **2017**, *46*, 5311–5346. [CrossRef]
190. Ates, H.C.; Dincer, C. Wearable Breath Analysis. *Nat. Rev. Bioeng.* **2023**, *1*, 80–82. [CrossRef]
191. Kaloumenou, M.; Skotadis, E.; Lagopati, N.; Efstathopoulos, E.; Tsoukalas, D. Breath Analysis: A Promising Tool for Disease Diagnosis—The Role of Sensors. *Sensors* **2022**, *22*, 1238. [CrossRef]
192. Pham, Y.L.; Beauchamp, J. Breath Biomarkers in Diagnostic Applications. *Molecules* **2021**, *26*, 5514. [CrossRef]
193. Choi, S.-J.; Jang, B.-H.; Lee, S.-J.; Min, B.K.; Rothschild, A.; Kim, I.-D. Selective Detection of Acetone and Hydrogen Sulfide for the Diagnosis of Diabetes and Halitosis Using SnO_2 Nanofibers Functionalized with Reduced Graphene Oxide Nanosheets. *ACS Appl. Mater. Interfaces* **2014**, *6*, 2588–2597. [CrossRef]
194. Shin, J.; Choi, S.-J.; Lee, I.; Youn, D.-Y.; Park, C.O.; Lee, J.-H.; Tuller, H.L.; Kim, I.-D. Thin-Wall Assembled SnO_2 Fibers Functionalized by Catalytic Pt Nanoparticles and Their Superior Exhaled-Breath-Sensing Properties for the Diagnosis of Diabetes. *Adv. Funct. Mater.* **2013**, *23*, 2357–2367. [CrossRef]
195. Choi, S.-J.; Lee, I.; Jang, B.-H.; Youn, D.-Y.; Ryu, W.-H.; Park, C.O.; Kim, I.-D. Selective Diagnosis of Diabetes Using Pt-Functionalized WO_3 Hemitube Networks As a Sensing Layer of Acetone in Exhaled Breath. *Anal. Chem.* **2013**, *85*, 1792–1796. [CrossRef]
196. Wang, L.; Cheng, Y.; Gopalan, S.; Luo, F.; Amreen, K.; Singh, R.K.; Goel, S.; Lin, Z.; Naidu, R. Review and Perspective: Gas Separation and Discrimination Technologies for Current Gas Sensors in Environmental Applications. *ACS Sens.* **2023**, *8*, 1373–1390. [CrossRef]
197. Zheng, F.; Jiang, H.-Y.; Yang, X.-T.; Guo, J.-H.; Sun, L.; Guo, Y.-Y.; Xu, H.; Yao, M.-S. Reviews of Wearable Healthcare Systems Based on Flexible Gas Sensors. *Chem. Eng. J.* **2024**, *490*, 151874. [CrossRef]

198. Zong, B.; Wu, S.; Yang, Y.; Li, Q.; Tao, T.; Mao, S. Smart Gas Sensors: Recent Developments and Future Prospective. *Nano-Micro Lett.* **2024**, *17*, 54. [CrossRef]
199. Li, W.; Guo, J.; Cai, L.; Qi, W.; Sun, Y.; Xu, J.-L.; Sun, M.; Zhu, H.; Xiang, L.; Xie, D.; et al. UV Light Irradiation Enhanced Gas Sensor Selectivity of NO_2 and SO_2 Using rGO Functionalized with Hollow SnO_2 Nanofibers. *Sens. Actuators B Chem.* **2019**, *290*, 443–452. [CrossRef]
200. Barry, T.I.; Stone, F.S.; Tompkins, F.C. The Reactions of Oxygen at Dark and Irradiated Zinc Oxide Surfaces. *Proc. R. Soc. A* **1997**, *255*, 124–144. [CrossRef]
201. Kim, J.-H.; Zheng, Y.; Mirzaei, A.; Kim, H.W.; Kim, S.S. Synthesis and Selective Sensing Properties of rGO/Metal-Coloaded SnO_2 Nanofibers. *J. Electron. Mater.* **2017**, *46*, 3531–3541. [CrossRef]
202. Kwon, Y.J.; Na, H.G.; Kang, S.Y.; Choi, S.-W.; Kim, S.S.; Kim, H.W. Selective Detection of Low Concentration Toluene Gas Using Pt-Decorated Carbon Nanotubes Sensors. *Sens. Actuators B Chem.* **2016**, *227*, 157–168. [CrossRef]
203. Orozco, J.M.; Webb, G. The Adsorption and Hydrogenation of Benzene and Toluene on Alumina- and Silica- Supported Palladium and Platinum Catalysts. *Appl. Catal.* **1983**, *6*, 67–84. [CrossRef]
204. Saeys, M.; Reyniers, M.-F.; Neurock, M.; Marin, G.B. Density Functional Theory Analysis of Benzene (De)Hydrogenation on Pt(111): Addition and Removal of the First Two H-Atoms. *J. Phys. Chem. B* **2003**, *107*, 3844–3855. [CrossRef]
205. Kim, J.-H.; Wu, P.; Kim, H.W.; Kim, S.S. Highly Selective Sensing of CO, C_6H_6, and C_7H_8 Gases by Catalytic Functionalization with Metal Nanoparticles. *ACS Appl. Mater. Interfaces* **2016**, *8*, 7173–7183. [CrossRef]
206. Abideen, Z.U.; Kim, J.-H.; Mirzaei, A.; Kim, H.W.; Kim, S.S. Sensing Behavior to Ppm-Level Gases and Synergistic Sensing Mechanism in Metal-Functionalized rGO-Loaded ZnO Nanofibers. *Sens. Actuators B Chem.* **2018**, *255*, 1884–1896. [CrossRef]
207. Sánchez-Vicente, C.; Santos, J.P.; Lozano, J.; Sayago, I.; Sanjurjo, J.L.; Azabal, A.; Ruiz-Valdepeñas, S. Graphene-Doped Tin Oxide Nanofibers and Nanoribbons as Gas Sensors to Detect Biomarkers of Different Diseases through the Breath. *Sensors* **2020**, *20*, 7223. [CrossRef]
208. Fedi, F.; Miglietta, M.L.; Polichetti, T.; Ricciardella, F.; Massera, E.; Ninno, D.; Di Francia, G. A Study on the Physicochemical Properties of Hydroalcoholic Solutions to Improve the Direct Exfoliation of Natural Graphite down to Few-Layers Graphene. *Mater. Res. Express* **2015**, *2*, 035601. [CrossRef]
209. Abideen, Z.U.; Park, J.Y.; Kim, H.W.; Kim, S.S. Graphene-Loaded Tin Oxide Nanofibers: Optimization and Sensing Performance. *Nanotechnology* **2016**, *28*, 035501. [CrossRef]
210. Burris, A.J.; Tran, K.; Cheng, Q. Tunable Enhancement of a Graphene/Polyaniline/Poly(Ethylene Oxide) Composite Electrospun Nanofiber Gas Sensor. *J. Anal. Test.* **2017**, *1*, 12. [CrossRef]
211. Zhang, Y.; Kim, J.J.; Chen, D.; Tuller, H.L.; Rutledge, G.C. Electrospun Polyaniline Fibers as Highly Sensitive Room Temperature Chemiresistive Sensors for Ammonia and Nitrogen Dioxide Gases. *Adv. Funct. Mater.* **2014**, *24*, 4005–4014. [CrossRef]
212. Niu, Y.; Wang, R.; Jiao, W.; Ding, G.; Hao, L.; Yang, F.; He, X. MoS_2 Graphene Fiber Based Gas Sensing Devices. *Carbon* **2015**, *95*, 34–41. [CrossRef]
213. Liu, X.; Cheng, S.; Liu, H.; Hu, S.; Zhang, D.; Ning, H. A Survey on Gas Sensing Technology. *Sensors* **2012**, *12*, 9635–9665. [CrossRef]
214. Kim, E.; Lee, S.; Kim, J.H.; Kim, C.; Byun, Y.T.; Kim, H.S.; Lee, T. Pattern Recognition for Selective Odor Detection with Gas Sensor Arrays. *Sensors* **2012**, *12*, 16262–16273. [CrossRef]
215. Ugale, A.D.; Umarji, G.G.; Jung, S.H.; Deshpande, N.G.; Lee, W.; Cho, H.K.; Yoo, J.B. ZnO Decorated Flexible and Strong Graphene Fibers for Sensing NO_2 and H_2S at Room Temperature. *Sens. Actuators B Chem.* **2020**, *308*, 127690. [CrossRef]
216. Lohmann, T.; von Klitzing, K.; Smet, J.H. Four-Terminal Magneto-Transport in Graphene p-n Junctions Created by Spatially Selective Doping. *Nano Lett.* **2009**, *9*, 1973–1979. [CrossRef]
217. Wang, D.H.; Hu, Y.; Zhao, J.J.; Zeng, L.L.; Tao, X.M.; Chen, W. Holey Reduced Graphene Oxide Nanosheets for High Performance Room Temperature Gas Sensing. *J. Mater. Chem. A* **2014**, *2*, 17415–17420. [CrossRef]
218. Wang, T.; Ouyang, Z.; Wang, F.; Liu, Y. A Review on Graphene Strain Sensors Based on Fiber Assemblies. *SN Appl. Sci.* **2020**, *2*, 862. [CrossRef]
219. Razaq, A.; Bibi, F.; Zheng, X.; Papadakis, R.; Jafri, S.H.M.; Li, H. Review on Graphene-, Graphene Oxide-, Reduced Graphene Oxide-Based Flexible Composites: From Fabrication to Applications. *Materials* **2022**, *15*, 1012. [CrossRef]
220. Wu, S. Recent Progress in Flexible Graphene-Based Composite Fiber Electrodes for Supercapacitors. *Crystals* **2021**, *11*, 1484. [CrossRef]

materials

MDPI

Communication

Visualizing Thermal Reduction in Graphene Oxide

Xiangrui Xu [1], Junjie Huang [1], Gesong Miao [1], Bo Yan [1], Yangbo Chen [1], Yinghui Zhou [1], Yufeng Zhang [1,2], Xueao Zhang [1,2,*] and Weiwei Cai [1,2,*]

[1] College of Physical Science and Technology, Xiamen University, Xiamen 361005, China
[2] Jiujiang Research Institute of Xiamen University, Jiujiang 360404, China
* Correspondence: xazhang@xmu.edu.cn (X.Z.); wwcai@xmu.edu.cn (W.C.)

Abstract: The reduction of graphene oxide (GO) is critical for tuning its properties. This study integrates optical contrast analysis with Raman spectroscopy and X-ray photoelectron spectroscopy (XPS) to investigate the structural and optical evolution of GO in thermal reduction. For GO on 100 nm SiO_2/Si, the R channel contrast exhibits superior sensitivity to structural changes, making it a reliable indicator of the reduction process. A theoretical model based on Fresnel equations reveals the role of SiO_2 thickness in modulating optical contrast, providing guidelines for substrate optimization and channel selection.

Keywords: graphene oxide; reduction; optical microscopy; X-ray photoelectron spectroscopy; Raman spectroscopy

1. Introduction

Due to its unique layered structure with oxygen-containing functional groups, GO has become a significant functional material for various applications, such as photonics [1–4], electronic [5–7], sensors [8,9], and energy technologies [10,11]. Thermal reduction is commonly used to remove oxygen-containing functional groups from GO and restore the sp^2-conjugated carbon framework. The reduction parameters, such as temperature and duration, exert a substantial influence on the reduction effect, thereby modulating the resulting properties [12–16]. Understanding reduction processes is crucial for optimizing the application potential of GO. Hence, various techniques have been employed to investigate the reduction [12–21].

Raman spectroscopy is commonly used to investigate the structural change in graphene-based material, with a focus on the characteristic features (i.e., the D and G modes). The D mode arises from out-of-plane vibrations induced by structural defects and disorders in the carbon lattice, including oxygenated functional groups, as well as double-resonant processes near the Brillouin Zone boundary. Meanwhile, the G mode originates from in-plane vibrations of the sp^2-hybridized carbon framework, corresponding to the first-order scattering of the E_{2g} mode [17,18,22–25]. Hence, the intensity ratio of these modes (I_D/I_G) and the full width at half maximum (FWHM) are widely deployed to reveal the structural changes during the GO reduction process [17,26–30]. On the other hand, XPS is employed to understand the electronic and chemical structure of materials. Hence, the GO reduction progress can be analyzed by XPS (e.g., the change in oxygen-containing functional groups). However, inconsistencies in peak selection and binding energy determination during spectral deconvolution have been widely reported [18,21,31–33]. The C 1s peak shift observed in GO relative to pristine graphene complicates charge correction procedures [32]. Moreover, the uncertainty of fitting parameters, especially for the asymmetric C-C sp^2 signal, affects the reliability of quantitative analysis [21]. While Raman spectroscopy is a powerful tool, it may require careful optimization of parameters (e.g., laser intensity, exposure time) to minimize sample damage and improve efficiency, especially for large-scale measurements.

In contrast, despite its inability to directly provide specific structural information, optical microscopy offers a rapid, cost-effective, and non-destructive approach for analyzing properties like thickness and oxidation degree in two-dimensional materials, including graphene [34–37], transition metal dichalcogenides [38–44], and others [45–50]. Changes in optical images and contrast changes extracted from them are used to evaluate the oxidation process of WTe_2 [51]. Changes in optical images are used to evaluate the oxidation process of $MoTe_2$ and $MoTe_2$ encapsulated by MoS_2 to evaluate the protective effect of MoS_2 [52]. Changes in optical images and contrast changes extracted from them are used to evaluate the oxidation process of $MoTe_2$ and $MoTe_2$ encapsulated by hexagonal boron nitride to evaluate the protective effect of hexagonal boron nitride [53]. Optical microscopy is used to identify the thickness of GO processed by vacuum heating to improve the optical contrast [54] and estimate the reduction degree of GO by correlating of XPS by optical contrast [27]. However, the evolution of optical contrast in various wavelength region during GO reduction and its physical mechanisms have not been comprehensively investigated.

In this work, an approach combining multi-channel optical contrast analysis with Raman spectroscopy and XPS is proposed to investigate the GO reduction process. It is found that optical contrast allows rapid assessment of reduction processes, while Raman spectroscopy and XPS provide detailed structural information. Different channels of optical contrast have different sensitivities to thermal reduction. For GO on 100 nm SiO_2/Si, the R channel contrast in optical images exhibits superior sensitivity to thermal reduction. A theoretical model based on Fresnel equations reveals the evolution of optical properties across distinct color channels and the modulation of these properties by SiO_2 layer thickness, providing guidelines for channel selection and substrate optimization for contrast analysis.

2. Method

A GO aqueous suspension (1.5 mg/mL) is prepared using a modified Hummers method [55] and spin-coated onto 100 nm SiO_2/Si substrates pre-treated with oxygen plasma (Harrick Plasma, PDC-32G-2, Ithaca, New York, NY, USA) for 10 min. The spin-coating process consists of two steps: 1500 rpm for 30 s and 3000 rpm for 60 s, ensuring uniform GO deposition. GO reduction is performed on 100 nm SiO_2/Si substrates in a tube furnace at temperatures of 200, 400, 600, and 800 °C under a vacuum of less than 1 Pa for 30 min without gas.

Optical images are acquired using a ZYJ-1000E optical microscope(China, Shanghai). Python code utilizing multiple libraries is employed to generate individual channel images and measure intensities. For individual channel image generation, Pillow (10.4.0) (PIL) is employed for RGB image reading and channel separation, generating pseudo-color images. For contrast quantification, OpenCV (4.10.0) (cv2) is used for image reading with interactive ROI selection to annotate substrate and material regions, NumPy (1.26.4) for calculating grayscale and R, G, and B channel contrasts. Raman measurements are performed at room temperature using a WITec Alpha-300 (Germany, Baden-Württemberg, Ulm) with a blue laser of 488 nm for stronger Raman signals. XPS measurements are performed using a Thermo Scientific K-Alpha (Waltham, MA, USA) with Al Kα radiation (1486.6 eV) as the excitation source. Chamber vacuum less than 2.0×10^{-7} mBar. The resolution of XPS spectra used is 0.1 eV. Atomic force microscopy (AFM) is performed using a NTEGRA Prima (Russia, Moscow).

3. Results and Discussion

Figure 1 presents optical and channel-separated images of GO before and after thermal reduction at different temperatures. The first column are optical mages, and the next three columns are pseudo-color images of the R, G, and B channels, respectively. While the visibility of GO on a 100 nm SiO_2/Si substrate initially appears poor, it progressively improves with increasing reduction temperature, consistent with previous reports [27,54]. However, distinct phenomena emerge in channel-separated images. R channel images show near-indistinguishable GO initially, followed by rapid visibility enhancement during reduction, surpassing optical image visibility. In G channel images, GO is always distinguishable and also shows a rapid increase in visibility. However, the visibility is poorer than that in R channel images. Although visibility in B channel images also increases with temperature, it remains significantly lower than in other channels. This channel-dependent visibility variation reveals wavelength-specific optical responses during GO reduction and offers an optical method for monitoring graphene oxide reduction.

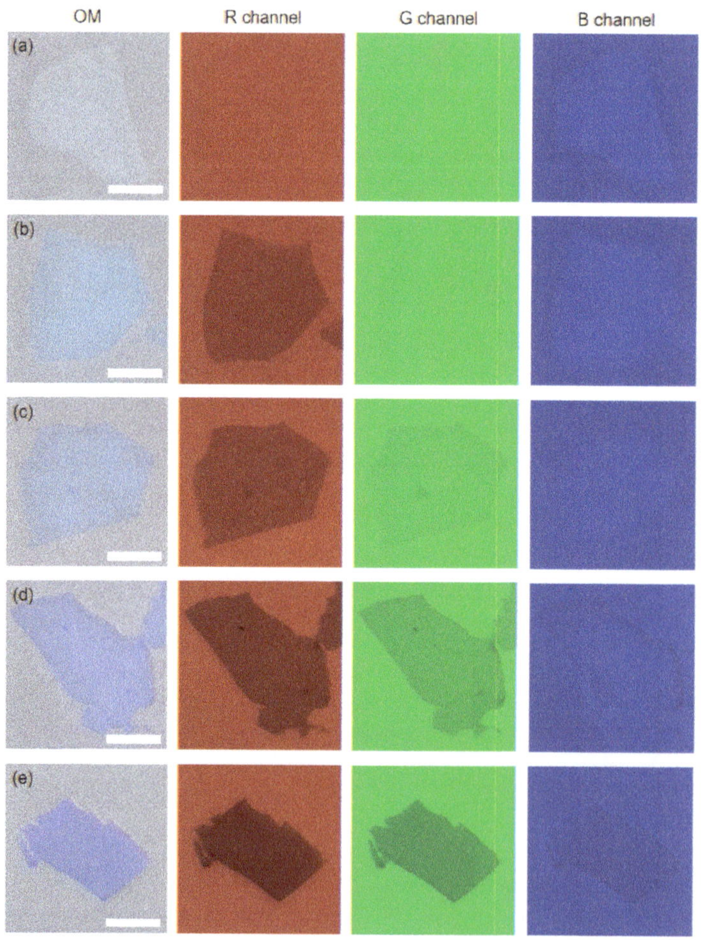

Figure 1. Optical and channel-separated images of GO before reduction (**a**) and after reduction at 200 °C (**b**), 400 °C (**c**), 600 °C (**d**), and 800 °C (**e**). The scale bars are 10 μm.

To complement the optical microscopy findings and provide a comprehensive understanding of the reduction process, Raman spectroscopy and XPS are applied for detailed chemical and structural characterization. These techniques enable examination of oxygen-containing functional group removal and sp^2-hybridized carbon framework restoration, offering critical insights into the reduction mechanism.

Figure 2a–e show the representative Raman spectra in the 1000–2000 cm^{-1} region with deconvoluted peaks of GO deposited on 100 nm SiO_2/Si substrates before and after reduction at various temperatures. Following spectral calibration with the silicon reference peak (520 cm^{-1}) and baseline subtraction using Asymmetric Least Squares Smoothing, peaks are carefully fitted with a Voigt function, which accounts for both Gaussian and Lorentzian contributions. The fitting process employs the Levenberg-Marquardt algorithm for optimization. The G and D′ peaks overlap, while the D, G, and D′ peaks alone are insufficient to accurately fit all Raman signals. Therefore, an additional f_1 peak located between the D and G peaks is included to improve the fitting results [56,57]. The complete Raman spectra are shown in Figure S1. The spectra exhibit characteristic D and G modes at ~1330 cm^{-1} and ~1600 cm^{-1}, respectively [58]. The intensity ratio of these modes (I_D/I_G) provides information on the defects [28,29,59], thereby accessing the structural changes during the reduction process. The I_D/I_G ratio decreases with increasing annealing temperature up to 600 °C, then rises with further increasing temperature, as depicted in Figure 2f. The initial decrease in the I_D/I_G ratio suggests the removal of oxygen-containing functional groups and the recovery of sp^2-hybridized carbon framework, accompanied by a reduction in defects, resulting in a corresponding decrease in the intensity of the D mode relative to the G mode. However, at 800 °C, the increase in the I_D/I_G ratio suggests the emergence of new structural defects.

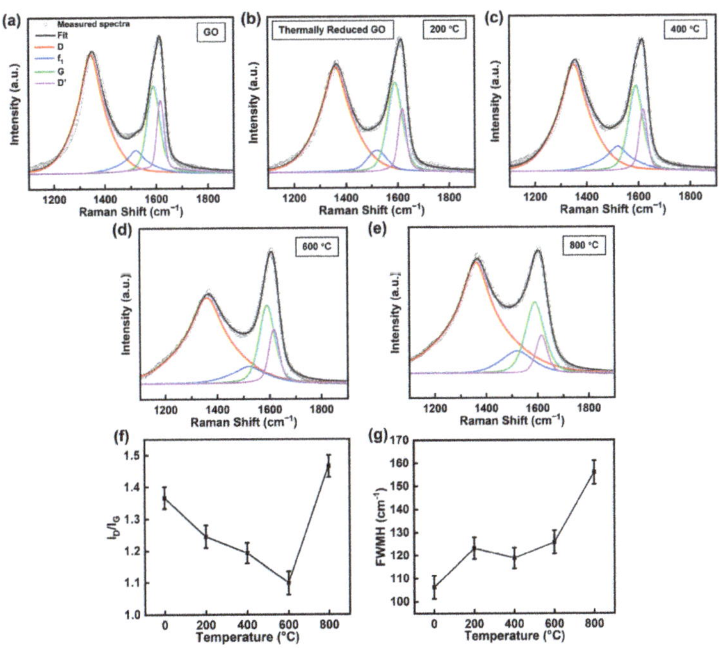

Figure 2. (**a**–**e**) Raman spectra in the 1000–2000 cm^{-1} region with deconvoluted peaks, (**f**) I_D/I_G ratio, and (**g**) FWHM of G mode for GO before and after reduction at various temperatures.

Figure 2g shows the temperature-dependent variation in the G mode FWHM of GO. The initial G mode FWHM of 106.2 cm^{-1} increases to 123.2 cm^{-1} at 200 °C, remains stable at 120.1 cm^{-1}, and 125.2 cm^{-1} for 400 °C and 600 °C respectively, and shows a significant increase to 156.1 cm^{-1} at 800 °C. The observed G mode broadening at 200 °C suggests a decrease in the crystalline quality of the sp^2-hybridized carbon framework. The nearly invariant FWHM values between 400–600 °C indicate that the defect density of the sp^2-hybridized carbon framework remains stable during intermediate thermal treatment. The FWHM increases at 800 °C, indicating another decrease in crystalline quality. This phenomenon is likely attributed to the formation of Stone-Wales defects, vacancies, and distortions during high-temperature annealing, leading to G mode broadening [60–63]. The non-monotonic behavior of the I_D/I_G ratio and G mode FWHM complicates the detection of the reduction of GO.

Figure 3 shows the representative C 1s XPS spectra of GO before and after reduction at various temperatures. The C 1s XPS spectra are deconvoluted using six Voigt functions for characteristic peaks after Shirley background subtraction [64]: C-C sp^2 (284.4 eV), C-C sp^3 (285 eV), C-O (285.7 eV), C-O-C (286.7 eV), C=O (288.0 eV), and O-C=O (290.1 eV) [21,64,65]. The mathematical reliability was assessed using the coefficient of determination, analysis of variance, and residual analysis to ensure statistically sound fitting results. The content of C 1S chemical groups is shown in Table S1. Thermal annealing progressively restores the graphitic structure in GO, as evidenced by the significant increase in C-C sp^2 content from 16% to 69% and the corresponding decrease in C-C sp^3 content from 24% to 10% after annealing at 800 °C. This transformation indicates effective recovery of the sp^2-hybridized carbon framework and defect healing. The dominant oxygen-containing functional C-O-C group decomposed rapidly at 200 °C and stabilized at higher temperatures, while the percentage of minor oxygen-containing functional groups (C-OH, C=O, and O-C=O) is consistently lower. XPS analysis effectively tracks chemical changes during the initial reduction stage (GO to 200 °C), where C-C sp^2 content increased sharply from 16% to 48% and sp^3 content decreased from 24% to 17%. However, its sensitivity diminished for detecting further structural changes during advanced thermal treatments (600 °C to 800 °C), as evidenced by the increase in C-C sp^2 content (63% to 69%) and decrease in C-C sp^3 content (12% to 10%). While they can still be distinguished, the discrepancy in C-C sp^2 content and C-C sp^3 content between the two temperatures is reduced.

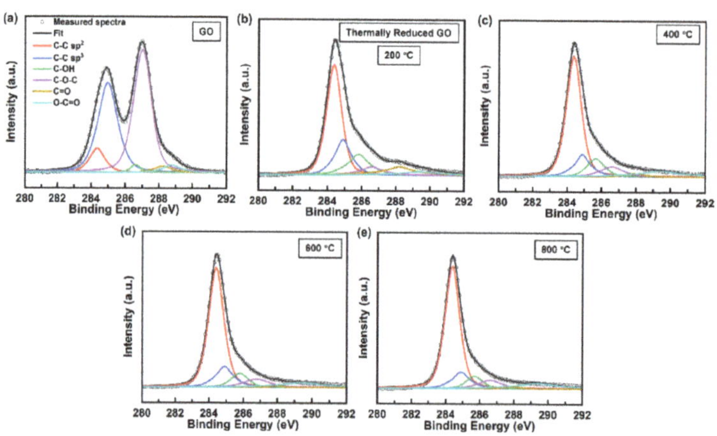

Figure 3. C 1s XPS spectra with deconvoluted peaks of GO before reduction (**a**) and after reduction at 200 °C (**b**), 400 °C (**c**), 600 °C (**d**), and 800 °C (**e**).

Clearly, the change in optical images of GO closely correlates with the reduction temperature as well as the structural change in GO. To quantitatively analyze the change, the contrast (C) of the GO flake is defined as:

$$C = \frac{I_{substrate} - I_{flake}}{I_{substrate}}. \quad (1)$$

where $I_{substrate}$ represents the substrate intensity and I_{flake} represents the GO flake intensity. A code programmed in Python (3.12.3) is employed for measuring intensities, as described in Method. This measure provides a numerical representation of the visual distinction between the GO and the underlying 100 nm SiO_2/Si substrate. Note that a higher contrast value indicates a more distinct boundary between the GO and the substrate.

Figure 4a demonstrates the statistical contrasts of optical images (RGB) and their channel-separated counterparts, illustrating a notable enhancement in contrast throughout the reduction process. The data points are the mean values obtained from measurements of three GO flakes in independent experiments conducted under the same conditions, and the error bars indicate the standard deviation. At different thermal reduction temperatures, the contrast of the different channels of GO changes differently, and this stems from the structural changes induced by the reduction process. The pristine GO exhibits poor visibility, as evidenced by its negative RGB contrast at −0.01. Specifically, the B and G channels show negative contrasts of −0.04 and −0.02, respectively, and a negative contrast means that the color of the GO is lighter than the substrate. While only the R channel displays a positive contrast of 0.02. The contrast of the B channel suggests that pristine GO exhibits smaller contrasts in the blue wavelength range, which will be confirmed in the following theoretical analysis. As the reduction temperature increases to 200 °C, the R channel contrast rises at 0.15, while the B channel remains negative at −0.06. The G channel contrast approaches 0. The RGB contrast becomes positive at 0.01. As the reduction temperature increases to 400 °C, the R channel contrast increases at 0.17, and the G channel increases at 0.03. The B channel contrast weakens at −0.03, which still suggests smaller contrasts in the blue wavelength range. The RGB contrast was enhanced at 0.04. As reduction temperatures rise above 600 °C, the optical properties continue to change. In the R channel, contrast surges from 0.26 at 600 °C to 0.33 at 800 °C. This increase is accompanied by increases in both G (from 0.12 to 0.20) and B (from 0.01 to 0.03) channels. The RGB contrast increases from 0.12 to 0.19. Note that the B channel contrast transitions from negative to positive values from 400 °C to 600 °C.

These results highlight the significant differences in contrast enhancement between color channels and reveals that the R channel exhibits higher contrast, suggesting superior sensitivity to structural changes in GO during reduction, making the R channel contrast the preferred optical indicator of the reduction progress of GO.

The enhanced contrast resulting from thermal reduction is primarily attributed to the increased refractive index [54,66]. A model is developed to investigate the optical properties of GO (and GO after reduction), analyzing the interaction of normal-incidence light with a triple-layer structure (GO/SiO_2/Si) as illustrated in Figure 4b. The optical contrast analysis is based on the Fresnel equations, which describe the reflection and transmission of light at the interface between two media with distinct refractive indices. The calculation requires the determination of the thickness d, and the refractive index $n(\lambda)$, of each layer. Note that the thickness of the Si layer d_3, is regarded as infinite. The thickness of the SiO_2 layer d_2 is 100 nm. The thickness of GO d_1 is 0.8 nm, as demonstrated by AFM analysis, as shown in Figure S2. The reflection indexes are derived and optimized from the relevant literature [54,66–70].

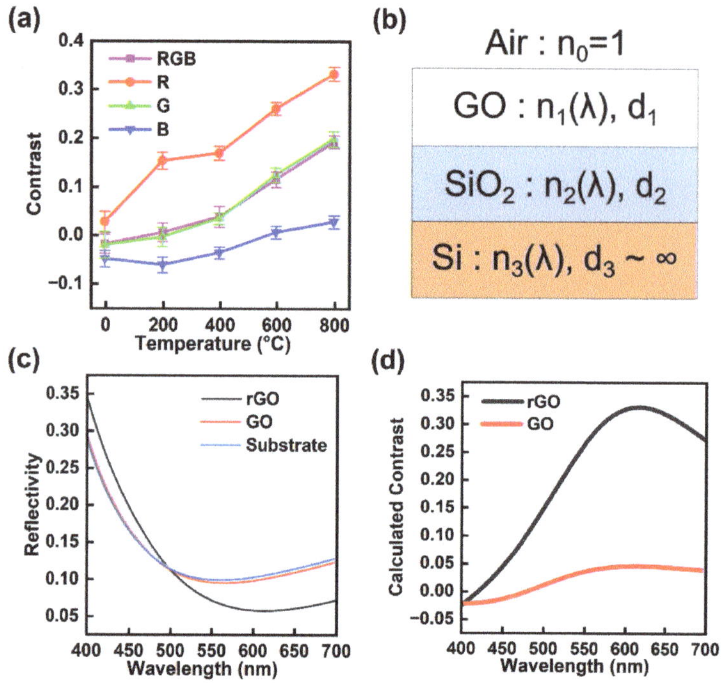

Figure 4. (**a**) Contrast of optical images (RGB) and channel-separated images before and after reduction at 200 °C, 400 °C, 600 °C, and 800 °C. (**b**) Schematic diagram of the GO/SiO_2/Si structure. (**c**) Calculated reflectivity spectra of rGO, GO, and the 100 nm SiO_2/Si substrate. (**d**) Calculated contrast spectra of rGO and GO.

As described in our previous work [71], the intensity of reflected light for normal incidence can be expressed as:

$$I = p_1^2 = \left| \frac{r_1 e^{i(\delta_1+\delta_2)} + r_2 e^{-i(\delta_1-\delta_2)} + r_3 e^{-i(\delta_1+\delta_2)} + r_1 r_2 r_3 e^{i(\delta_1-\delta_2)}}{e^{i(\delta_1+\delta_2)} + r_1 r_2 e^{-i(\delta_1-\delta_2)} + r_1 r_3 e^{-i(\delta_1+\delta_2)} + r_2 r_3 e^{i(\delta_1-\delta_2)}} \right|^2, \quad (2)$$

where

$$r_1 = \frac{n_0 - n_1}{n_0 + n_1}, r_2 = \frac{n_1 - n_2}{n_1 + n_2}, r_3 = \frac{n_2 - n_3}{n_2 + n_3}, \quad (3)$$

are the reflection coefficients at each interface determined by the Fresnel formula, while

$$\delta_1 = \frac{2\pi n_1 d_1}{\lambda}; \delta_2 = \frac{2\pi n_2 d_2}{\lambda}, \quad (4)$$

are the phase thickness.

The reflection of the entire structure (p_1), is calculated by the recursive method beginning with the bottom layer:

$$p_k e^{i\varphi_k} = \frac{r_k + r_{k+1} e^{-2i\delta_k}}{1 + r_k r_{k+1} e^{-2i\delta_k}}; p_{k-1} e^{i\varphi_{k-1}} = \frac{r_{k-1} + p_k e^{i\varphi_k} e^{-2i\delta_{k-1}}}{1 + r_{k-1} p_k e^{i\varphi_k} e^{-2i\varphi_{k-1}}}; \ldots; p_1. \quad (5)$$

Figure 4c shows the reflectivity spectra of GO, GO after reduction at 800 °C (rGO), and the 100 nm SiO_2/Si substrate. The R, G, and B channels are defined as light in the 580–700, 480–600, and 400–500 nm ranges, respectively. The enhanced reflectivity in the B channel, accompanied by low overall reflectivity, produces a grey-blue tint in optical images. The reflectivity difference is significant in the R channel. The calculated contrast spectra of GO and rGO on 100 nm SiO_2/Si are shown in Figure 4d, enabling quantitative analysis across different channels. The calculated contrast differences in RGB, R, G, and B channels are 0.19, 0.27, 0.22, and 0.007. These distinct contrast characteristics across wavelength ranges account for the differential channel responses observed in optical images and the superiority of the R channel.

Note that the optical properties are sensitive to SiO_2 thickness variations. Figure 5a,b show the calculated reflectivity contour plots of the GO and rGO on the 100 nm SiO_2/Si substrate, exhibiting similar reflectance variation patterns with increasing SiO_2 thickness, characterized by periodic intensity variations manifested as alternating light–dark bands with linear slopes and high reflectivity, which mostly occur in the short wavelength range, characterized by deep yellow in the lower part. Figure 5c shows the corresponding contrast difference contour plot, which demonstrates analogous periodic intensity modulations. SiO_2 thickness variations induce distinct optical properties across color channels. While the R channel demonstrates the greatest contrast difference on 100 nm SiO_2/Si substrates, specific SiO_2 thickness ranges favor RGB, B, or G channels, characterized by yellow lines cutting across the plot. As demonstrated in Figure 5d–g, all channels exhibit periodic oscillatory behavior in contrast difference with SiO_2 thickness, showing distinct periodicities and extremum values. The RGB channel demonstrates the shortest oscillation period, followed by the B, G, and R channels, as calculated by the difference between the wavelengths of the last and first extreme point divided by the number of extreme points. The ratio of the last extreme point to the first extreme point indicates that the R channel undergoes the most significant decay in contrast, followed by the G, RGB, and B channels.

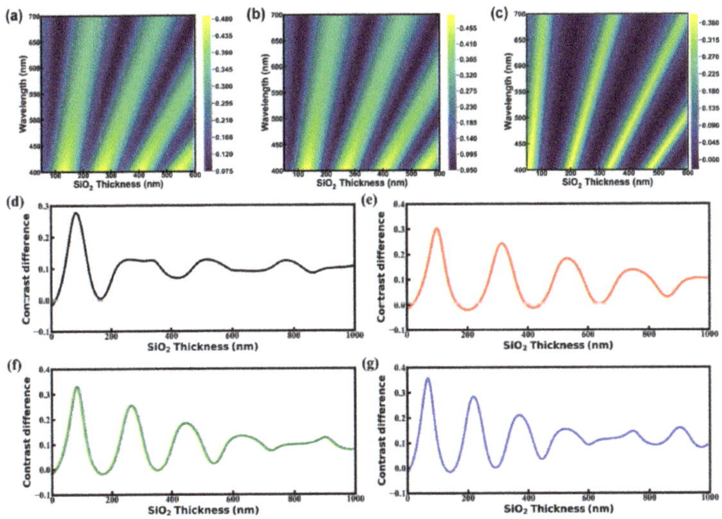

Figure 5. Contour plot of calculated reflectivity versus wavelength and SiO_2 thickness of rGO (**a**) and GO (**b**) on 100 nm SiO_2/Si substrate. (**c**) Contour plot of calculated contrast differences versus wavelength and SiO_2 thickness. Calculated contrast differences versus SiO_2 thickness in the RGB (**d**), R (**e**), G (**f**), and B (**g**) channels.

4. Conclusions

In summary, an integrated approach combining optical contrast analysis with Raman spectroscopy and XPS is presented to investigate the reduction of GO. The capability of optical microscopy as a rapid and effective tool for characterizing the reduction process of GO on 100 nm SiO_2/Si is demonstrated. The results show that the R channel contrast for this sample is highly sensitive to structural changes during thermal reduction, making it a reliable indicator of reduction progress. While Raman spectroscopy and XPS provide detailed structural insights, the observed non-monotonic behavior in Raman spectra and weakened but still effective discrimination of XPS data at high temperatures demonstrate the utility of optical contrast analysis for fast, large-area assessment of reduction progress. The theoretical model based on Fresnel equations provides insights into differences across color channels and how SiO_2 layer thickness modulates optical properties, offering guidelines for substrate optimization and channel selection. In industrial settings, the selection of color channels sensitive to thermal reduction based on substrate materials enables rapid, non-destructive quality control. This approach proves particularly valuable for efficient material monitoring while preserving product integrity. Future research directions should focus on extending this methodology to other 2D materials like transition metal dichalcogenides to streamline post-synthesis characterization. Additionally, integrating automated optical contrast analysis with machine learning could significantly improve throughput and accuracy for industrial-scale material screening.

Supplementary Materials: The following supporting information can be downloaded at: https://www.mdpi.com/article/10.3390/ma18102222/s1, Figure S1: Raman spectra of GO before and after reduction at various temperatures; Figure S2: AFM height image of GO on 100 nm SiO_2/Si. The scale bar is 10 μm; Table S1: Content of C chemical groups.

Author Contributions: Conceptualization, X.X., Y.C. and W.C.; Methodology, X.X., J.H., B.Y. and Y.Z. (Yinghui Zhou); Software, X.X. and G.M.; Validation, X.X., Formal analysis, X.X. and Y.Z. (Yangbo Chen); Investigation, X.X.; Data curation, X.X., J.H., G.M. and B.Y.; Writing—original draft, X.X.; Writing—review & editing, Y.Z. (Yufeng Zhang) and W.C.; Supervision, Y.Z. (Yinghui Zhou), Y.Z. (Yufeng Zhang), X.Z. and W.C.; Project administration, W.C.; Funding acquisition, X.Z. All authors have read and agreed to the published version of the manuscript.

Funding: The authors appreciate the financial support from the National Natural Science Foundation of China (Nos. 12374193 and 12174321).

Institutional Review Board Statement: Not applicable.

Informed Consent Statement: Not applicable.

Data Availability Statement: All data that support the findings of this study are included within the article (and any Supplementary Files).

Conflicts of Interest: The authors declare no conflict of interest.

References

1. Zheng, X.; Jia, B.; Lin, H.; Qiu, L.; Li, D.; Gu, M. Highly efficient and ultra-broadband graphene oxide ultrathin lenses with three-dimensional subwavelength focusing. *Nat. Commun.* **2015**, *6*, 8433. [CrossRef] [PubMed]
2. Wu, J.; Yang, Y.; Qu, Y.; Xu, X.; Liang, Y.; Chu, S.T.; Little, B.E.; Morandotti, R.; Jia, B.; Moss, D.J. Graphene Oxide Waveguide and Micro-Ring Resonator Polarizers. *Laser Photonics Rev.* **2019**, *13*, 1900056. [CrossRef]
3. Zheng, X.; Xu, B.; Li, S.; Lin, H.; Qiu, L.; Li, D.; Jia, B. Free-standing graphene oxide mid-infrared polarizers. *Nanoscale* **2020**, *12*, 11480–11488. [CrossRef]
4. Gan, S.X.; Chew, J.W.; Ng, K.B.; Tey, L.S.; Chong, W.Y.; Goh, B.T.; Lai, C.K.; Choi, D.-Y.; Madden, S.; Ahmad, H. Single-mode fiber multi-level all-optical switching using GSST-graphene oxide hybrid thin film structure. *J. Appl. Phys.* **2024**, *136*, 063101. [CrossRef]
5. Gao, W.; Singh, N.; Song, L.; Liu, Z.; Reddy, A.L.M.; Ci, L.; Vajtai, R.; Zhang, Q.; Wei, B.; Ajayan, P.M. Direct laser writing of micro-supercapacitors on hydrated graphite oxide films. *Nat. Nanotechnol.* **2011**, *6*, 496–500. [CrossRef]
6. Afroj, S.; Tan, S.; Abdelkader, A.M.; Novoselov, K.S.; Karim, N. Highly Conductive, Scalable, and Machine Washable Graphene-Based E-Textiles for Multifunctional Wearable Electronic Applications. *Adv. Funct. Mater.* **2020**, *30*, 2000293. [CrossRef]
7. Wang, Z.; Yang, X.; Wang, G.; Yang, X.; Qiao, L.; Lu, M. MXene enhanced reduced graphene oxide aerogel for high-performance supercapacitors. *J. Chem. Phys.* **2024**, *161*, 074704. [CrossRef]

8. Alves, L.S.M.; Neves, M.F.F.d.; Benatto, L.; Ramos, M.K.; Eising, M.; de Oliveira, C.K.B.Q.M.; Zarbin, A.J.G.; Roman, L.S. Influence of Nanostructuring Sensors Based on Graphene Oxide and PEDOT:PSS for Methanol Detection. *IEEE Sens. J.* **2023**, *23*, 1845–1853. [CrossRef]
9. Park, S.-J.; Kim, J.; Kang, S.; Cha, H.J.; Shin, H.; Park, J.; Jang, Y.-S.; Woo, J.-S.; Won, C.; Min, D.-H. Discovery of direct-acting antiviral agents with a graphene-based fluorescent nanosensor. *Sci. Adv.* **2020**, *6*, eaaz8201. [CrossRef]
10. Kweon, D.H.; Baek, J.-B. Edge-Functionalized Graphene Nanoplatelets as Metal-Free Electrocatalysts for Dye-Sensitized Solar Cells. *Adv. Mater.* **2019**, *31*, 1804440. [CrossRef]
11. Vaqueiro-Contreras, M.; Bartlam, C.; Bonilla, R.S.; Markevich, V.P.; Halsall, M.P.; Vijayaraghavan, A.; Peaker, A.R. Graphene oxide films for field effect surface passivation of silicon for solar cells. *Sol. Energy Mater. Sol. Cells* **2018**, *187*, 189–193. [CrossRef]
12. Gao, X.; Jang, J.; Nagase, S. Hydrazine and Thermal Reduction of Graphene Oxide: Reaction Mechanisms, Product Structures, and Reaction Design. *J. Phys. Chem. C* **2010**, *114*, 832–842. [CrossRef]
13. Alam, S.N.; Sharma, N.; Kumar, L. Synthesis of Graphene Oxide (GO) by Modified Hummers Method and Its Thermal Reduction to Obtain Reduced Graphene Oxide (rGO)*. *Graphene* **2017**, *6*, 1–18. [CrossRef]
14. Sengupta, I.; Chakraborty, S.; Talukdar, M.; Pal, S.K.; Chakraborty, S. Thermal reduction of graphene oxide: How temperature influences purity. *J. Mater. Res.* **2018**, *33*, 4113–4122. [CrossRef]
15. Sengupta, I.; Kumar, S.S.S.S.; Pal, S.K.; Chakraborty, S. Characterization of structural transformation of graphene oxide to reduced graphene oxide during thermal annealing. *J. Mater. Res.* **2020**, *35*, 1197–1204. [CrossRef]
16. Pelaez-Fernandez, M.; Bermejo, A.; Benito, A.M.; Maser, W.K.; Arenal, R. Detailed thermal reduction analyses of graphene oxide via in-situ TEM/EELS studies. *Carbon* **2021**, *178*, 477–487. [CrossRef]
17. Lee, A.Y.; Yang, K.; Anh, N.D.; Park, C.; Lee, S.M.; Lee, T.G.; Jeong, M.S. Raman study of D* band in graphene oxide and its correlation with reduction. *Appl. Surf. Sci.* **2021**, *536*, 147990. [CrossRef]
18. Yang, D.; Velamakanni, A.; Bozoklu, G.; Park, S.; Stoller, M.; Piner, R.D.; Stankovich, S.; Jung, I.; Field, D.A.; Ventrice, C.A.; et al. Chemical analysis of graphene oxide films after heat and chemical treatments by X-ray photoelectron and Micro-Raman spectroscopy. *Carbon* **2009**, *47*, 145–152. [CrossRef]
19. Ganguly, A.; Sharma, S.; Papakonstantinou, P.; Hamilton, J. Probing the Thermal Deoxygenation of Graphene Oxide Using High-Resolution In Situ X-ray-Based Spectroscopies. *J. Phys. Chem. C* **2011**, *115*, 17009–17019. [CrossRef]
20. Erickson, K.; Erni, R.; Lee, Z.; Alem, N.; Gannett, W.; Zettl, A. Determination of the Local Chemical Structure of Graphene Oxide and Reduced Graphene Oxide. *Adv. Mater.* **2010**, *22*, 4467–4472. [CrossRef]
21. Kovtun, A.; Jones, D.; Dell'Elce, S.; Treossi, E.; Liscio, A.; Palermo, V. Accurate chemical analysis of oxygenated graphene-based materials using X-ray photoelectron spectroscopy. *Carbon* **2019**, *143*, 268–275. [CrossRef]
22. Ferrari, A.C. Raman spectroscopy of graphene and graphite: Disorder, electron–phonon coupling, doping and nonadiabatic effects. *Solid State Commun.* **2007**, *143*, 47–57. [CrossRef]
23. Vidano, R.; Fischbach, D.B. New Lines in the Raman Spectra of Carbons and Graphite. *J. Am. Ceram. Soc.* **1978**, *61*, 13–17. [CrossRef]
24. Ammar, M.R.; Galy, N.; Rouzaud, J.N.; Toulhoat, N.; Vaudey, C.E.; Simon, P.; Moncoffre, N. Characterizing various types of defects in nuclear graphite using Raman scattering: Heat treatment, ion irradiation and polishing. *Carbon* **2015**, *95*, 364–373. [CrossRef]
25. Eckmann, A.; Felten, A.; Verzhbitskiy, I.; Davey, R.; Casiraghi, C. Raman study on defective graphene: Effect of the excitation energy, type, and amount of defects. *Phys. Rev. B* **2013**, *88*, 035426. [CrossRef]
26. Dresselhaus, M.S.; Jorio, A.; Hofmann, M.; Dresselhaus, G.; Saito, R. Perspectives on Carbon Nanotubes and Graphene Raman Spectroscopy. *Nano Lett.* **2010**, *10*, 751–758. [CrossRef] [PubMed]
27. Perrozzi, F.; Prezioso, S.; Donarelli, M.; Bisti, F.; De Marco, P.; Santucci, S.; Nardone, M.; Treossi, E.; Palermo, V.; Ottaviano, L. Use of Optical Contrast To Estimate the Degree of Reduction of Graphene Oxide. *J. Phys. Chem. C* **2013**, *117*, 620–625. [CrossRef]
28. Lucchese, M.M.; Stavale, F.; Ferreira, E.H.M.; Vilani, C.; Moutinho, M.V.O.; Capaz, R.B.; Achete, C.A.; Jorio, A. Quantifying ion-induced defects and Raman relaxation length in graphene. *Carbon* **2010**, *48*, 1592–1597. [CrossRef]
29. Cançado, L.G.; Jorio, A.; Ferreira, E.H.M.; Stavale, F.; Achete, C.A.; Capaz, R.B.; Moutinho, M.V.O.; Lombardo, A.; Kulmala, T.S.; Ferrari, A.C. Quantifying Defects in Graphene via Raman Spectroscopy at Different Excitation Energies. *Nano Lett.* **2011**, *11*, 3190–3196. [CrossRef]
30. Eigler, S.; Dotzer, C.; Hirsch, A. Visualization of defect densities in reduced graphene oxide. *Carbon* **2012**, *50*, 3666–3673. [CrossRef]
31. Stobinski, L.; Lesiak, B.; Malolepszy, A.; Mazurkiewicz, M.; Mierzwa, B.; Zemek, J.; Jiricek, P.; Bieloshapka, I. Graphene oxide and reduced graphene oxide studied by the XRD, TEM and electron spectroscopy methods. *J. Electron. Spectrosc. Relat. Phenom.* **2014**, *195*, 145–154. [CrossRef]
32. Carvalho, A.; Costa, M.C.F.; Marangoni, V.S.; Ng, P.R.; Nguyen, T.L.H.; Castro Neto, A.H. The Degree of Oxidation of Graphene Oxide. *Nanomaterials* **2021**, *11*, 560. [CrossRef]
33. Pei, S.; Cheng, H.-M. The reduction of graphene oxide. *Carbon* **2012**, *50*, 3210–3228. [CrossRef]

34. Wang, P.; Liu, Z.-B.; Chen, X.-D.; Xing, F.; Jiang, W.-S.; Dong, B.; Xin, W.; Tian, J.-G. Accurate layers determination of graphene on transparent substrate based on polarization-sensitive absorption effect. *Appl. Phys. Lett.* **2013**, *103*, 181902. [CrossRef]
35. Ni, Z.H.; Wang, H.M.; Kasim, J.; Fan, H.M.; Yu, T.; Wu, Y.H.; Feng, Y.P.; Shen, Z.X. Graphene thickness determination using reflection and contrast spectroscopy. *Nano Lett.* **2007**, *7*, 2758–2763. [CrossRef]
36. Lu, Y.; Li, X.-L.; Zhang, X.; Wu, J.-B.; Tan, P.-H. Optical contrast determination of the thickness of SiO_2 film on Si substrate partially covered by two-dimensional crystal flakes. *Sci. Bull.* **2015**, *60*, 806–811. [CrossRef]
37. Bayle, M.; Reckinger, N.; Felten, A.; Landois, P.; Lancry, O.; Dutertre, B.; Colomer, J.-F.; Zahab, A.-A.; Henrard, L.; Sauvajol, J.-L.; et al. Determining the number of layers in few-layer graphene by combining raman spectroscopy and optical contrast. *J. Raman Spectrosc.* **2018**, *49*, 36–45. [CrossRef]
38. Dong, X.; Yetisen, A.K.; Tian, H.; Güler, İ.; Stier, A.V.; Li, Z.; Köhler, M.H.; Dong, J.; Jakobi, M.; Finley, J.J.; et al. Line-scan hyperspectral imaging microscopy with linear unmixing for automated two-dimensional crystals identification. *ACS Photonics* **2020**, *7*, 1216–1225. [CrossRef]
39. Li, Y.; Dong, N.; Zhang, S.; Wang, K.; Zhang, L.; Wang, J. Optical identification of layered MoS_2 via the characteristic matrix method. *Nanoscale* **2015**, *8*, 1210–1215. [CrossRef]
40. Li, X.; Shi, Y.; Li, S.; Shi, W.; Han, W.; Zhou, C.; Zhao, X.; Liang, B. Layer-number dependent reflection spectra of MoS_2 flakes on SiO_2/Si substrate. *Opt. Mater. Express* **2018**, *8*, 3082–3091. [CrossRef]
41. Late, D.J.; Liu, B.; Matte, H.S.S.R.; Rao, C.N.R.; Dravid, V.P. Rapid characterization of ultrathin layers of chalcogenides on SiO_2/Si substrates. *Adv. Funct. Mater.* **2012**, *22*, 1894–1905. [CrossRef]
42. Castellanos-Gomez, A.; Quereda, J.; Meulen, H.P.v.d.; Agraït, N.; Rubio-Bollinger, G. Spatially resolved optical absorption spectroscopy of single- and few-layer MoS2 by hyperspectral imaging. *Nanotechnology* **2016**, *27*, 115705. [CrossRef]
43. Castellanos-Gomez, A.; Agraït, N.; Rubio-Bollinger, G. Optical identification of atomically thin dichalcogenide crystals. *Appl. Phys. Lett.* **2010**, *96*, 213116. [CrossRef]
44. Aslan, B.; Chenet, D.A.; van der Zande, A.M.; Hone, J.C.; Heinz, T.F. Linearly polarized excitons in single- and few-layer ReS_2 crystals. *ACS Photonics* **2016**, *3*, 96–101. [CrossRef]
45. Golla, D.; Chattrakun, K.; Watanabe, K.; Taniguchi, T.; LeRoy, B.J.; Sandhu, A. Optical thickness determination of hexagonal boron nitride flakes. *Appl. Phys. Lett.* **2013**, *102*, 161906. [CrossRef]
46. Zhao, Q.; Puebla, S.; Zhang, W.; Wang, T.; Frisenda, R.; Castellanos-Gomez, A. Thickness identification of thin InSe by optical microscopy methods. *Adv. Photonics Res.* **2020**, *1*, 2000025. [CrossRef]
47. Puebla, S.; Mariscal-Jiménez, A.; Galán, R.S.; Munuera, C.; Castellanos-Gomez, A. Optical-based thickness measurement of MoO_3 nanosheets. *Nanomaterials* **2020**, *10*, 1272. [CrossRef]
48. Huang, L.; Shang, Z.; Gao, M.; Miao, C.; Cheng, Y.; Huang, W. Optimized parameters for identifying the layer number of few layer chromium tri-iodide from a theoretical perspective: Implications for two-dimensional spintronics. *ACS Appl. Nano Mater.* **2020**, *3*, 8382–8388. [CrossRef]
49. Mao, N.; Tang, J.; Xie, L.; Wu, J.; Han, B.; Lin, J.; Deng, S.; Ji, W.; Xu, H.; Liu, K.; et al. Optical anisotropy of black phosphorus in the visible regime. *J. Am. Chem. Soc.* **2016**, *138*, 300–305. [CrossRef]
50. Chen, H.; Fei, W.; Zhou, J.; Miao, C.; Guo, W. Layer identification of colorful black phosphorus. *Small* **2017**, *13*, 1602336. [CrossRef]
51. Naylor, C.H.; Parkin, W.M.; Gao, Z.; Kang, H.; Noyan, M.; Wexler, R.B.; Tan, L.Z.; Kim, Y.; Kehayias, C.E.; Streller, F.; et al. Large-Area Synthesis of High-Quality Monolayer 1T′-WTe_2 Flakes. *2D Mater.* **2017**, *4*, 021008. [CrossRef] [PubMed]
52. Xu, T.; Li, S.; Li, A.; Yu, Y.; Zhang, H.; Hu, P.; Zhou, W.; Sheng, L.; Jiang, T.; Cheng, H.; et al. Structural Evolution of Atomically Thin 1T′-$MoTe_2$ Alloyed in Chalcogen Atmosphere. *Small Struct.* **2022**, *3*, 2200025. [CrossRef]
53. Simona, P.; Leonardo, M.; Domenica, C.; Dong, K.; Stiven, F.; Sergio, P.; Filippo, F.; Vaidotas, M.; Camilla, C. Synthesis of Large-Scale Monolayer 1T′-$MoTe_2$ and Its Stabilization via Scalable hBN Encapsulation. *ACS Nano.* **2021**, *15*, 4213–4225.
54. Guo, X.; Li, J.; Yu, Y.; Zafar, A.; Ni, Z. Thermally enhanced optical contrast of graphene oxide for thickness identification. *Nanotechnology* **2019**, *30*, 295704. [CrossRef]
55. Marcano, D.C.; Kosynkin, D.V.; Berlin, J.M.; Sinitskii, A.; Sun, Z.; Slesarev, A.; Alemany, L.B.; Lu, W.; Tour, J.M. Improved Synthesis of Graphene Oxide. *ACS Nano.* **2010**, *4*, 4806–4814. [CrossRef]
56. Mathew, S.; Chan, T.; Zhan, D.; Gopinadhan, K.; Barman, A.-R.; Breese, M.; Dhar, S.; Shen, Z.; Venkatesan, T.; Thong, J.T. The Effect of Layer Number and Substrate on the Stability of Graphene under MeV Proton Beam Irradiation. *Carbon* **2011**, *49*, 1720–1726. [CrossRef]
57. Xu, L.; Cheng, L. Graphite Oxide under High Pressure: A Raman Spectroscopic Study. *J. Nanomater.* **2013**, *2013*, 731875. [CrossRef]
58. Ferrari, A.C.; Robertson, J. Interpretation of Raman spectra of disordered and amorphous carbon. *Phys. Rev. B* **2000**, *61*, 14095–14107. [CrossRef]
59. Ioni, Y.; Timur, K.; Ivan, S.; Vasiliy, B.; Ayrat, M.D. Revealing the Effect of Graphite Source on the Properties of Synthesized Graphene Oxide. *Carbon Lett.* **2024**, *34*, 1219–1228. [CrossRef]

60. Kudin, K.N.; Ozbas, B.; Schniepp, H.C.; Prud'homme, R.K.; Aksay, I.A.; Car, R. Raman Spectra of Graphite Oxide and Functionalized Graphene Sheets. *Nano Lett.* **2008**, *8*, 36–41. [CrossRef]
61. Park, H.; Lim, S.; Nguyen, D.D.; Suk, J.W. Electrical Measurements of Thermally Reduced Graphene Oxide Powders under Pressure. *Nanomaterials* **2019**, *9*, 1387. [CrossRef] [PubMed]
62. Chen, C.-M.; Huang, J.-Q.; Zhang, Q.; Gong, W.-Z.; Yang, Q.-H.; Wang, M.-Z.; Yang, Y.-G. Annealing a graphene oxide film to produce a free standing high conductive graphene film. *Carbon* **2012**, *50*, 659–667. [CrossRef]
63. Wang, M.; Duong, L.D.; Oh, J.-S.; Mai, N.T.; Kim, S.; Hong, S.; Hwang, T.; Lee, Y.; Nam, J.-D. Large-Area, Conductive and Flexible Reduced Graphene Oxide (RGO) Membrane Fabricated by Electrophoretic Deposition (EPD). *ACS Appl. Mater. Interfaces* **2014**, *6*, 1747–1753. [CrossRef]
64. Perrozzi, F.; Croce, S.; Treossi, E.; Palermo, V.; Santucci, S.; Fioravanti, G.; Ottaviano, L. Reduction dependent wetting properties of graphene oxide. *Carbon* **2014**, *77*, 473–480. [CrossRef]
65. Larciprete, R.; Fabris, S.; Sun, T.; Lacovig, P.; Baraldi, A.; Lizzit, S. Dual Path Mechanism in the Thermal Reduction of Graphene Oxide. *J. Am. Chem. Soc.* **2011**, *133*, 17315–17321. [CrossRef]
66. Jung, I.; Vaupel, M.; Pelton, M.; Piner, R.; Dikin, D.A.; Stankovich, S.; An, J.; Ruoff, R.S. Characterization of Thermally Reduced Graphene Oxide by Imaging Ellipsometry. *J. Phys. Chem. C* **2008**, *112*, 8499–8506. [CrossRef]
67. Yang, H.; Hu, H.; Wang, Y.; Yu, T. Rapid and non-destructive identification of graphene oxide thickness using white light contrast spectroscopy. *Carbon* **2013**, *52*, 528–534. [CrossRef]
68. Li, Y.; Singh, A.; Krylyuk, S.; Davydov, A.; Jaramillo, R. Near-infrared photonic phase-change properties of transition metal ditellurides. In *Low-Dimensional Materials and Devices 2019*; Kobayashi, N.P., Talin, A.A., Davydov, A.V., Eds.; SPIE: San Diego, CA, USA, 2019; p. 28. [CrossRef]
69. Palik, E.D. *Handbook of Optical Constants of Solids*, 1st ed.; Academic Press: San Diego, CA, USA, 1997; ISBN 978-0-12-544423-1.
70. Blake, P.; Hill, E.W.; Castro Neto, A.H.; Novoselov, K.S.; Jiang, D.; Yang, R.; Booth, T.J.; Geim, A.K. Making graphene visible. *Appl. Phys. Lett.* **2007**, *91*, 063124. [CrossRef]
71. Xu, X.; Su, Y.; Miao, G.; Huang, J.; Lin, G.; Zhu, T.; Xu, Y.; Huang, C.; Zhou, Y.; Zhang, Y.; et al. Visualizing Oxidation in Monolayer 1T′-$MoTe_2$. *J. Phys. D Appl. Phys.* **2025**, *58*, 175302. [CrossRef]

materials

MDPI

Article

Sustainable Eco-Friendly Synthesis of Gold Nanoparticles Anchored on Graphene Oxide: Influence of Reductant Concentration on Nanoparticle Morphology

Mariano Palomba [1], Gianfranco Carotenuto [1], Maria Grazia Raucci [1], Antonio Ruotolo [2] and Angela Longo [1,*]

[1] Institute for Polymers, Composites, and Biomaterials, National Research Council, SS Napoli/Portici, 80125 Napoli, Italy; mariano.palomba@cnr.it (M.P.); giancaro@unina.it (G.C.); mariagrazia.raucci@cnr.it (M.G.R.)

[2] Department of Engineering, College of Charleston, Charleston, SC 29424, USA; ruotoloa@cofc.edu

* Correspondence: angela.longo@cnr.it; Tel.: +39-0817758824

Abstract: Gold nanoparticles (AuNPs) anchored on graphene oxide (GO) have had a significant interest for their unique optical, electrical, and catalytic properties. This study presents an eco-friendly and sustainable synthesis of AuNPs on GO sheets using L-ascorbic acid (L-aa) as a green reducing agent and polyvinylpyrrolidone (PVP) as a stabilizer. The effect of reductant concentration on nanoparticle morphology was systematically investigated using UV–Visible spectroscopy and transmission electron microscopy (TEM). Results indicate the formation of AuNPs anchored on GO sheets and that an increase in the L-aa amount leads to both an increase in nanoparticle size and a morphological transition from spherical to irregular structures. The simultaneous nucleation and growth processes result in the formation of multiple families of nanostructures, as confirmed by TEM analysis, which reveals two distinct size distributions. At higher L-aa concentrations, the nanoparticles shape evolves into irregular morphologies due to selective growth along a preferential facet. This approach not only enables precise control over AuNP size and shape but also aligns with green chemistry principles, making it a promising route for applications in plasmonics, sensors, and photothermal therapy.

Keywords: gold nanoparticles; graphene oxide; green reducing agent; surface stabilizer

1. Introduction

The sustainable synthesis of gold nanoparticles (AuNPs) anchored on graphene oxide (GO) sheets has garnered significant attention due to their combined functional properties [1] which hold promise for electronics [2], optoelectronics [2], catalysis [3], and particularly for biomedical applications [4].

AuNPs are highly efficient at absorbing light and converting it into heat, making them ideal for photothermal therapy, where near-infrared light is used to generate localized heat and destroy cancer cells [5]. In addition, AuNPs possess biocompatibility, antioxidant properties, and anti-inflammatory effects, which make them suitable for enhancing bone regeneration. The inclusion of GO provides mechanical strength, increased surface area, and bioactivity, all of which are beneficial in improving cellular interactions and tissue regeneration [6]. GO also acts as a biocompatible carrier and stabilizer for AuNPs, due to its large surface area, thereby enhancing their stability, dispersibility, and overall efficacy in photothermal therapy [7]. Moreover, GO absorbs near-infrared radiation, further contributing to heat generation and amplifying treatment effectiveness [8]. By controlling the morphology of nanoparticles, influenced by the reductant concentration, these particles can be optimized for use in biological systems.

When it comes to osteoporotic bone regeneration, AuNPs can promote the differentiation of stem cells into osteoblasts, stimulate bone matrix production, and enhance bone mineralization [9]. They can also facilitate the delivery of growth factors or drugs directly to the bone tissue, helping to repair damaged bone and regenerate osteoporotic defects. Moreover, the eco-friendly synthesis method ensures that these nanoparticles are safe and sustainable, reducing the risk of toxicity when used in clinical applications. Thus, these gold nanoparticle-functionalized graphene oxide composites can play a vital role in the development of advanced therapeutic strategies for bone regeneration, offering promising prospects for the treatment of osteoporosis and other bone-related diseases.

Furthermore, as mentioned, the integration of AuNPs and GO is revolutionizing electronics. Reduced graphene oxide (rGO) offers excellent electrical conductivity, which is further improved by the anchored AuNPs. These nanoparticles form conductive bridges between GO sheets, reducing electron scattering and enhancing composite performance [10]. The surface plasmon resonance (SPR) of AuNPs amplifies light–matter interactions, enhancing the efficiency of high-performance optoelectronic devices, such as photodetectors and solar cells, offering a unique combination of conductivity, flexibility, and sensitivity. In addition, applications range from flexible, transparent electronics to energy storage systems and wearable devices [1,11,12].

Nowadays, the control of the plasmonic absorption and morphology of AuNPs remains a critical focus for advancing next-generation technologies.

In addition, the growing attention towards sustainability emphasizes the use of eco-friendly materials, minimizing toxic chemicals, and developing methods of synthesis reducing energy consumption, maximizing efficiency [13,14]. A widely used method for synthesizing AuNPs anchored on GO is simultaneous reduction in an aqueous solution. Typically, a gold precursor (i.e., $HAuCl_4$) is mixed with a GO suspension, and a green reducing agent, like plant extracts, microorganisms [15], or rose water [16], etc., is added to reduce Au^{3+} to Au^0, forming AuNPs directly on GO sheets. This approach replaces toxic chemicals like hydrazine and sodium borohydride with green alternatives, adhering to the principles of green chemistry. Among green reductants, L-ascorbic acid (L-aa) stands out for its effectiveness [17–19]. L-aa facilitates the controlled synthesis of gold and silver nanoparticles and efficiently reduces GO to rGO, yielding materials with enhanced electrical conductivity. Key advantages of using L-ascorbic acid include the following:

- Environmentally friendly and non-toxic properties.
- High efficiency in reducing metal ions, enabling the synthesis of nanoparticles with controlled size and morphology.
- Effective reduction of GO to rGO with improved properties.
- Low risk of introducing impurities, as L-aa contains only carbon, oxygen, and hydrogen atoms.

This approach not only leverages the advanced properties of AuNPs and GO but also aligns with green chemistry principles, ensuring a sustainable and efficient synthesis process.

Here, a simple method to obtain colloid solutions of AuNPs anchored on GO by reduction of $HAuCl_4$ using L-aa is presented. The stability of colloid solutions is ensured by the presence of polyvinylpyrrolidone (PVP) as a stabilizer [20]. In addition, the influence of reductant concentration on nanoparticle morphology was analyzed through UV–Visible spectroscopy and transmission electron microscopy (TEM).

2. Materials and Methods

2.1. Materials

The precursors used for the synthesis of the AuNPs/graphene colloidal solution were as follows: (i) graphene oxide (GO) nanosheets as the graphene-based support; (ii) tetrachloroauric acid ($HAuCl_4$, Sigma-Aldrich, St. Louis, MO, USA, 99.9%) as the metal nanoparticle precursor; (iii) L-ascorbic acid (L-aa, $C_6H_8O_6$, Sigma-Aldrich, SSt. Louis, MO, USA, reagent grade) as a green reducing agent; and (iv) poly(N-vinylpyrrolidone) (PVP, Sigma-Aldrich, St. Louis, MO, USA, Mw = 10,000 a.m.u.) as a protective agent.

For the synthesis of GO nanosheets, the following reagents were used without further purification: sulfuric acid (H_2SO_4, Sigma-Aldrich, St. Louis, MO, USA, 99.999%), potassium nitrate (KNO_3, Sigma-Aldrich, >99.0%), potassium permanganate ($KMnO_4$, Sigma-Aldrich, St. Louis, MO, USA, >99.0%), hydrogen peroxide (H_2O_2, Sigma-Aldrich, St. Louis, MO, USA, 30%), and anhydrous ethanol (C_2H_6O, Carlo Erba, Milano, Italy, ACS reagent).

Additionally, distilled water was used for all experiments.

2.2. GO Nanosheet Synthesis

GO nanosheets were synthesized using a modified Hummers' method applied to graphite nanoplatelets [21]. A mixture of graphite nanoplatelets and KNO_3 in H_2SO_4 was stirred for a few minutes in an ice bath. $KMnO_4$ was then slowly added under stirring to prevent the temperature from exceeding 20 °C. The mixture was subsequently stirred for 1 h at 35 °C. A color change from dark purplish green to dark brown confirmed the oxidation of the nanoplatelets.

To stop the oxidation process, distilled water was gradually added to the solution, followed by the addition of a diluted aqueous H_2O_2 solution (≈4% v/v). The resulting GO nanosheets were separated from the reaction mixture by filtration and purified through successive cycles of centrifugation, ultrasonication, and re-dispersion in water until a pH of 5–6 was achieved. Further details on this synthesis method can be found in previous publications [18] and references therein.

2.3. Synthesis of AuNPs/Graphene Colloid Solutions

The preparation of AuNPs/graphene colloid solutions involved three steps. Firstly, a solution was prepared by dissolving 1 g of PVP into 10 g of distillate water. To ensure complete dissolution of PVP in water and obtain a stable solution, the mixture was stirred continuously for 48 h at room temperature [22]. After that, 1 mL of aqueous solution of GO (40% in weight) was added. This main solution contains PVP, as a passivating agent, and GO, as a graphene-like substrate for gold nanoparticle deposition. A second solution was freshly prepared by dissolving 12.5 mg of tetrachloroauric acid in distilled water at room temperature. This solution containing the metallic precursor was added to the main solution under continuous stirring. Then, a freshly aqueous solution of L-aa, as a reducing agent, was quickly injected under stirring and made to react for 15 min.

To investigate the effect of L-ascorbic acid (L-aa) on the size and morphology of gold nanoparticles (AuNPs), the weight ratio between the metallic acid and the reducing agent varied, starting from stoichiometric conditions.

For this, the balanced chemical equation between $HAuCl_4$ and L-aa is as follows:

$$2\,HAuCl_4 + 3\,C_6H_8O_6 = 2\,Au + 3\,C_6H_6O_6 + 8\,HCl \quad (1)$$

In stoichiometric conditions, where molar ratio R = 2/3 and for a fixed amount of gold produced (i.e., 12 mg of $HAuCl_4$), the corresponding weight ratio of the reactants is 1.29.

This value was used as the starting point, from which the amount of L-aa was progressively increased to explore its effect on nanoparticle formation.

The amount of L-aa added influenced the nucleation process, which in turn affected the size and morphology of the obtained AuNPs and promoted the partial reduction of GO. Table 1 summarizes the amounts used for the preparation of the different samples.

Table 1. The amounts of the reducing agent and the weight ratio $AuHCl_4$/L-aa for the preparation of the AuNPs/graphene hybrid structures.

Name	[$HAuCl_4$] mg	[L-aa]	Ratio [$HAuCl_4$]/[L-aa]
Sample 1	12.52	9.72	1.29
Sample 2	12.36	12.75	0.96
Sample 3	12.49	14.66	0.85
Sample 4	12.59	16.72	0.75
Sample 5	12.10	18.77	0.64
Sample 6	12.84	20.22	0.60
Sample 7	12.31	22.66	0.54
Sample 8	12.57	24.52	0.51

As visible in Figure 1, eight different samples were obtained, with a color shift from red to blue, by varying the weight ratio, R, from 1.29 to 0.51. It is well known that this typical optical property of metal nanoparticles, named Surface Plasmon Resonance (SPR), is strictly dependent on the shape and the size of the nanoparticles.

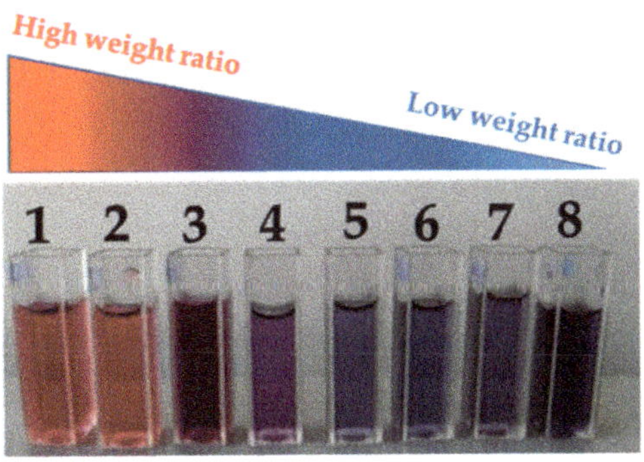

Figure 1. Different samples of AuNPs/graphene colloid solution obtained by varying the weight ratio of $HAuCl_4$/L-aa.

2.4. Instrumentations

The spectra of precursors and the SPR of colloid solutions were acquired with a Perkin-Elmer Lambda 850 spectrophotometer (PerkinElmer, Waltham, MA, USA), at room temperature (25 °C) between 200 nm and 900 nm with a scan speed of 800 nm/min and a read interval of 1 nm. For measurements, an aliquot of the sample was placed into a 1 cm path length quartz cuvette in the sample light path, while the distilled water was placed into a quartz cuvette in the reference light path.

Qualitative and quantitative elemental analysis measurements were performed by scanning electron microscope (SEM) (FEI Quanta 200 FEG, Hillsboro, OR, USA) equipped with an Oxford Inca Energy Dispersive X-ray (EDX) microanalyzer (Inca Oxford 250, High Wycombe, UK).

The synthesized colloid solutions were morphologically characterized by TEM to investigate the sheet-like morphology of rGO, and the formation of AuNPs and their size and dispersion patterns. A few drops of each sample were air-dried on a copper grid coated with carbon film for TEM analysis. The measurements were carried out by an FEI Tecnai G2 Spirit TWIN (Philips, Amsterdam, The Netherlands) microscope, operating at 120 kV and equipped with a LaB6 filament. The sizes of AuNPs from TEM images were analyzed using the National Institutes of Health (NIH) Image J1.48i. This software is an open-source image processing software designed to analyze multidimensional scientific images, such as TEM micrographs. Measurements needed to analyze the changes in AuNP morphology were constructed using the Feret diameter. This diameter is a measure of an object's size along a specified direction. In general, it can be defined as the distance between the two parallel planes restricting the object perpendicular to that direction. This measure is used in the analysis of particle sizes, for example, in microscopy, where it is applied to projections of a three-dimensional (3D) object on a 2D plane [23].

The effect of the L-aa amount on sample morphology was analyzed by comparing optical results with morphological characterization.

3. Results and Discussion

This section is divided into two subparagraphs, providing a detailed description of the experimental results, their interpretation, and the corresponding discussion. To ensure a clearer and more concise presentation, selected experimental results from only three samples, i.e., sample 1, sample 4, and sample 8, are presented, as they exhibit distinct and specific properties.

3.1. Optical Characterization

The optical spectra of all precursors were acquired and a comparison of these spectra with those of the obtained samples confirms the complete conversion of the precursors through the reduction process.

Figure 2 illustrates UV–Visible spectra of the obtained colloid solution of AuNPs/graphene obtained using different weight ratio, R, from 1.29 to 0.51 characterized by a presence of a typical SPR peak. As previously mentioned, the SPR peak position (λ_m) of Au nanoparticles is strictly dependent on size and shape [24]. As shown by the experimental spectra, the wavelength (λ_m) of the SPR peak shifts to a higher wavelength, indicating that the samples contain Au nanoparticles with variations in size and shape.

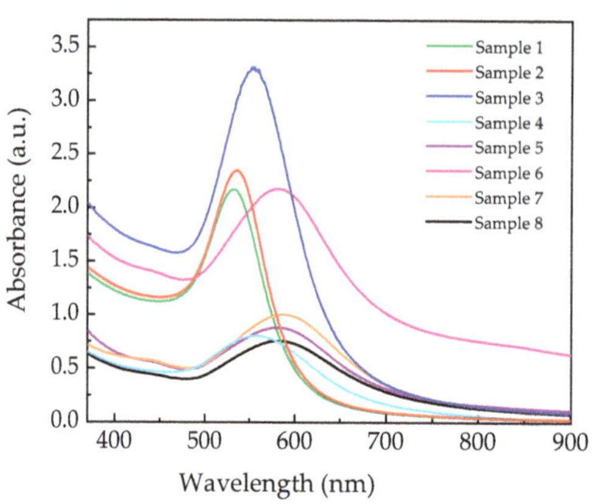

Figure 2. UV–Visible spectra of AuNPs/graphene colloid solution obtained by varying the weight ratio of $HAuCl_4$/L-aa.

All spectra were normalized to maximum values in the range from 450 to 900 nm and fitted with a Gaussian curve to better evaluate the λ_m position. In Figure 3, for greater clarity, only spectra of sample 1, sample 4, and sample 8 are shown.

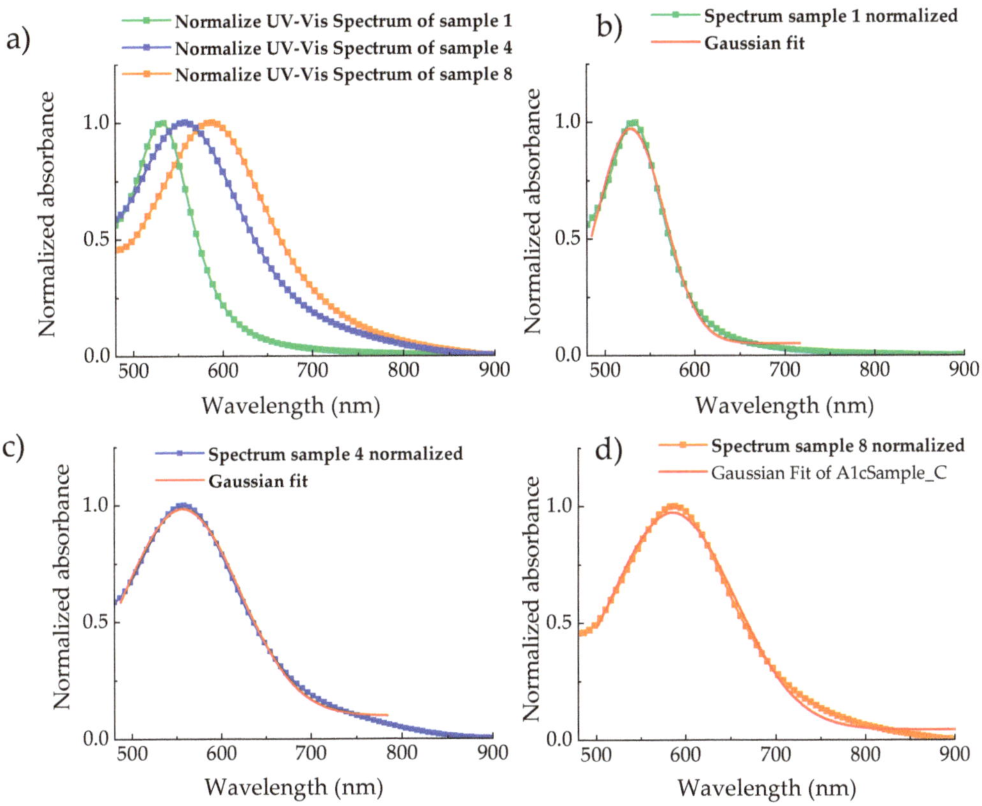

Figure 3. UV–Visible spectra of AuNPs/graphene colloid solution of sample 1, sample 4, and sample 8 normalized (**a**), and the fit of experimental data by Gaussian curve of AuNPs/graphene colloid solution of sample 1, sample 4, and sample 8 (**b**), (**c**), and (**d**), respectively.

Accordingly, Figure 4 summarizes the behavior of the peak position of SPR, λ_m, and the full width at half maximum (FWHM), obtained through the Gaussian fit of the optical spectra. As the weight ratio R decreases from sample 1 to sample 8, λ_m shifts to higher wavelengths, from 530 nm to 590 nm (see black squares and line in Figure 5). Similarly, the FWHM increases from 25 to 140 as R decreases from sample 1 to sample 4, then stabilizes around 130 (see red circles and line in Figure 5). These trends indicate that small, monodisperse Au nanoparticles were obtained with a stoichiometric amount of L-aa, whereas larger nanoparticles with irregular shapes formed at higher concentrations.

The main observation is thus that, as R decreases, the main SPR band red-shifts and becomes broader, resulting in a band/shoulder corresponding to the nanostructure asperities mode becoming more intense. Such changes in the optical response can be easily justified in terms of the morphology of the particles.

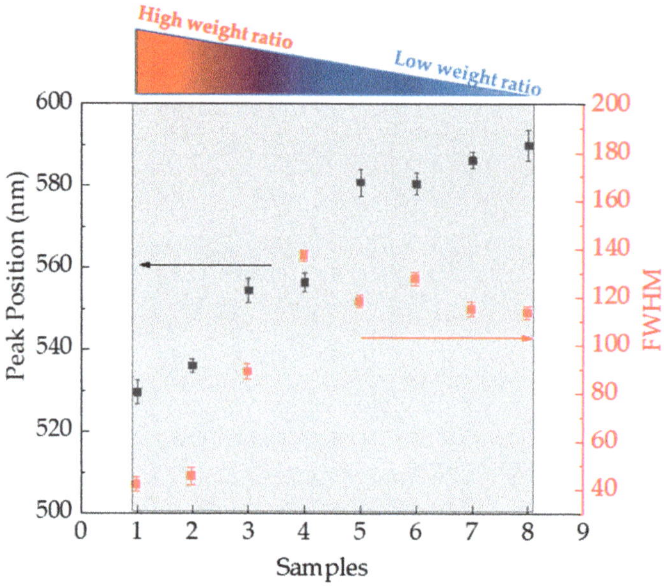

Figure 4. Peak position of SPR, λ_m, and the full width at half maximum (FWHM) obtained by UV–Visible spectra of all AuNPs/graphene colloid solutions.

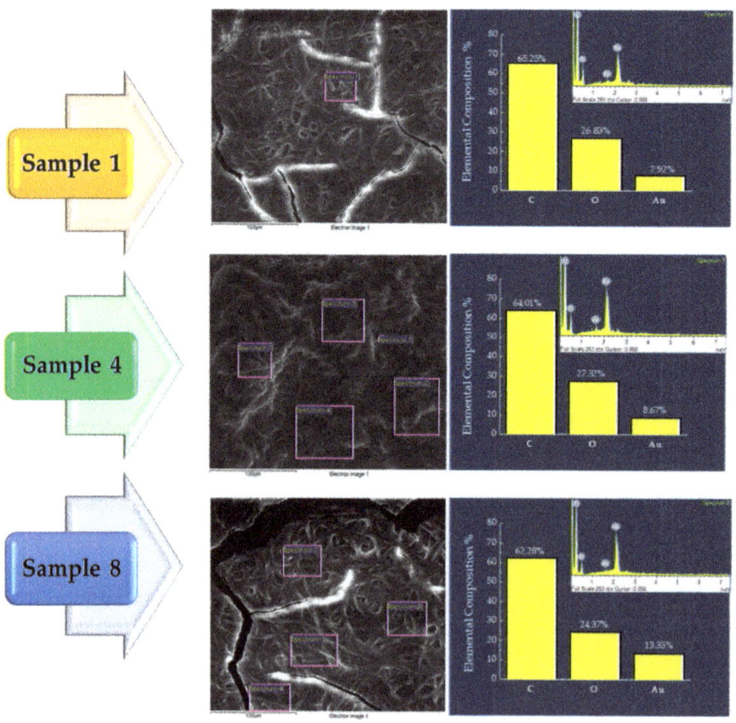

Figure 5. SEM micrographs and corresponding EDS elemental analysis of samples 1, 4, and 8. Each image indicates the area where EDS was performed. The elemental composition is presented as bar charts showing weight percentages of C, O, and Au. Inset: representative EDS spectra for each sample.

3.2. Morphological and Structural Characterization

EDS analysis was performed on the samples to determine the chemical composition and identify potential impurities. SEM micrographs of sample 1, sample 4, and sample 8 at 10,000× magnification and the EDS analysis, are shown in Figure 5, along with histograms displaying the element composition percentages. SEM images acquired at low magnification were used solely to indicate the region where EDS elemental analysis was performed. These images do not aim to provide morphological information. The element Au constitutes, on average, around 9 wt% (percentage in weight) of the total composition for all samples. Additionally, due to the presence of PVP and GO, the amounts of carbon and oxygen remain relatively constant across all samples, at approximately 60 wt% and 25 wt%, respectively. The inset in Figure 5 displays the elemental spectra obtained for all samples.

Representative SEM images of sample 3 (see Figure 6) are shown to highlight the composite nature of the hybrid system. The SEM images clearly reveal the presence of spherical nanoparticles, attributable to AuNPs, embedded within a polymeric matrix. Unfortunately, due to their similar morphology and low contrast in SEM, the GO sheets cannot be distinguished from the surrounding PVP matrix.

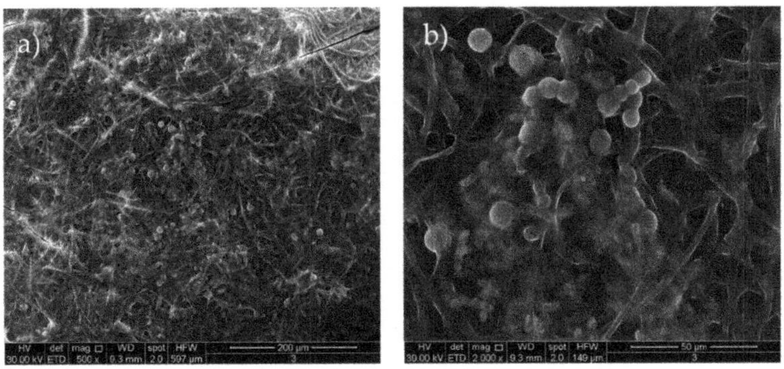

Figure 6. SEM of AuNPs distributed in a stabilizing PVP of typical sample 3 at two different magnifications (**a**,**b**).

For this purpose, the TEM analysis was carried out. Figure 7 shows TEM images of sample 1, sample 3, sample 4, and sample 8. The micrographs clearly display the successful synthesis of AuNPs anchored on graphene oxide sheets (indicated in figure by yellow line), and reveal that the morphology of AuNPs anchored to the surface differs significantly and is strongly related to decrease in weight ratio R from 1.29 to 0.51. As shown by the micrographs at different magnifications of sample 1 (see Figure 7a–c) at high weight ratio, the AuNPs are well-anchored on the GO surface and are nearly uniform and homogenous in their spherical shape and size. In addition, the histograms (see Figure 7d) show that the average size of the nanoparticles is 11 nm, with a relatively narrow distribution. The micrographs of sample 3 (Figure 7e–g) show nanoparticles that are anchored on the GO sheet, predominantly spherical, and comparable in size and distribution to those in sample 1 (Figure 7a–c) and sample 2 (data not shown). A comparison of the Feret diameter distributions (see histograms in Figure 7d,h) shows a similar Gaussian-like profile in samples 1 and 3, with mean values centered around ~11 nm and ~13 nm, respectively. These results confirm that, in both cases, the nucleation and growth processes occur under comparable kinetic regimes, resulting in spherical nanoparticles with narrow size distributions. These observations suggest that no significant morphological changes occur between samples 1 and 3, supporting the idea that nanoparticle aggregation and morphology alteration begin to be prominent only from sample 4 onward. Decreasing R ratio, the sample 4 micrographs (see Figure 7i,l,m) reveal the formation, on GO sheet, of two different AuNPs classes. It is possible to identify a class of uniform and spherical nanoparticles characterized by the smallest size, and a second group characterized by irregular shape and biggest size. The distribution shows two average sizes centered at 6 nm and 22 nm (see Figure 7n). In the case of lowest R, the sample 8 micrographs, see Figure 7o–q, clearly reveal a formation of AuNPs anchored on the GO sheet characterized by irregular shapes and multiple protrusions radiating from a central core. This structure results from anisotropic growth mechanisms triggered by the high L-aa concentration, which alters both nucleation kinetics and the capping efficiency of PVP. The number of particles is reduced, consistent with fewer nucleation events and extensive growth, possibly driven by Ostwald ripening. The broad Feret diameter distribution (Figure 7r), spanning from ~30 to 65 nm, reflects the heterogeneous population of branched nanostructures formed under these conditions.

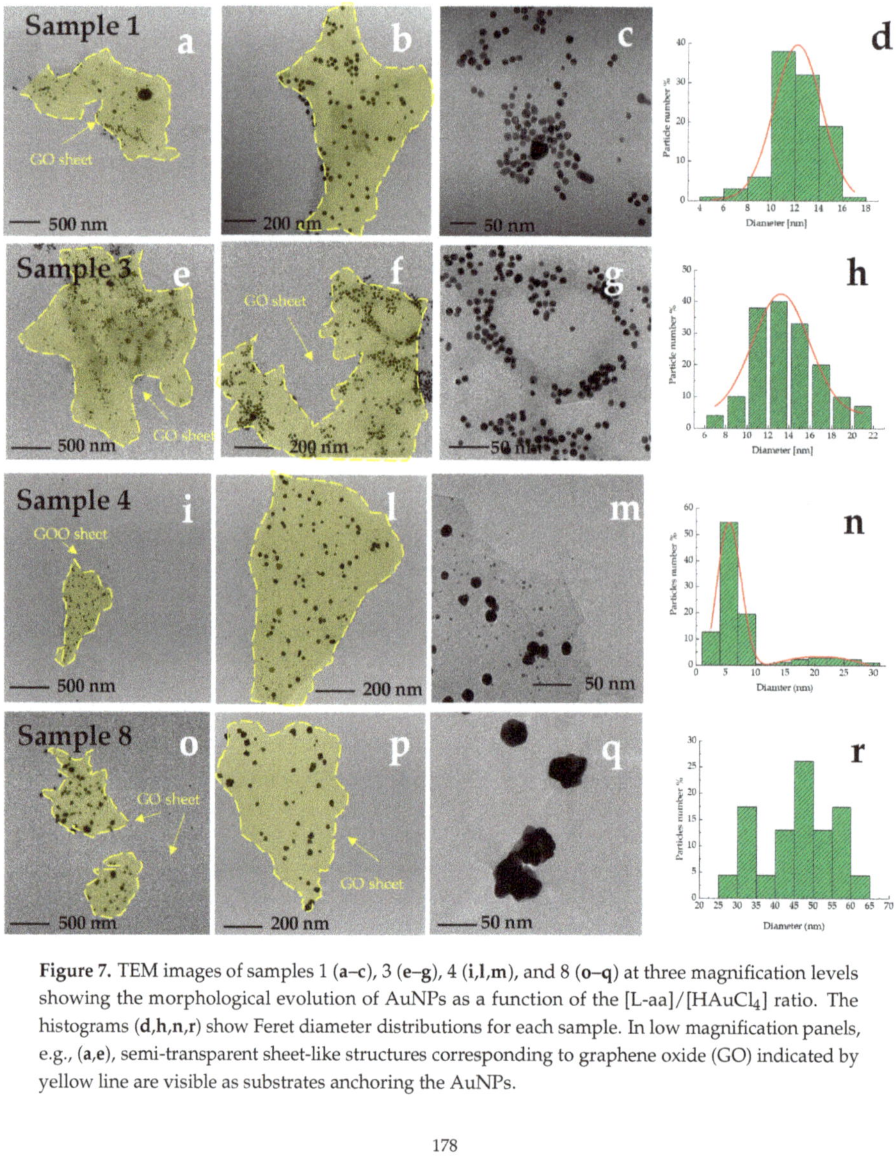

Figure 7. TEM images of samples 1 (**a–c**), 3 (**e–g**), 4 (**i**,**l**,**m**), and 8 (**o–q**) at three magnification levels showing the morphological evolution of AuNPs as a function of the [L-aa]/[$HAuCl_4$] ratio. The histograms (**d**,**h**,**n**,**r**) show Feret diameter distributions for each sample. In low magnification panels, e.g., (**a**,**e**), semi-transparent sheet-like structures corresponding to graphene oxide (GO) indicated by yellow line are visible as substrates anchoring the AuNPs.

In other words, at lower magnification (e.g., panels Figure 7a,e,i,p), the presence of thin, wrinkled GO sheets (see yellow line in Figure 7) can be clearly identified as semi-transparent layers supporting the AuNPs. These sheets are typical of exfoliated GO and provide evidence that the nanoparticles are anchored on the GO surface, not freely dispersed. Although the polymeric matrix of PVP is a fundamental stabilizer in the colloid system, it is not detectable in TEM images due to its low electron density and high transparency to the electron beam. Consequently, the uniform background observed in several micrographs does not indicate an absence of GO but rather the non-observable nature of PVP in these imaging conditions. In addition, significant changes in morphology and size distribution appear only from sample 4 onward, where bimodal size populations and irregular shapes become evident. As shown in Figure 7c, AuNPs in Sample 1 exhibit a predominantly spherical morphology, with narrow size distribution and good dispersion on GO sheets. In contrast, AuNPs in Sample 7 display (see Figure 7m) a branched and irregular structure, consistent with uncontrolled growth and preferential facet development. These observations confirm the morphological evolution induced by L-aa concentration, as also reflected in the shift and broadening of the size distribution histograms.

This marks the onset of morphology transition due to increased L-aa concentration and simultaneous growth mechanisms, as also confirmed by the UV-Vis spectral red-shift and FWHM broadening.

These results confirm the eco-friendly effectiveness of L-aa as a reducing agent for both $HAuCl_4$ and highlight the role of PVP as a passivating and stabilizing agent. PVP is well known for its ability to enhance the dispersion of nanoparticles in aqueous systems and is frequently employed in combination with graphene-based materials to improve colloidal stability [22]. Furthermore, it has been reported to assist in transporting gold nanoparticles into the interlaminar regions of GO, while simultaneously stabilizing their structure [25]. In this configuration, PVP not only prevents nanoparticle aggregation, but also promotes effective anchoring of AuNPs onto the GO surface. This structural arrangement has been widely described in the literature for its role in improving both nanoparticle distribution and overall composite stability [20,25].

At low concentrations of L-aa, the shape of the AuNPs is thermodynamically controlled, leading to spherical structures due to the adsorption and stabilizing effect of PVP. In this case, the formation of the particles follows a two-step process. The first step is nucleation, during which the average particle radius is less than a few nanometers. Subsequently, nanoparticle growth is believed to occur through either Ostwald ripening or coalescence. With increasing L-aa concentration, smaller spherical AuNP seeds form due to a reduction in nanoparticle size. In this scenario, nucleation and growth occur simultaneously, resulting in multiple families of nanostructures. This phenomenon is particularly evident in the TEM micrographs, which reveal two distinct families with different size distributions. As the L-aa concentration continues to increase, this aggregation process further evolves, ultimately leading to the formation of irregular AuNPs. Additionally, a notable decrease in particle number is observed, accompanied by an increase in particle size, which can be directly attributed to Ostwald ripening.

These morphological evolutions are consistent with UV–Visible observations, particularly the progressive red-shift and broadening of the SPR band (Figure 5), indicating increased size and shape anisotropy. Such trends are attributed to the combined effect of L-aa concentration on nucleation kinetics, partial reduction of GO, and local pH variations, which modulate the balance between thermodynamic and kinetic growth regimes.

In other words, beyond the observed effects of the reducing agent concentration on nanoparticle size and morphology, our investigation reveals that increasing the concentration of L-ascorbic acid induces a slight decrease in the pH of the reaction medium.

This pH shift plays a critical role in modulating the nucleation and growth mechanisms of AuNPs. Under near-neutral conditions, the efficient passivation provided by PVP leads to the formation of small, monodisperse, and predominantly spherical nanoparticles. However, when higher amounts of L-ascorbic acid are used, the resulting lower pH can diminish the capping efficiency of PVP. This change promotes more rapid nucleation and anisotropic growth, giving rise to irregular or irregular, branched nanoparticle morphologies. Therefore, careful control of both the reducing agent concentration and pH is essential for fine-tuning the structural properties of the nanoparticles, ultimately broadening their applicability in plasmonic, sensors, and photothermal therapy.

4. Conclusions

In conclusion, a sustainability and green chemistry approach is presented. The synthesis of AuNPs anchored on GO is performed using an environmentally friendly reducing agent (L-ascorbic acid), replacing toxic chemicals like hydrazine and sodium borohydride. In addition, enhanced stability and better dispersibility are obtained thanks to the use of polyvinylpyrrolidone (PVP) as a stabilizer that improves the stability of colloidal solutions and prevents nanoparticle aggregation. The study systematically investigates how the concentration of the reducing agent influences nanoparticle size and shape, providing valuable insights for tuning material properties.

The UV–Visible spectroscopy, TEM, and SEM-EDS analyses confirm the successful synthesis, composition, and morphology of the hybrid structures. Depending on the reducing agent concentration, the synthesized AuNPs exhibit spherical, irregular, and star-like morphologies, which significantly influence their potential applications in plasmonic, sensors, and photothermal therapy. While the morphological and optical results confirm the successful formation of AuNPs/GO composites stabilized by PVP, additional structural and chemical analyses such as XRD, XPS, and TGA would further validate the incorporation and interaction of each component. These characterizations are planned for future studies to deepen the understanding of the hybrid system. The method is simple, cost-effective, and scalable, making it suitable for industrial applications in nanoelectronics and sustainable materials development.

Author Contributions: Conceptualization, A.L.; software, A.L.; investigation, A.L.; resources, M.G.R. and A.L.; data curation, A.L. and M.P.; writing—original draft preparation, A.L., M.P., G.C., A.R. and M.G.R.; writing—review and editing, A.L., M.P., G.C., A.R. and M.G.R.; visualization, A.L. All authors have read and agreed to the published version of the manuscript.

Funding: This study was funded by Unione Europea Next Generation EU PRIN2022-2022W8SLMW—Smart Injectable Scaffolds for Sclerostin Based Bone Resorption Treatment—SMART4SCLERO [CUP: B53D23008940006].

Institutional Review Board Statement: Not applicable.

Informed Consent Statement: Not applicable.

Data Availability Statement: The data presented in this study are available on request from the corresponding author.

Acknowledgments: The authors are grateful to Maria Cristina Del Barone of LAMEST laboratory (IPCB-CNR) for SEM and TEM analysis.

Conflicts of Interest: The authors declare no conflicts of interest.

References

1. Idisi, D.O.; Oke, J.A.; Bello, I.T. *Graphene Oxide/Au Nanoparticles: Synthesis, Properties, and Application: A Mini-Review*; John Wiley and Sons Ltd.: Hoboken, NJ, USA, 2021. [CrossRef]

2. Wu, J.; Lin, H.; Moss, D.J.; Loh, K.P.; Jia, B. Graphene oxide for photonics, electronics and optoelectronics. *Nat. Rev. Chem.* **2023**, *7*, 162–183. [CrossRef]
3. Nancy, P.; Nair, A.K.; Antoine, R.; Thomas, S.; Kalarikkal, N. In situ decoration of gold nanoparticles on graphene oxide via nanosecond laser ablation for remarkable chemical sensing and catalysis. *Nanomaterials* **2019**, *9*, 1201. [CrossRef]
4. Turcheniuk, K.; Boukherroub, R.; Szunerits, S. Gold–graphene nanocomposites for sensing and biomedical applications. *J. Mater. Chem. B* **2015**, *3*, 4301–4324. [CrossRef]
5. Riley, R.S.; Day, E.S. Gold nanoparticle-mediated photothermal therapy: Applications and opportunities for multimodal cancer treatment. *WIREs Nanomed. Nanobiotechnol.* **2017**, *9*, e1449. [CrossRef]
6. Raucci, M.G.; Giugliano, D.; Longo, A.; Zeppetelli, S.; Carotenuto, G.; Ambrosio, L. Comparative facile methods for preparing graphene oxide–hydroxyapatite for bone tissue engineering. *J. Tissue Eng. Regen. Med.* **2017**, *11*, 2204–2216. [CrossRef]
7. Delgado-Corrales, B.J.; Chopra, V.; Chauhan, G. Gold nanostars and nanourchins for enhanced photothermal therapy, bioimaging, and theranostics. *J. Mater. Chem. B* **2025**, *13*, 399–428. [CrossRef]
8. Robinson, J.T.; Tabakman, S.M.; Liang, Y.; Wang, H.; Sanchez Casalongue, H.; Vinh, D.; Dai, H. Ultrasmall Reduced Graphene Oxide with High Near-Infrared Absorbance for Photothermal Therapy. *J. Am. Chem. Soc.* **2011**, *133*, 6825–6831. [CrossRef]
9. Pissuwan, D.; Poomrattanangoon, S.; Chungchaiyart, P. Trends in Using Gold Nanoparticles for Inducing Cell Differentiation: A Review. *Appl. Nano Mater.* **2022**, *5*, 3110–3120. [CrossRef]
10. Zhang, J.; Zhang, X.; Shen, J.; Pan, H.; Chen, Z.; Li, Y.; Zhu, S. Nanoarchitectonics of graphene oxide with functionalized cellulose nanocrystals achieving simultaneous dual connections and defect repair through catalytic graphitization for high thermal conductivity. *Carbon* **2023**, *201*, 295–306. [CrossRef]
11. Su, P.-G.; Shiu, W.-L.; Tsai, M.-S. Flexible humidity sensor based on Au nanoparticles/graphene oxide/thiolated silica sol–gel film. *Sens. Actuators B Chem.* **2015**, *216*, 467–475. [CrossRef]
12. Chuang, M.-K.; Lin, S.-W.; Chen, F.-C.; Chu, C.-W.; Hsu, C.-S. Gold nanoparticle-decorated graphene oxides for plasmonic-enhanced polymer photovoltaic devices. *Nanoscale* **2014**, *6*, 1573–1579. [CrossRef]
13. Khosravi, A.; Zarepour, A.; Iravani, S.; Varma, R.S.; Zarrabi, A. Sustainable synthesis: Natural processes shaping the nanocircular economy. *Environ. Sci. Nano* **2024**, *11*, 688–707. [CrossRef]
14. Gupta, D.; Boora, A.; Thakur, A.; Gupta, T.K. Green and sustainable synthesis of nanomaterials: Recent advancements and limitations. *Environ. Res.* **2023**, *231*, 116316. [CrossRef]
15. El-Maghrabi, N.; El-Borady, O.M.; Hosny, M.; Fawzy, M. Catalytic and Medical Potential of a Phyto-Functionalized Reduced Graphene Oxide–Gold Nanocomposite Using Willow-Leaved Knotgrass. *ACS Omega* **2021**, *6*, 34954–34966. [CrossRef]
16. Tabrizi, M.A.; Varkani, J.N. Green synthesis of reduced graphene oxide decorated with gold nanoparticles and its glucose sensing application. *Sens. Actuators B Chem.* **2014**, *202*, 475–482. [CrossRef]
17. Umer, A.; Naveed, S.; Ramzan, N.; Rafique, M.S.; Imran, M. A green method for the synthesis of Copper Nanoparticles using L-ascorbic acid. *Matéria* **2014**, *19*, 197–203. [CrossRef]
18. Palomba, M.; Carotenuto, G.; Longo, A. A Brief Review: The Use of L-Ascorbic Acid as a Green Reducing Agent of Graphene Oxide. *Materials* **2022**, *15*, 6456. [CrossRef]
19. Kulkarni, S.R.; Borse, D.B.; Agarwal, A.D., Saptale, S.P. Green synthesis of Ag nanoparticles using Vitamin C (Ascorbic Acid) in a batch process. In Proceedings of the International Conference on Nanoscience, Engineering and Technology (ICONSET 2011), Chennai, India, 28–30 November 2011; pp. 88–90. [CrossRef]
20. Koczkur, K.M.; Mourdikoudis, S.; Polavarapu, L.; Skrabalak, S.E. Polyvinylpyrrolidone (PVP) in nanoparticle synthesis. *Dalton Trans.* **2015**, *44*, 17883–17905. [CrossRef]
21. Carotenuto, G.; Longo, A.; Nicolais, L.; De Nicola, S.; Pugliese, E.; Ciofini, M.; Locatelli, M.; Lapucci, A.; Meucci, R. Laser-Induced Thermal Expansion of H_2SO_4-Intercalated Graphite Lattice. *J. Phys. Chem. C* **2015**, *119*, 15942–15947. [CrossRef]
22. Volpe, M.V.; Longo, A.; Pasquini, L.; Casuscelli, V.; Carotenuto, G. Synthesis and characterization of gold-based quantum dots. *J. Mater. Sci. Lett.* **2003**, *22*, 1697–1699. [CrossRef]
23. Merkus, H.G. Particle Size Measurements. In *Particle Technology Series*; Springer: Dordrecht, The Netherlands, 2009; Volume 17. [CrossRef]
24. Amendola, V.; Pilot, R.; Frasconi, M.; Maragò, O.M.; Iatì, M.A. Surface plasmon resonance in gold nanoparticles: A review. *J. Phys. Condens. Matter.* **2017**, *29*, 203002. [CrossRef]
25. Jayan, J.S.; Pal, K.; Saritha, A.; Deeraj, B.D.S.; Joseph, K. Graphene oxide as multi-functional initiator and effective molecular reinforcement in PVP/epoxy composites. *J. Mol. Struct.* **2021**, *1230*, 129873. [CrossRef]

MDPI AG
Grosspeteranlage 5
4052 Basel
Switzerland
Tel.: +41 61 683 77 34

Materials Editorial Office
E-mail: materials@mdpi.com
www.mdpi.com/journal/materials

www.ingramcontent.com/pod-product-compliance
Lightning Source LLC
Chambersburg PA
CBHW061546210926
51850CB00009B/2018